Oxford International Resources

5

Maths
Teacher's Guide

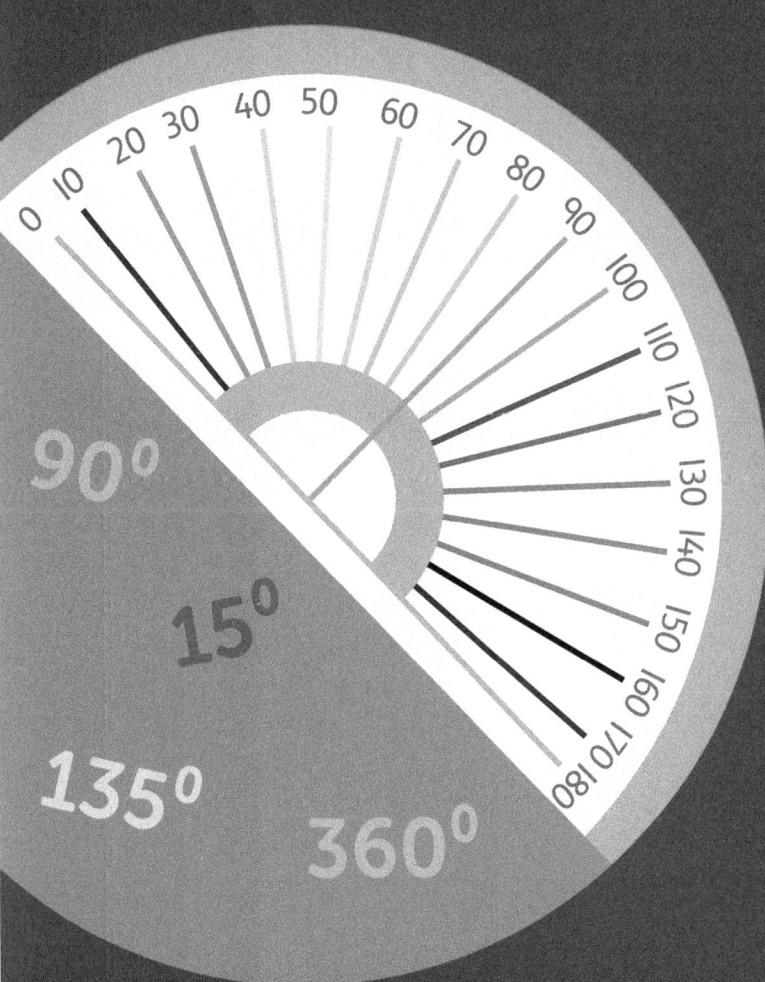

Tony Cotton
Caroline Clissold
Ray Huntley

OXFORD
UNIVERSITY PRESS

Great Clarendon Street, Oxford, OX2 6DP, United Kingdom

Oxford University Press is a department of the University of Oxford. It furthers the University's objective of excellence in research, scholarship, and education by publishing worldwide. Oxford is a registered trade mark of Oxford University Press in the UK and in certain other countries.

© Ray Huntley, Caroline Clissold and Tony Cotton 2021

The moral rights of the author have been asserted.

First published in 2014

All rights reserved. No part of this publication may be reproduced, stored in a retrieval system, or transmitted, in any form or by any means, without the prior permission in writing of Oxford University Press, or as expressly permitted by law, by licence or under terms agreed with the appropriate reprographics rights organization. Enquiries concerning reproduction outside the scope of the above should be sent to the Rights Department, Oxford University Press, at the address above.

You must not circulate this work in any other form and you must impose this same condition on any acquirer.

British Library Cataloguing in Publication Data

Data available

ISBN 9781382017305

11

Paper used in the production of this book is a natural, recyclable product made from wood grown in sustainable forests. The manufacturing process conforms to the environmental regulations of the country of origin.

Printed and bound by CPI Group (UK) Ltd, Croydon, CR0 4YY

Acknowledgements

The publisher and authors would like to thank the following for permission to use photographs and other copyright material:

Cover: Artwork by Peskimo. **Photos: pv(t):** foto-bee/Alamy Stock Photo; **pv(m):** Alistair McDonald/Shutterstock; **pv(b):** 2R fotografia/Shutterstock; **px:** Monkey Business Images/Shutterstock; **pxviii:** monkeybusinessimages/iStockphoto.

Artwork by Q2A Media Services Pvt. Ltd.

Every effort has been made to contact copyright holders of material reproduced in this book. Any omissions will be rectified in subsequent printings if notice is given to the publisher.

The manufacturer's authorised representative in the EU for product safety is Oxford University Press España S.A. of El Parque Empresarial San Fernando de Henares, Avenida de Castilla, 2 – 28830 Madrid (www.oup.es/en or product.safety@oup.com). OUP España S.A. also acts as importer into Spain of products made by the manufacturer.

Contents

Introduction	iv
1 Number and place value	**1**
Overview	1
Engage	2
1A Place value	3
1B Rounding	6
1C Ordering and comparing	9
1D Number sequences	13
1E Odd and even numbers	19
1F Roman numerals	22
1G Number problems	25
Connect	28
Review	30
2 Addition and subtraction	**31**
Overview	31
Engage	32
2A Partitioning to add or subtract	33
2B Adding and subtracting near multiples	35
2C Which strategy?	39
2D Written methods of adding and subtracting	42
2E Adding and subtracting to solve problems	48
Connect	51
Review	53
3 Multiplication and division	**54**
Overview	54
Engage	56
3A Multiplication and division facts	57
3B Factors and multiples	60
3C Using known facts to multiply	63
3D Doubling and halving	67
3E Written methods for multiplying	70
3F Multiplying and dividing by 10, 100 and 1000	74
3G Written methods for dividing	77
3H Prime numbers	80
3I Square and cube numbers	83
3J Multiplying and dividing to solve problems	87
Connect	90
Review	91
4 Fractions, decimals and percentages	**93**
Overview	93
Engage	95
4A Equivalent fractions	96
4B Fraction and decimal equivalents	102
4C Improper fractions and mixed numbers	105
4D Adding and subtracting fractions	108
4E Multiplying fractions	111
4F Ordering fractions	115
4G Thousandths	118
4H Rounding decimals	121
4I Percentages	125
4J Proportion	130
4K Ratio	133
Connect	136
Review	137
5 Length, mass, capacity and volume	**139**
Overview	139
Engage	141
5A Units of measure	142
5B Measuring length	145
5C Centimetres and millimetres	148
5D Measuring mass	151
5E Measuring capacity	154
5F Imperial units	157
5G Volume	160
5H Problem solving with measures	163
Connect	166
Review	168
6 Area and perimeter	**169**
Overview	169
Engage	170
6A Understanding perimeter	171
6B Understanding area	174
6C Calculating area and perimeter	177
Connect	180
Review	181
7 Time	**183**
Overview	183
Engage	184
7A Converting between units of time	185
7B Calculating time intervals	188
7C Using calendars	192
Connect	195
Review	196
8 Geometry – properties of shapes	**198**
Overview	198
Engage	200
8A Regular and irregular polygons	201
8B Symmetry in polygons	204
8C Identifying 3D shapes	207
8D Angles	210
8E Angle sums	214
Connect	217
Review	218
9 Geometry – position and direction	**219**
Overview	219
Engage	220
9A Coordinates	221
9B Reflection	224
9C Translations	229
Connect	232
Review	233
10 Statistics	**234**
Overview	234
Engage	235
10A Frequency tables and bar charts	236
10B Line graphs	239
10C Timetables	242
10D Probability	246
Connect	249
Review	250
Glossary	**252**

Introduction

The joy of learning maths

We are living in an ever-changing world, where the way we work, live, learn, communicate and relate to one another is constantly shifting. In this climate, we need to instill in our learners the skills to equip them for every eventuality so they are able to overcome challenges, adapt to change and have the best chance of success. To do this, we need to evolve beyond traditional teaching approaches and foster an environment where students can start to build lifelong learning skills for success. Students need to learn how to learn, how to problem solve, be agile and work flexibly. Going hand-in-hand with this is the development of self-awareness and mindfulness through the promotion of wellbeing to ensure that students learn the socio-emotional skills to succeed.

With *Oxford International Primary Maths*, students develop lifelong learning skills as well as mathematical skills. The course promotes the development of real-world skills including financial literacy. The activities in the Student Books and Practice Books offer numerous opportunities to think creatively and develop interpersonal skills. Fundamentally, *Oxford International Primary Maths* promotes students' self-development as critical thinking and motivation are at the heart of the problem-solving approach in the course.

This series is based on the English National Curriculum Programme of Study for Primary Maths. The *Oxford International Primary Maths* books for each stage meet all the learning objectives from the curriculum. Each lesson includes the learning objectives and a summary of the key teaching points. A full mapping grid identifying the unit and lesson where each objective can be found is available online at www.oxfordowl.co.uk

Oxford International Primary Maths: A problem-solving approach

In this second edition of *Oxford International Primary Maths*, there is a strong focus on using a problem-solving approach. While mathematical facts are important, it is unlikely that simply giving students the information they need will result in them understanding the mathematics and being able to apply their learning in new problem-solving situations. This is often described as a move from 'surface learning' to 'deep learning'.

Many people remember mathematics lessons as places where the teacher stood at the front of the class writing on the board. Students wrote down the information, maybe worked through a couple of examples with the teacher and then proceeded to complete a series of exercises to practise the skill that they had been taught. This can be described as a *didactic* approach and it relies on the idea that direct instruction is the appropriate strategy to adopt. The authors of this series would argue that *heuristic* strategies encourage students to explore the mathematics for themselves supported by the teacher. 'Heuristic' derives from the Greek word meaning to discover, and in mathematics learning, heuristic strategies are ones where students engage in exploration and discovery to solve a problem. Heuristic strategies include making a visual representation of a problem, making a calculated guess or estimate, simplifying a problem or following a known method. This results in a deeper understanding.

When faced with any problem in mathematics, there are recognised stages to go through in order to solve the problem, and these have been developed and agreed by many researchers. One version that summaries the problem-solving process comes from Georg Polya.

1. Understand the problem.

2. Devise a plan.

3. Carry out the plan.

4. Check the reasoning.

In following these stages, students use a number of skills that support problem solving, such as using trial and improvement, working systematically, pattern spotting, visualising, conjecturing and generalising.

Embedding a mastery approach

In recent years, the term 'mastery' has been used in conjunction with mathematics learning. It has been drawn from teaching approaches in countries where mathematics performance is deemed to be very high. The essence of mastery is to produce students who have deep conceptual understanding and procedural fluency through learning in a collaborative and problem-solving context. Mastery learning incorporates use of manipulatives, exposure to different methods of solving a problem, dialogue and explanation.

Following a Concrete Pictorial Abstract (CPA) approach

One of the more successful approaches to learning was provided by Jerome Bruner in his model of enactive, iconic and symbolic modes. This has been developed in recent years to form the CPA approach. CPA stands for concrete, pictorial and abstract, each of which aligns with Bruner's modes. The concrete phase involves students making use of physical manipulatives to help understand the learning, before moving to record the learning in pictorial form as individuals. As the learning develops, students will begin to recognise how to record their learning in a more general and abstract way. The CPA approach is not necessarily sequential, and students might move between the different modes as they work through a problem.

Oxford International Primary Maths and the use of manipulatives

Throughout the series, students are encouraged to use manipulatives, or concrete objects, to model addition, subtraction, multiplication and division. These manipulatives include:

- base-ten equipment (ones-cubes, tens-rods, hundreds-flats and thousands-cubes)

- place-value counters

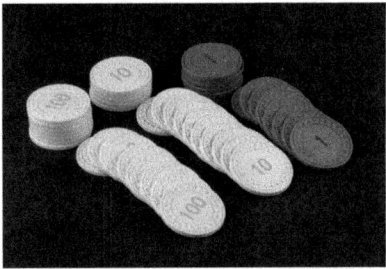

- number rods.

Such manipulatives are used to explain to students how the written methods 'work', for example by modelling exchanging 10 ones-cubes for 1 tens-rod in an addition.

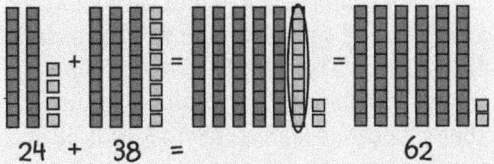

Differentiation

There are several ways that you can differentiate learning in the classroom. These include differentiation by:

- task
- outcome
- support
- grouping.

It has been traditional in some schools to offer up to three different levels of tasks for each lesson. This is differentiation by task. It is important that all students are exploring the same area of mathematics as they can collaborate and discuss their mathematics in a way that is not possible if students are engaged on different activities. This approach has been extensively researched and published by Jo Boaler of Stanford University, California. For example, she has outlined projects that gave students in different schools either a differentiated approach in lessons, or lessons where everyone worked on the same task (Boaler, J., 2005. *The Elephant in the Classroom.* Souvenir Press). Where all abilities worked on the same task, every student made and sustained 'better than expected' progress, and performed better on statutory tests and exams. The Education Endowment Foundation teacher's toolkit suggests that collaborative learning can result in a five-month acceleration in students' learning. (See https://educationendowmentfoundation.org.uk/resources/teaching-learning-toolkit.)

The expectation in this series is that all students will be offered the same starting point. The activities are carefully designed to be accessible to all students in your class and the teacher's notes for the activity offer differentiated outcomes for students. It is also important that you offer differentiated support to different students. You will mainly do this through the sort of questioning that you engage in and support you offer. You will ask challenging questions and supporting questions to help all students access the task. For example, when engaging in a simple counting activity with some students you might model the action of counting by placing a finger on each object as you count and emphasise the last number you say to model that the last number you say gives the number of objects in the set. You might ask other students engaged in the same activity to compare two sets, or to find one more or one less than the set they are counting.

Grouping students to promote a growth mindset

When engaging in learning mathematics, it is expected that you will use a variety of student groupings. This may be a change for some teachers who have previously grouped students by prior attainment in their classroom. Research has shown that grouping students 'by ability', which usually means grouping students using test results, can have a negative impact on their future attainment. It is more effective to use a range of ways of grouping students. You will decide on the most appropriate way of grouping depending on the activity. You are also given advice in the teacher's notes. It is important that the teacher is active in deciding which form of grouping is appropriate. It is also important that students learn how to operate in a range of different groups and with a range of different students so that they get used to working in a variety of ways and with different people.

The three main ways of grouping students are based on:

- friendship
- ability/prior experience
- mixed attainment.

Friendship groups, are most appropriate for activities in which the students have been given some element of choice. Perhaps they are carrying out some research for a data handling project or exploring data on animals to develop their understanding of measurement. This grouping is the default if teachers do not actively group students.

Ability groups, or groups based on students' prior experience, may be helpful if the lesson requires a very specific prior knowledge. You can group together the students you know have this knowledge and they can then work with minimal guidance from you, which allows you to focus on groups who need additional support.

Mixed-attainment groups are encouraged for the majority of the activities. This form of grouping is also favoured by those following a mastery approach. Working in collaborative, all-attainment groups also supports students' wellbeing and promotes a growth mindset, as described in research by Carol Dweck. She found that students who were grouped by ability tended to stay in those groupings throughout their school life, and regard themselves as having a fixed ability that could not be changed. This has dire consequences for students in middle or lower sets. When placed in mixed-ability groups, all students can develop a growth mindset which enables them to believe they can learn and improve, whatever their starting point (Dweck, C., 2007. 'The Perils and Promise of Praise'. *Educational Leadership*. October 2007, 65(2), 34–39). A growth mindset is promoted when students do not feel that their future success is predicated on prior achievement. This kind of grouping is particularly helpful for students new to English. Mixed-attainment groups allow students who are less confident in English to hear more-confident peers using mathematical vocabulary. Research has shown that mixed-attainment groups benefit both high attainers, who become more secure in their mathematics knowledge through explaining their thinking to peers, and those less secure in their mathematical knowledge as peer teaching has been shown to be effective.

Whatever form of grouping you choose, it is helpful to assign roles to individuals in the group. Some teachers use 'role cards' to remind members of the group of the role they should play. Here are some examples of roles.

- Leader: You should make sure everyone has a chance to speak and focus the discussion around the task.
- Time keeper: You should encourage the group to stay on task. Announce when the time is half way through and when time is nearly up.
- Recorder: You should write down group members' ideas or draw a collective graphic. You will write on the board during the presentation.
- Presenter: You will present the group's findings to the whole class at the end of the session.
- Resource organiser: You will make sure that group members have all the resources they need during the task.

Assessment

Assessment is the process of establishing how individual students are progressing and what they have achieved, or a means of measuring their learning. Assessment is usually carried out in two main ways – assessment of learning and assessment for learning.

Assessment of learning is sometimes called summative assessment, and takes place at the end of a lesson, a unit, a term or even a year. It measures what students know at that point as a summary of their learning to that point. In *Oxford International Primary Maths*, summative assessment opportunities are provided in the Review lesson at the end of each unit in the Student Book, while half-termly summative assessment opportunities are provided through printable resources, available online.

Assessment for learning is an approach brought to prominence by Paul Black and Dylan Wiliam and is based on the notion that students have a full, clear sense of what they are learning, where they have reached in their learning and what they need to do to improve further. It is carried out during lessons and gives teachers continuous data on each student's learning, as well as allowing students to track their own learning, which provides greater motivation. (Black, P., Harrison, C., Lee, C., Marshall, B., and Wiliam, D. 2004. 'Inside the Black Box: Assessment for Learning in the Classroom'. *Phi Delta Kappan*. (86)1, 8–21.)

It is suggested that there are five key strategies for assessment for learning. These are outlined below with suggestions of how you can do this in your classroom.

1. Being clear about learning objectives and success criteria with the students.

Each activity has at least one learning objective. At the beginning of a lesson, share the activity's learning objective with students. This should be more than simply stating the objective. You should make sure that students understand the objective and how you will measure success. For example, you might say: *I know that you can all count 10 objects* and all count to 10 as a class. Then you point to 20 on a number line and ask: *Does anyone know what this number is?* If a student knows it is 20 praise them, if no-one knows, tell them it is 20 and say: *By the end of the lesson I will be able to listen to you count to 20.*

2. Planning student discussions that give you evidence of their learning.

Every activity plan in the Teacher's Guide offers the opportunity for small-group or whole-class discussion. There are also examples of probing questions that you can ask to assess students' current understanding. For example, if a group has been counting two sets of objects you can ask: *Were there more or less in the second group? How do you know?*

3. Giving students feedback that helps them move forward.

This allows students to know whether or not they are meeting the success criteria and what they can do next to move their learning on. Developing the example above, if a group has been comparing two sets and understands the concept of 'more' and 'less' you could ask them to make sets that are one more and one less, or even two more and two less.

4   Activating students to act as instructional resources for each other.

Collaborative group work in mixed-attainment groups, as described by Jo Boaler in her research (see under Differentiation earlier), gives students the opportunity to operate both as learners and teachers, with peer learning being highly effective. Not only is understanding of the mathematics enhanced, but students can support each other in assessing their progress.

5   Activating students as owners of their own learning.

The key point here is to listen carefully to the students and adapt your questioning to support individual development and to follow individual interests.

Questioning is key

The most skilled mathematics teachers can ask open questions to elicit students' current understandings. Skilful open questioning also allows students to articulate their current understanding carefully and though this process either consolidate their understanding or come to realise where they have made a mistake. The list below offers a series of open questions that can be used whatever mathematics you are teaching.

- *How are these the same/different?*
- *About how many/how long/many more ... do you think there will be?*
- *What would happen if ...?*
- *How else could you have done that?*
- *Why did you ...?*
- *How did you ...?*
- *How do you know that is correct?*

If you want students to check their solutions and consolidate their learning it is helpful to ask them to explain how they reached their solution to a friend. Similarly, to support students in reflecting on their learning you might ask the following.

- *What mathematics did you use to solve the problem?*
- *What new mathematics did you learn?*
- *What key words did you use?*
- *What was the most challenging part of the activity?*
- *What did you do when you got stuck?*
- *What other questions could you ask?*
- *Did this remind you of any other areas of mathematics?*

In *Oxford International Primary Maths*, there is an opportunity to ask these reflective questions, and for students to reflect on their learning, at the end of each unit in the Review lesson of the Practice Book.

Word problems

Word problems are useful as an assessment of children's understanding of the correct mathematics to use in any given situation. In *Oxford International Primary Maths* word problems are included throughout the units and on every Student Book Review page as part of the end-of-unit assessment. Many teachers find teaching word problems a challenge. This area is particularly challenging for students with a limited English vocabulary as word problems are tightly bound to linguistic ability. We have to decode and understand what the problem is asking us to do before we can begin to apply our mathematical knowledge. Some teachers have found the following acronym helpful when working with students on solving word problems.

R: Read the problem carefully.

U: Understand what the problem is asking you to do.

C: Choose the mathematics or arithmetical operations that you need to use to solve the problem.

S: Solve the problem.

A: Answer the problem.

C: Check that the answer is accurate and reasonable.

It is often helpful for students to underline key facts and write down the operations they are going to use before they solve the problem. For example:

> Tony rode his bicycle 7 miles to school with his friend. On his way home he took a short cut which was only 5 miles. How far did he cycle altogether?
>
> *This will be an addition calculation.*

It is a useful activity for students to annotate word problems and write down the operation(s) they will use without carrying out the calculation as this focuses on the skill of understanding the problem and choosing the operations appropriately.

Another activity that helps students to become skilled at solving word problems is asking them to write their own word problems based on a picture or a set of objects. Here is an example.

- How many black cubes are there? (3)
- Two friends took three cubes each. How many were left? (2)
- If I take out the black cubes, how many are left? (5)
- If I share the cubes equally between two people, how many do they each get? (4)

Wellbeing and *Oxford International Primary Maths*

It is thought that students learn more and feel more connected to their learning when they are active in their lessons. *Oxford International Primary Maths* has active learning at its heart. Most lessons start with a whole-class session that usually includes a range of physical or active activites. You will see this signified by a 'star-jump' icon in the Teacher's Guide.

Many adults and children have felt anxious about their learning of mathematics at some stage. This anxiety is reduced by working collaboratively in all-attainment groups. There is also a reflective session at the end of each lesson and the formative assessment activity in the Practice Book asks students to reflect on their learning across the unit.

Wellbeing is also supported by effective questioning to support and stretch students and by planning group work carefully. These areas have already been discussed above.

Language support

The challenges

Ministries of Education at both local and national level are increasingly adopting the policy of English Medium Instruction (EMI), for either one or two subjects or across the whole curriculum. The rationale for doing so varies according to the local context, but improving the levels of achievement in English is an important factor.

In international schools an additional reason is likely to be that students do not share a mother tongue with each other or perhaps the teacher. English is, therefore, chosen as the medium for instruction so that all students are in the same position and to provide the opportunity to develop proficiency in an international language.

This does not mean that the mathematics teacher is now being asked to replace the English teacher, or to have the same skills or knowledge of English (though in many primary schools one teacher may indeed teach both). What it does mean, however, is that mathematics teachers have to view their role differently: they have to become much more language aware. It is this recognition of the need to ensure that the delivery of the content is not negatively impacted by the use of the second language that informs the planning and methodology of EMI.

This raises significant challenges, including:

- the teacher's knowledge of English
- students' level of English (which may vary considerably in international schools)
- resources that provide appropriate language support
- assessment tools which ensure that it is the content and not the language that is being tested
- differentiation that acknowledges different levels of proficiency in both language and content.

Meeting the challenges positively

Perhaps lack of confidence in their own English proficiency is one of the most common concerns among teachers. However, while it is a factor, success in EMI is not necessarily linked to teachers' proficiency in English. Teachers who have English as their mother tongue may well lack the sensitivity to, or awareness of, the language that a non-native speaker has acquired through learning and studying the second language. Developing this awareness and demonstrating it in both materials and method is the key to effective EMI.

Classroom language/Teacher Talk

Often non-native-speaker teachers are more concerned about their ability to run and manage the whole class in English than they are about the teaching of the mathematics concepts, as the resources or textbook should help them with the latter. However, this use of English in the class is very important as it provides exposure to the second language, which plays a valuable role in language acquisition. It is also true that the Teacher Talk for purposes such as checking attendance and collecting homework does not have to be totally accurate or accessible to students. When teaching the mathematics concepts, however, it is essential that the Teacher Talk is comprehensible. Some basic strategies to ensure this include:

- simplify your language
- use short, simple sentences and project your voice
- paraphrase (say in a different way) as necessary
- use visuals, write or draw on the board, gestures and body language to clarify meaning
- repeat as necessary
- plan before the lesson
- prepare clear, simple instructions and check understanding.

Creating a language-rich environment

Primary teachers often excel at providing a colourful and engaging physical environment for students. In the EMI classroom, this becomes even more important. Posters, 'word walls', lists of key structures, students' work, English signs and notices all provide a backdrop that provides the opportunity for language exposure and language acquisition.

Planning

When planning, look carefully at each stage of the unit and identify the language demands. This means thinking about what language students will need to understand or produce, and deciding how best to scaffold the learning to ensure that language does not become an obstacle to understanding the concept. This involves providing language support and goes beyond the familiar strategy of identifying key vocabulary.

Support for listening and reading

Listening and reading are receptive skills, requiring understanding rather than production of language. If you are asking students to listen to or read texts in English, ask yourself the following questions when you are planning the unit.

- Do I need to teach any vocabulary before they listen/read?
- How can I prepare them for the content of the text so that they are not listening 'cold'?

- Can I provide visual support to help them understand the key content?
- How many times should I ask them to read/listen?
- What simple question can I set before they listen/read for the first time to focus their attention?
- How can I check more detailed understanding of the text? Can I use a graphic organiser (e.g. tables, charts and diagrams) or gap-fill task to reduce the language demands?
- Do I need to differentiate the task for those students who find reading/listening difficult?
- Could I make the tasks interactive (e.g. jigsaw reading, when students access different information before coming together then share information)?
- How am I going to check their answers and give feedback?

Support for speaking and writing

Speaking and writing are productive skills because students doing these need to produce language. They are different from the receptive skills of listening and reading where students receive language from other sources. These skills may require more input from the teacher.

When you plan to use a task that requires students to *produce* English (speak or write), you need to think about how to help them do this.

This means that you have to think in detail about what language the task requires (Language Demands, LD) and what strategies you will use to help them use English to perform the task (Language Support, LS).

You need to ask yourself the following questions.

- What *vocabulary* does the task require? (LD)
- Do I need to teach this before they start? How? (LS)
- What *phrases/sentences* will they need? Think about the language for learning mathematics (e.g. predicting and comparing). What structures do they need for these language functions? (LD)
- Will they be able to produce these sentences or should I provide some *scaffolding* [e.g. sentence starters/sentence frames/gapped sentences (see below)]? (LS)

    A square has ____ sides.

    A triangle has ____ sides.

    A quadrilateral has ____ sides.

    A pentagon has ____ sides.

- While I am *monitoring* this task is there any way I can provide further support for their use of English (especially for the less-confident students)? (LS)
- What language will students need to use at the *feedback* stage (e.g. when they present their task)? Do I need to scaffold this? (LD, LS)

Teaching vocabulary and structures

Vocabulary

Learning the key mathematics vocabulary is central to EMI and 'learning' means more than simply understanding the meaning. Knowing a word also involves being able to *pronounce* it accurately and *use* it appropriately. Below is a list of strategies that could be useful.

- Avoid writing the list of vocabulary on the board at the start of the unit and 'explaining' it. The vocabulary should be introduced as and when it arises. Word boxes are provided on each page of the Student Books and Practice Books with the key words for the lesson. This helps students associate the word or phrase with the concept and context.
- Before the lesson, check that you are confident with the pronunciation and spelling of the vocabulary that will be used. Write the vocabulary clearly on the board when you first introduce it in the lesson. If you think students may struggle to pronounce words, decide how best to model the pronunciation.
- Give students a chance to say a word once they have understood it. The most efficient way to do this is through repetition drilling.
- Use visuals whenever possible to reinforce students' understanding of the word.
- Ensure that students are recording the vocabulary systematically in their glossaries at the back of their Student Books, and, if possible, use a word wall that lists the vocabulary under unit or topic headings.
- Remember to use and revise the vocabulary.

Structures

In order for students to talk or write about their mathematics, they will need to go beyond vocabulary: they will also need to use those phrases and sentence frames that a particular task requires.

For example, they may need the following expressions in mathematics.

> X is the same as Y.
>
> The sides are the same length.
>
> The next number in the sequence.
>
> I predict that X will happen.
>
> If X happens, then Y happens.
>
> The next step is …

You need to build up banks of common mathematics phrases and encourage students to record them. This is an important part of identifying the language demands and providing the necessary support. You do not have to focus on grammar as the language can be taught as phrases rather than specific grammatical structures.

Using this Teacher's Guide

Every unit of the Teacher's Guide begins with useful background information that includes the following.

The Big idea: The main mathematical concept covered in the unit is outlined.

Look out for: This section focuses on tricky concepts that may need explaining prior to any learning taking place.

Common misconceptions: Common errors that students make, or misunderstandings that students have, are identified. This section offers advice on how to deal with these misconceptions.

Key vocabulary: This is a list of the key mathematical words used in the unit.

Coverage in lessons: The English National Curriculum objectives covered in the unit are listed.

Every lesson in the Student Book and Practice Book has corresponding lesson notes in the Teacher's Guide. These comprehensive lesson notes include the following.

A mini reproduction: This shows the relevant pages from the Student Book.

Global skills: These are the skills that aim to foster a classroom environment where students develop the skills for success. The skills are: *creative skills* where students are problem solving, investigating or exploring new maths content; *real-world skills* where students are taking part in research, or presenting and interpreting information, or if they are dealing with money and developing their financial literacy; *interpersonal skills* where students are practising their teamwork and communication, often through working in pairs or larger groups; and *self-development skills* where students have the opportunity to reflect on their learning and talk about what went well and what they are still uncertain about.

The key vocabulary and resources: Key vocabulary used in the lesson, and the concrete resources required for the activity, are listed.

Language support: This includes a range of strategies, including card sorts and card games, word walls, team games to define or explain words, use of similar words to explain meaning and exploration of the origins of words.

The key principles underpinning the language support are listed below.

Words should be introduced and explained carefully.
Words should be explained in context.
Repetition is vital.
Words should be linked to pictures or actions.
Students should develop their own glossaries.
The learning of mathematics vocabulary should be fun.
Language should not be a barrier to effective learning of mathematics.

Detailed lesson notes: Comprehensive lesson notes include an introduction activity and main activity. These notes refer to the Student Book and Practice Book, where relevant. The notes include probing questions for formative assessment, which are italicised. Icons are used to suggest the groupings that should be used at each point of the activity (whole class, small group, pairs, individual). A separate 'star-jump' icon indicates that the activities give students an opportunity for physical movement (standing up, jumping, moving around) rather than doing activities sitting down.

Differentiation: The Teacher's Guide offers strategies for you to *support* those students who may have difficulty accessing the task; to *consolidate* the learning for those students who need a little more practice; and to *extend* the learning for those who need more challenge.

The Teacher's Guide also offers differentiated outcomes. These outcomes are listed in the form of:

All students
Most students
Some students

Stretch zone: Each activity in the Student Book and the Practice Book has a Stretch zone question to support deeper learning. The Teacher's Guide provides additional notes on these activities.

Reflection time: Suggestions are made on how to bring the class back together to reflect on the learning and share ideas.

Answers: Answers to all the Student Book and Practice Book activities are provided.

Review pages: The Teacher's Guide has notes on the Review pages of the Student Book (summative assessment), with answers to the assessment questions, and the Practice Book (a formative, reflective review).

Digital resources: Where it is appropriate to use digital resources in a lesson, such as sharing the interactive Student eBook page on an interactive whiteboard (IWB), suggestions are embedded in the lesson plan.

Resources sheets: These photocopiable resources can be used with some of the main activities. They are referenced in the resources section of the lesson plan and are available on the Oxford Owl website (www.oxfordowl.co.uk).

Tour of a typical unit

Engage lesson

> The 'Big question' provides a discussion stimulus about the key idea of the unit.

1 Numbers and counting

? How do we use numbers?

In this unit you will:
- count, read and write numbers to 100
- count in twos, fives and tens
- know and make numbers using objects and pictures
- use words such as equal to, more than, less than (fewer), most, least
- read and write numbers from 1 to 20 in words.

> Learning objectives are stated clearly at the beginning of every unit.

Engage

Which numbers can you see in the classroom?

Which numbers can you see on your way to school?

What is the biggest number you have ever seen?

> Further questions allow students to develop communication skills.

> The Engage spread is bright and colourful, with artwork or photos to spark interest in young students and provide discussion points.

xi

Student Book Discover and Explore

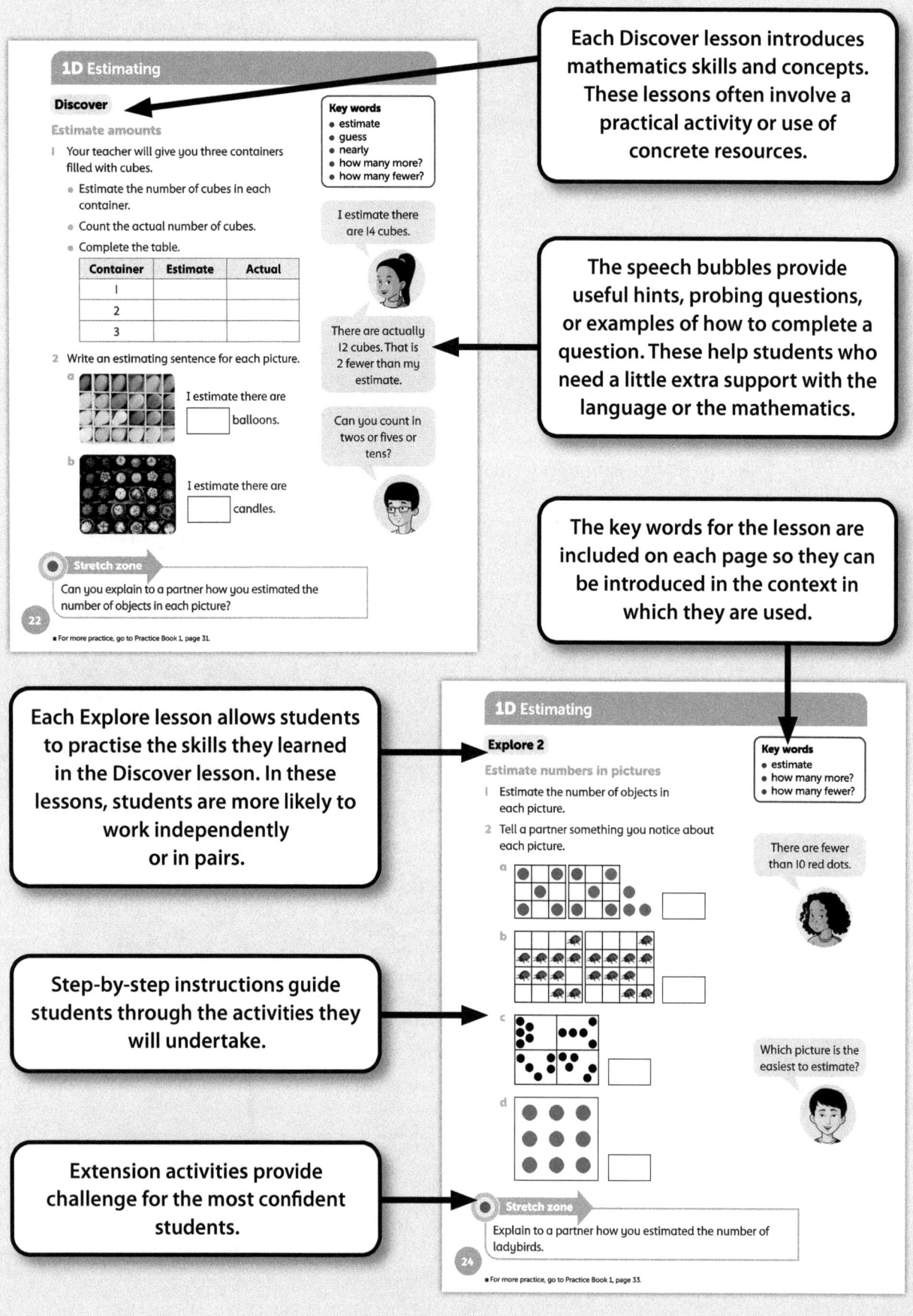

Connect lesson

The Connect lesson makes links between the different areas of mathematics in the unit.

1 Numbers and counting

Connect

Make a number poster

Work as a group.

1. Collect some magazines. Talk about which magazines might have numbers in them. What do the numbers tell us?

 We use numbers to count or to say how many of something there are.

The 'Big idea' sums up what students have discovered in the unit. It answers the Big question on the Engage page.

Connect activities are often set in real-life contexts to make the link between mathematics and the real world.

2. Cut out pictures that have numbers.

What is the biggest number on your poster?

What is the smallest number on your poster?

3. Make a poster to display in class.
4. Talk in your group about the numbers you have found.

Stretch zone

Take photographs of numbers on the way home from school. What job are the numbers doing? Explain your ideas to a partner.

A further extension activity provides a challenge for the most confident students.

Review lesson

1 Numbers and counting

Review

1 Draw the beads and write the numbers in the spaces.

Beads	Numbers	Words
	5	
 (16 beads)		sixteen
 (6 beads)		
		three
 (2 beads)		
	12	
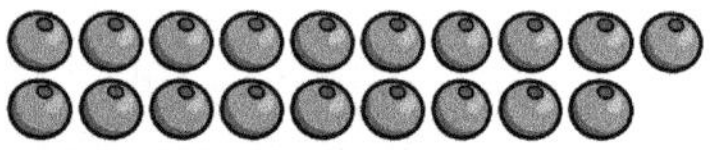 (19 beads)		nineteen
	1	
 (4 beads)		four
	14	
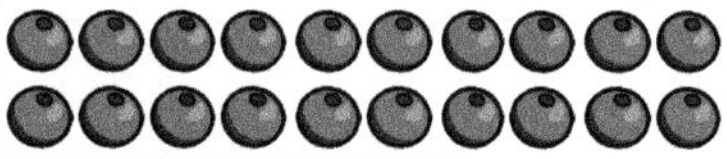 (20 beads)		twenty

2 Samir has a bracelet with 19 beads. Lina's bracelet has one more bead than Samir's. How many beads are on Lina's bracelet? ☐

Celine's bracelet has 10 more beads than Lina's. How many beads are on Celine's bracelet? ☐

26

> Students' progress is assessed through the questions and tasks at the end of each unit. In Student Books 2 and 6, these questions reflect the style of the SATs (national Standard Assessment Tests).

> A word problem is always included on the Review page.

Practice Book Discover and Explore

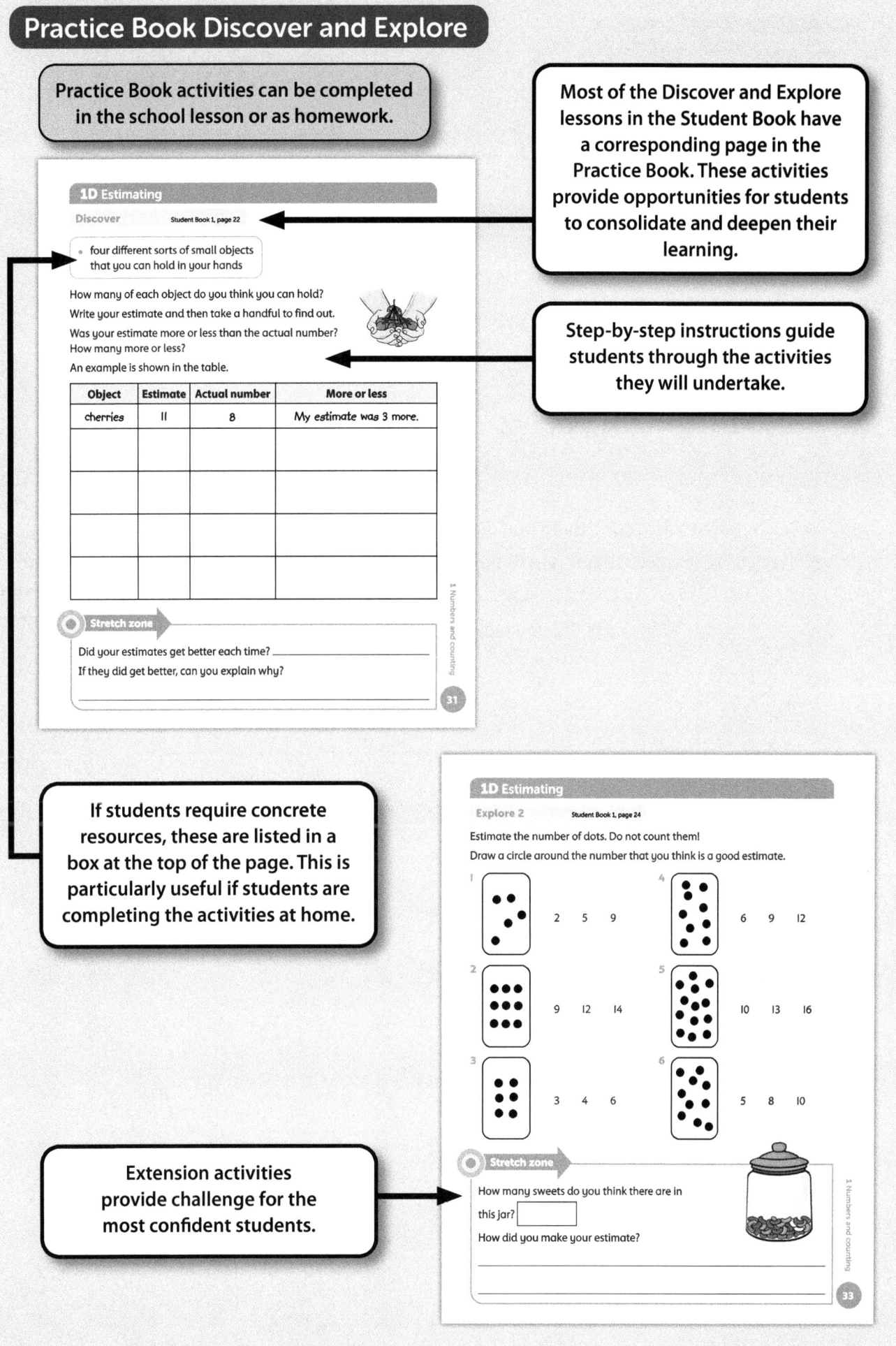

Practice Book activities can be completed in the school lesson or as homework.

Most of the Discover and Explore lessons in the Student Book have a corresponding page in the Practice Book. These activities provide opportunities for students to consolidate and deepen their learning.

Step-by-step instructions guide students through the activities they will undertake.

If students require concrete resources, these are listed in a box at the top of the page. This is particularly useful if students are completing the activities at home.

Extension activities provide challenge for the most confident students.

Practice Book Review

> Each Review page in the Practice Book includes a reminder of all the topics learned in the unit.

1 Numbers and counting

Review

 1 Draw a face next to each bubble to show how you feel about your learning.

- counting objects
- reading and writing numbers
- counting in twos, fives and tens
- estimating quantities

 2 Tell a partner about one thing you did really well in this unit.

3 Draw or write about things you found easy, challenging or really hard.

> Self-assessment activities help students to reflect on their learning.

What work did you feel confident doing?

What work was challenging?

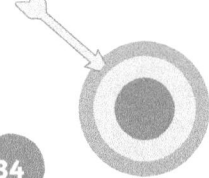
Is there any work you might need some extra help with?

Component overview

The Student Books

The Student Books are write-in textbooks for students to read and use. There are six Student Books: one for each school year at primary school. The Student Books introduce learning through a mixture of practical, discussion and independent activities.

Student Book	Typical student age range
Student Book 1	Age 5–6
Student Book 2	Age 6–7
Student Book 3	Age 7–8
Student Book 4	Age 8–9
Student Book 5	Age 9–10
Student Book 6	Age 10–11

The Practice Books

The Practice Books are write-in workbooks for students to read and use. There are six Practice Books: one for each school year at primary school. The Practice Books provide deeper learning opportunities through a range of independent activities, which can be completed in school or at home.

Practice Book	Typical student age range
Practice Book 1	Age 5–6
Practice Book 2	Age 6–7
Practice Book 3	Age 7–8
Practice Book 4	Age 8–9
Practice Book 5	Age 9–10
Practice Book 6	Age 10–11

 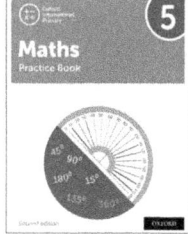 

The Teacher's Guides

There are six Teacher's Guides: one for each school year at primary school. Each Teacher's Guide includes:

- an introduction with advice about delivering mathematics in primary schools using *Oxford International Primary Mathematics*
- a unit overview, giving advice on teaching each unit, including common misconceptions and how to deal with them
- a lesson plan for every lesson in the Student Book and corresponding pages in the Practice Book
- model answers to each question in the Student Book and Practice Book.

Digital resources
Interactive eBooks

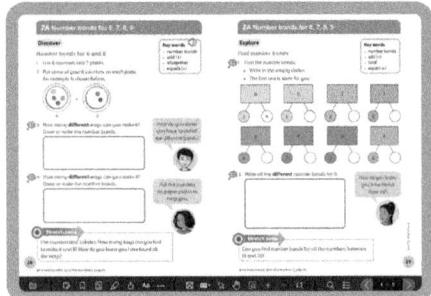

For the teacher

Teachers can access the Student Books, Practice Books and Teacher's Guides online in eBook format, on the Oxford Owl website (www.oxfordowl.co.uk).

The enhanced eBooks show the course content on screen, making it easier for teachers to deliver engaging lessons.

For the students

Teachers can allocate an eBook version of the Student Books to students for use at home. The Student eBooks include interactive activities, worksheets and audio of all the key vocabulary,

Assessment resources

The downloadable assessment materials offer you additional opportunities to assess students' progress. The materials include:

- end-of-unit summative assessment
- end-of-year summative assessment.

Every test comes with everything you need to assess and record progress including:

- answers
- mark schemes and guidance on assessment.

Oxford Primary Illustrated Maths Dictionary

The *Oxford Primary Illustrated Maths Dictionary* gives comprehensive coverage of the key maths terminology students use in the course. Entries are in alphabetical order, and each includes a clear and straightforward definition along with a fun and informative colour illustration or diagram to help explain the meaning. The dictionary is suitable for students with English as an Additional Language.

The curriculum

The Oxford International Curriculum offers a new approach to teaching and learning focused on wellbeing, which places joy at the heart of the curriculum and develops the global skills students need for their future academic, personal and career success.

Through six subjects – English, Maths, Science, Computing, Wellbeing and Global Skills Projects – the Oxford International Curriculum offers a coherent and holistic approach to ensure continuity and progression across every student's educational journey, equipping them with the skills to shape their own future. Through this approach, we can help your students discover the joy of learning and develop the global skills they need to thrive in a changing world.

1 Number and place value

Overview

Big idea

The main Big idea for this unit is place value. In the Hindu-Arabic numeral system we use, all numbers can be represented using just the ten digits from 0 to 9, so we count in base 10 – derived from the ten fingers on our hands. Numbers larger than 9 are made using powers of the base: tens, hundreds, thousands and so on.

This unit continues to develop students' understanding of the place-value system whereby the value of a digit depends on which position within a number it is written. When a digit moves one position to the left, its value increases ten times. When it moves one position to the right, its value decreases ten times. Using these conventions lies behind our counting and calculation methods. A key idea is that of exchange, so in calculations we want students to use the language of 'exchange one of these for ten of those' and so on. Place value is supported concretely using base-10 equipment or counters.

Look out for

- **Students who partition incorrectly because they do not understand the place value of digits**, for example a student who partitions 5042 as 5000 + 400 + 20. Using place-value cards to distinguish each digit can help with this, allowing them to see that the 4 represents 40 and the 2 represent 2 ones.
- **Students who struggle to build sequences where the numbers do not increase or decrease by the same amount from each number to the next.** Help them to calculate the differences between terms and look for a constant pattern in the differences.

Possible misconceptions

- **Students may think that to multiply by 10 you simply 'add a zero'.** This strategy only works for whole numbers. For example, when multiplying 12.7 by 10, simply 'adding a zero' gives the answer 12.70, which is the same number as the original. The correct answer (127) is ten times bigger. Encourage students to practise and develop their understanding. Say, *Multiplying by 10 makes the number ten times larger. Each digit moves to the left.* For example, for 21 × 10, 1 becomes 10 and 20 becomes 200.

Key vocabulary

- thousand, ten thousand, hundred thousand, 6/5/4-digit number, partition
- place holder, multiple, count on/back
- odd number, even number, negative number, positive number, below freezing
- decimal numbers, tenths, hundredths, thousandths, ten thousandths, hundred thousandths
- >, greater than, <, less than, round, rounding to the nearest …, approximately, ascending order, descending, order
- step size, sequence, number sequence, identify the rule, extend the sequence, rule of the sequence, pattern, generalisation

Coverage in lessons

Learning objective	E	1A	1B	1C	1D	1E	1F	1G	C	R
Read, write, order and compare numbers to at least 1 000 000 and determine the value of each digit.		✓		✓		✓		✓	✓	✓
Count forwards or backwards in steps of powers of 10 for any given number up to 1 000 000.	✓				✓					
Interpret negative numbers in context, count forwards and backwards with positive and negative whole numbers, including through zero.				✓	✓					
Round any number up to 1 000 000 to the nearest 10, 100, 1000, 10 000 and 100 000.			✓					✓	✓	✓
Solve number problems and practical problems that involve all of the above.				✓		✓		✓	✓	✓
Read Roman numerals to 1000 (M) and recognise years written in Roman numerals.							✓			

1 Number and place value

Engage Student Book page 6

Big question
- How can I extend my knowledge of place value so that I understand numbers up to 1 million?

Global skills
- **Creative skills:** problem solving
- **Interpersonal skills:** communication/teamwork
- **Self-development skills:** reflecting on learning

Key vocabulary
- hundred thousand, ten thousand, positive numbers, negative numbers, tenths, hundredths, greater than >, less than <, partition, rounding to nearest ... , sequence, multiple

Resources
- counting stick
- mini whiteboards and markers

Language support
Support students with the necessary language, providing definitions as necessary. Use phrases and questions that include the key vocabulary to reinforce it, for example:
- *A sequence is ...*
- *What sequence did you make?*
- *Continue my sequence: 0.25, 0.5 ...*
- *What is the next number in this sequence: 1.2, 1, 0.8?*

 **Introductory activity**

Say, *Choose a step size that you can count in. Count in this step size with your partner.* Students might choose numbers they have practised previously. Encourage them to think of others, for example: fifties, tens of thousands, millions. Ask pairs to share their ideas.

Show them a counting stick and say, *Zero is at this end. Counting in steps of 150, what number goes at the other end?* Together, count from zero to the tenth multiple to confirm and then back to zero. *What happens when we count back from zero?* Agree that when you count back from zero you count in negative numbers. Together, count back from zero to the 10th multiple and then forward to zero again.

 **Main activity**

Look together at page 6 of the Student Book. If you have access to an IWB, display the page. *What do you notice about the number labels on the counting stick? In what size steps are they counting in on this counting stick?* (twenty fives) *If you labelled the interval between each of these numbers what would they be?* (175, 225, 275, 325, 375) *What number label could you put at either end of the counting stick?* (125 and 425)

Ask students to work in pairs to complete each of the questions on page 6 of the Student Book. Encourage them to use whiteboards to explore the sequences. They should not expect to be able to find the sequences immediately. Discuss Vihaan's and Blaine's **sequences**:
- Vihaan started on ⁻140.
- Blaine's sequence was 150, 200, 250, 300, 350, 400. His step size was 50.

Differentiation
Supporting: Choose a start number and step size to get students started on making a number sequence.

Consolidating: Ask students to explain their thinking as they work out the sequences in the Student Book.

Extending: Set challenges that involve students creating sequences by counting forwards and backwards in fractions and decimals, including below zero, and Roman numerals.

 Reflection time

Ask pairs to write one of their sequences on the board. Discuss how to find the step sizes. Then invite the rest of the class to continue the sequence. They should also work in pairs to complete the sequences on their whiteboards.

Unit 1 Number and place value

1A Place value

Discover Student Book page 7 • Practice Book page 14

Specific learning focus
- Know what each digit represents in 5- and 6-digit numbers.
- Be able to pronounce the number names correctly.

Global skills
- **Creative skills:** exploring
- **Real-world skills:** interpreting information
- **Interpersonal skills:** communication
- **Self-development skills:** reflecting on learning

Key vocabulary
- hundred thousand, ten thousand, thousands, hundreds, tens, ones

Resources
- Resource sheet 1.1: place-value table

Language support

Listen to students say the numbers that they made up. Encourage them to say them in full and not to say each digit separately, for example:

456 341 is 'four hundred and fifty-six thousand, three hundred and forty-one'.

Model this in discussions with students.

Introductory activity

Ask five students to give you a digit 0–9. As they do, write the digits on the board, writing each new digit on the right-hand end of the previous number, building to a 5-digit number. Ask the class to tell you what the number is. For example: starting with the digit 4 – *four*, 42 – *forty-two*, 425 – *four hundred and twenty-five*, 4253 – *four **thousand** two hundred and fifty-three* and so on. Point to each digit in turn and ask students to write down what its value is. For example, for the 2 in 42 537 in the thousands place, they write 2000.

Now repeat the activity but 'grow' the number from the left-hand end, for example 46, 846, 1846, 51 846. Complete with saying each number aloud and writing the value of each digit.

Main activity

Ask students to count in hundreds from the number you have made at the start of the Introductory activity until they cross the hundreds boundary into the next thousand. Record the count on the board so that students can see and explain the pattern.

Do this for counting in thousands and tens of thousands. Repeat this process several times. Include 6-digit numbers.

Look at page 7 of the Student Book together. Ask questions to prompt students to reflect and share what they know about place-value tables, for example:

- *How much do the numbers increase as you move down the first column? How much do they increase as you move down the second column?*
- *What happens to the numbers in a row as you move from right to left?*

Students should work in pairs to complete the questions on page 7 of the Student Book. They can use Resource sheet 1.1 as they work.

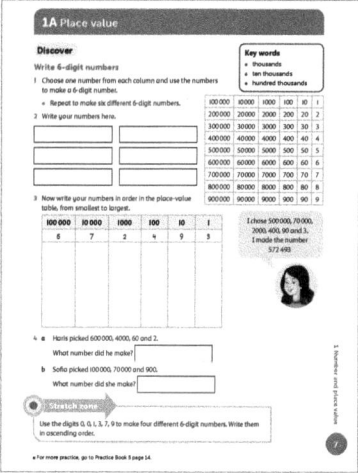

Differentiation

Supporting: Ask students to read their numbers to you so that you can support them in saying the numbers correctly.

Consolidating: Encourage students to support other students who are less confident.

Extending: Ask students to count on in hundreds and thousands aloud from the numbers they are making.

Stretch zone: *Use the digits 0, 0, 1, 3, 7, 9 to make four different 6-digit numbers. Write them in ascending order.*

Check that students are clear on what ascending means before they begin. You could also ask students to make up 5-digit numbers. Then ask them to count on 12 times in steps of 100, 1000 or 10 000 and to write the new number.

Unit 1 Number and place value

Reflection time

Ask pairs to say one of the numbers that they made up. Write them in a random order on the board. Ask them to tell you a way to find out which is the highest and which is the lowest. Accept any solutions that are correct. Ask students to do this on their whiteboards or paper. This will help you to assess their understanding.

Practice Book: Students complete Practice Book page 14. They can do this directly after the Main activity, as homework, or as the focus of a separate mathematics session to help students consolidate their learning and build fluency. Confirm that children can say how many digits there are in 1 million (7).

Differentiated outcomes	
All students	should create and say 6-digit numbers correctly with support.
Most students	will create and say 6-digit numbers correctly and be able to explain how they know they are correct.
Some students	may understand place value in numbers beyond 6-digits.

Answers

Student Book page 7

Answers will vary because students make their own numbers and order them. For example:

1 and 2   247 315

819 637

593 728

725 491

461 826

3   247 315, 461 826, 593 728, 725 491, 819 637

4   Haris: 604 062

Sofia: 170 900

Practice Book page 14

Students choose digit cards to make their own 6-digit numbers, so answers will vary. Check that they have written the numbers in order correctly, starting with the smallest, and that they have written the correct number names in words.

Stretch zone: 1 023 456 – one million, twenty-three thousand, four hundred and fifty-six.

1A Place value

Explore   Student Book pages 8–9 • Practice Book page 15

Specific learning focus

- Partition any number up to one million into thousands, hundreds, tens and ones.

Global skills

- **Creative skills:** investigating
- **Real-world skills** presenting information
- **Interpersonal skills:** communication

Key vocabulary

- hundred thousand, ten thousand, thousand, hundred, tens, ones, partition, >, <

Resources

- 0–9 digit cards
- place-value counters
- place-value tables (optional)

Language support

Start a working wall to display examples of 6-digit numbers students have made, with the words for how to say them aloud next to them. For example, display, 926 043: nine hundred and twenty -six thousand and forty-three.

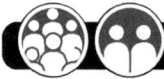 **Introductory activity**

Use a set of digit cards. Ask a student to choose and place six digit cards on the table. Write on the board three different numbers that can be made with the six digits. Ask students to look at the numbers and take turns to say them aloud to a partner. Ask different students to tell you, for example, digits that show the hundreds and ten thousands. Ask them to write these in numbers, for example 6000. Next, ask them to swap different digits and to read the new number.

Ask, *Is it bigger or smaller than the previous number? By approximately how much?*

For example:

- if they swap the 3 and 0 (926 043)

the number is smaller by approximately 300

- if they swap the 2 and 6 (962 043)

the number is bigger by approximately 40 000.

Ask them to explain their thinking at each step.

4   Unit 1 Number and place value

Discuss the role of the zero as a place holder.

Ask, *Remove the zero. What is the number without the zero?*

Repeat this process with different students.

 **Main activity**

Point out the speech bubbles on page 8 of the Student Book. Can students see how the digits chosen have been listed for their values in the table and then set out with place-value counters in question 2? Ask students to work in pairs and to make a 5-digit number using their digit cards. Ask them to write this down and then to **partition** it, for example:

86 457 = 80 000 + 6000 + 400 + 50 + 7

For each number made, pairs set it out using place-value counters and say the number aloud. The example above would be *eighty-six thousand, four hundred and fifty-seven* represented with eight 10 000 counters, six 1000 counters, four 100 counters, five 10 counters and seven 1 counters.

Students should record how many of each counter they used in the table. When they have made three or four more examples, ask them to do the same for five 6-digit numbers. At each stage, students should read their numbers to their partner.

Ask questions such as:
- How can you partition your 5-digit number?
- Which is the hundreds digit?
- The tens of thousands digit is …
- Is this number bigger or smaller than your previous number? By approximately how much?

For the second part of the activity in the Student Book, students need to write numbers in words. Encourage them to read the numbers aloud and to check with their partner. Remind them of the conventions of writing numbers in words, for example placing a comma after the thousands and hyphenating any 2-digit numbers greater than 20 that we say, such as eighty-two thousand, six hundred and twenty-one.

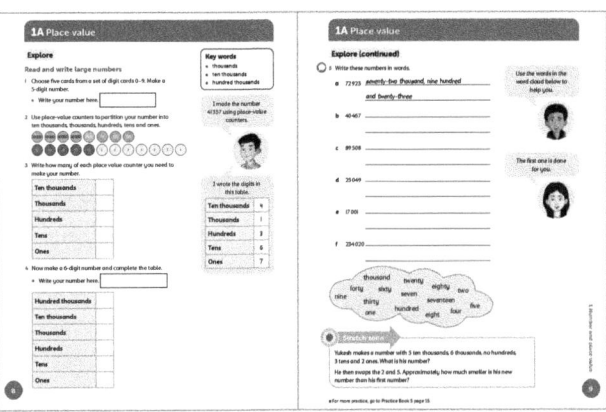

Differentiation

Supporting: Provide students with a place-value table.

Consolidating: Ask students to explain the values of the digits in their numbers.

Extending: Challenge students to think of the values in different ways. For example, in 41 357 there are 5 tens, but there are also 35 tens, or 135 tens and so on.

Stretch zone: *Yukesh makes a number with 5* **ten thousands***, 6 thousands, no hundreds, 3 tens and 2 ones. What is his number? He then swaps the 2 and 5. Approximately how much smaller is his new number than his first number?*

Check that students record the number as 56 032 and that if you swap the 2 and the 5 the new number is 26 035, which is approximately 30 000 smaller.

Ask students to explore the effect of swapping two digits in a 5-digit number.

Reflection time

Invite individual students to write one of their numbers on the board. Ask them to point to each digit in turn starting from the left-hand digit. The class says what the digits are. Then ask the class to partition that number. Finally, discuss the Stretch zone question. Ask, *What happened when Yukesh swapped the two digits? Approximately how much smaller/bigger are they? How did you work that out?*

Practice Book: Students complete Practice Book page 15. They can do this directly after the Main activity, as homework, or as the focus of a separate mathematics session to help students consolidate their learning and build fluency. Ask students to tell you what the symbols < and > represent. Discuss how you complete a statement that includes more than one inequality symbol.

Differentiated outcomes	
All students	should create and say 6-digit numbers using a place-value table and counters, for example, for support.
Most students	will create, say, write in words and partition 6-digit numbers.
Some students	may be able to partition in more than one way and be able to describe how a number changes when the place of its digits are changed.

Unit 1 Number and place value

Answers

Student Book pages 8–9

Answers will vary because students make their own 5- and 6-digit numbers. For example, for page 8:

1. 63 281

2. (Students use place-value counters to partition their number.)

3.
Ten thousands	6	Tens	8
Thousands	3	Ones	1
Hundreds	2		

4. 804 953

Hundred thousands	8	Hundreds	9
Tens of thousands	0	Tens	5
Thousands	4	Ones	3

5. **b** forty thousand, four hundred and sixty-seven
   c eighty-nine thousand, five hundred and eight
   d twenty-five thousand and forty-nine
   e seventeen thousand and one
   f two hundred and thirty-four thousand and twenty

Practice Book page 15

Answers will vary because students choose digits 0–9 to make their own decimal numbers to use for number sentences. They then use the digits to find specified numbers.

Stretch zone: One hundred and twenty-three thousand, four hundred and fifty-six.

Answers will vary depending which digits are swapped.

1B Rounding

Discover Student Book page 10 • Practice Book page 16

Specific learning focus
- Round 4-digit numbers to the nearest 10, 100, 1000, 10 000 or 100 000.

Global skills
- **Creative skills:** problem solving/exploring
- **Real-world skills** presenting information
- **Interpersonal skills:** communication/teamwork

Key vocabulary
- hundred thousand, ten thousand, greater than >, less than <, rounding to the nearest …

Resources
- 0–9 digit cards
- Resource sheet 1.2: number lines

Language support

Provide students with sentence frames to support them in their discussions of rounding, for example:
- ___ rounded to the nearest ___ is ___.
- I rounded down because ___ was closer to ___.
- I rounded up because ___ was closer to ___.

Introductory activity

Draw a number line on the board with 0 at one end and 100 at the other. Mark the tens divisions.

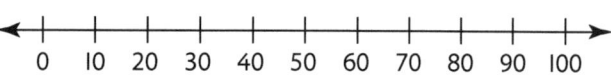

Invite individual students to position different numbers on the line, starting with a two-digit number. As they place their number ask, *Which ten is nearest to your number? How do you know?* Ask students to mark the midpoint between the tens for their number to show how they can tell which ten is nearer. For example, if the student chooses 57, ask, *Is 57 nearer to 50 or to 60?* (60) *Can you tell me why?* (Students mark 55 as the midpoint between 50 and 60, then say that 57 is nearer to 60 as it is past the midpoint.)

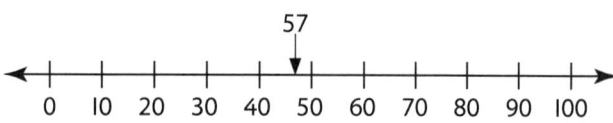

Main activity

Extend the Introductory activity with 5-digit numbers, with students placing them on a 0 to 100 000 number line with the ten thousand divisions marked. Build up to this using number lines to 1000 and 10 000 first, if appropriate. Ask similar questions to enable students to describe the positions of the numbers, for example: *Which pair of ten thousand numbers does your number fall between? Which ten thousand is nearest your number?*

Unit 1 Number and place value

Students should then work in pairs on the Student Book activities. When answering the word problems, they take it in turns to round a number and explain to their partner why they are rounding up or down.

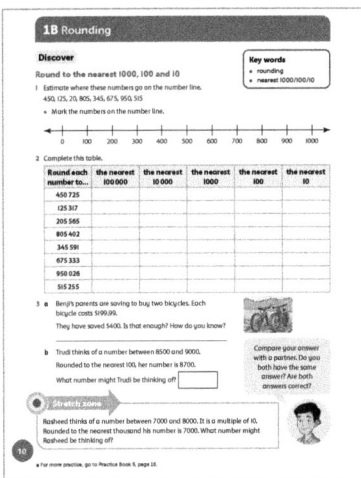

Differentiation

Supporting: Support students to come to an understanding of rounding 6-digit numbers by using number lines such as those on Resource sheet 1.2.

Consolidating: Ask students to explain their strategies for creating and placing each number on a number line and rounding.

Extending: Ask students to create their own rounding word problems.

Stretch zone: *Rasheed thinks of a number between 7000 and 8000. It is a multiple of 10. Rounded to the nearest thousand his number is 7000. What number might Rasheed be thinking of?*

Students should identify possible number (that is, multiples of 10 from 7010 to 7490).

 Reflection time

Invite students to share their answers. Ask: *How did you round the numbers in the table?* The rest of the class check their work to make sure that they agree. Discuss the possible answers to the word problems. To finish the lesson, choose some pairs who completed the extension activity to share their word problems. Students should work on these in pairs and the pair who created the problem should model the answer at the front of the class.

Practice Book: Students complete Practice Book page 16. They can do this directly after the Main activity, as homework, or as the focus of a separate mathematics session to help students consolidate their learning and build fluency.

Tell students that it is fine for them to draw additional number lines on a separate sheet of paper for each number to support them in rounding to the nearest 10, 100 and 1000.

Differentiated outcomes	
All students	should round accurately by using a marked number line.
Most students	should round accurately by using a number line.
Some students	may begin to round without a number line.

Answers

Student Book page 10

1 Check that students have written the numbers on the number line in the correct places.

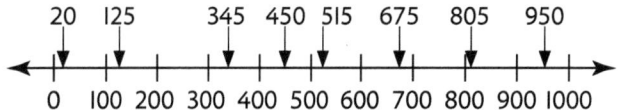

2	Round to nearest 100 000	Round to nearest 10 000	Round to nearest 1000	Round to nearest 100	Round to nearest 10
450 725	500 000	450 000	451 000	450 700	450 730
125 317	100 000	130 000	125 000	125 300	125 320
205 565	200 000	210 000	206 000	205 600	205 570
805 402	800 000	810 000	805 000	805 400	805 400
345 591	300 000	350 000	346 000	345 600	345 590
675 333	700 000	680 000	675 000	675 300	675 330
950 026	1 000 000	950 000	950 000	950 000	950 030
515 255	500 000	520 000	515 000	515 300	515 260

3 a Benji's parents have enough money because $199.99 rounded is $200 and two lots of $200 is $400. $199.99 is just less than $200 so they have enough.

b Trudi's number could be between 8650 and 8749.

Practice Book page 16

Answers will vary because students choose digit cards to make their own 4-digit numbers. Check that students have written their numbers in the table in order, starting with the smallest, and that they have rounded them to the nearest 10, 100 or 1000 correctly.

Stretch zone: Check that students have marked their chosen numbers correctly on the number line.

Unit 1 Number and place value

1B Rounding

Explore Student Book page 11 • Practice Book page 17

Specific learning focus
- Round a number with up to 6 digits to the nearest 100 000.

Global skills
- **Creative skills:** problem solving/exploring
- **Interpersonal skills:** communication/teamwork

Key vocabulary
- tenths, hundredths, thousandths, ten thousandths, hundred thousandths, rounding to the nearest ten/hundred/thousand/ten thousand/hundred thousand

Resources
- 0–9 digit cards, two sets per pair
- counters of two different colours per pair
- mini whiteboards and markers

Language support
Display examples of how to use a number line to round to various multiples of 10. Include thought and speech bubbles with text explaining the thinking. Ask, for example, *How do I round 722 to the nearest 100? Which hundreds numbers does 722 come between? 700 and 800. Next I look at the tens. Are they closer to 700 or 800?*

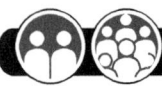 **Introductory activity**

Set the following problem.

Sam wants to buy a car for $9599 and a set of wheels for $1895. Approximately how much money does he need?

Ask students to discuss with a partner how to find the approximate amount and ask them to calculate an approximate answer. Ask pairs to share their strategies, for example rounding to the largest value digits ($10 000 and $2000) and use these responses to bring the discussion back to rounding to make estimates.

 **Main activity**

Make a 4-digit number, using digit cards. Ask pairs to round these to the nearest 1000 and write their answers on their whiteboards.

Draw a number line from 0 to 10 000 on the board, marked in thousands. Invite students to say their 4-digit number. You can then point to the line and ask questions to pinpoint the number's location. For example, if you have 3751, point to 2000 and 3000 and ask, *Does it go between these? Why not?* (It is greater than 3000.) Then point to 3000 and 4000 and ask the same again. (Yes, it is greater than 3000 but less than 4000.) Ask, *Which is it nearer to?* (4000) *How do you know?* (It is larger than 3500, which is the midpoint.) Repeat with other numbers.

Ask students questions to prompt the appropriate language, for example:
- *What does it mean to round? Can you explain in a different way?*
- *Why is rounding useful?*
- *What is the approximate answer to this calculation: 319 + 293?*
- *How did you come to that solution?*

Model the rules for the game on page 11 of the Student Book by playing three rounds with one of the more-confident students. After each turn, students explain why they are able to place a counter on the number. Students play the game in pairs. While students are working, ask them to explain how they choose the star numbers for a particular whole number. Ask: *What are the rules for rounding a number with 3 digits to the nearest 10?*

When they have successfully played this game, ask each pair to join with another pair. They create a 'board' to play the same game but with numbers having 6 digits and a choice of rounding to the nearest 10, 100, 1000 or 10 000, determined by choosing a card labelled with these powers of 10 or perhaps a spinner. Select one of these game boards to use for reflection time.

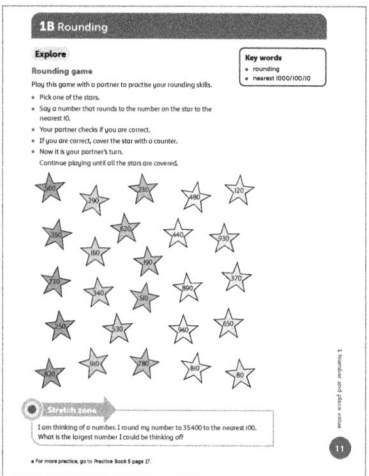

Differentiation

Supporting: Encourage students to continue to use a number line to support them in rounding to the nearest multiple of 10. Students may benefit from focusing on rounding to multiples of 10 for a time. Allow them to adapt the game to, for example, round all their numbers to the nearest 10 000 rather than changing each at each turn.

Consolidating: Ask students to explain their strategies for rounding.

Extending: Challenge students to design a points system for the board game to reward correct rounding.

Stretch zone: *I am thinking of a number. I round my number to 35 400 to the nearest 100. What is the largest number I could be thinking of?*

Students should use their knowledge of rounding to identify the largest possible number as 35 449. They could also make up similar problems, work out the possible answers and then share with another student to solve.

 Reflection time

Using a board designed by one of the small groups, play a game involving rounding numbers with 5 digits. Groups then exchange game boards and play the game. They should give feedback on the game boards using the 'two stars and a wish' strategy, saying two things they liked about the game and one thing they would like to see done differently. They may wish for easier or harder numbers, or more or fewer numbers. Choose one board and then play together, encouraging students to explain their thinking.

Practice Book: Students complete Practice Book page 17. They can do this directly after the Main activity, as homework, or as the focus of a separate mathematics session to help students consolidate their learning and build fluency. If possible, encourage students to locate the cities on a map and do some research on one or more cities, focusing in particular on numerical data. For example, Kuching has an area of 450 km², whereas Mataram has an area 61.3 km².

Differentiated outcomes	
All students	should be able to round 5-digit numbers to the nearest 10 000 with support.
Most students	should be able to round 5-digit numbers to the nearest 10 000.
Some students	may begin to round 6-digit numbers to the nearest 1 000 000.

Answers

Student Book page 11

Answers will vary because students choose different numbers to round to in the game.

Practice Book page 17

	City	Population	Nearest 1000	Nearest 10 000	Nearest 100 000
1	Mataram	506 635	507 000	510 000	500 000
2	Hue	401 189	401 000	400 000	400 000
3	Kuching	613 486	613 000	610 000	600 000
4	Lampang	449 892	450 000	450 000	400 000
5	Bacolod	614 635	615 000	610 000	600 000
6	Vientiane	684 486	684 000	680 000	700 000
7	Padang	982 378	982 000	980 000	1 000 000
8	Nonthaburi	964 933	965 000	960 000	1 000 000
9	Vung Tau	419 784	420 000	420 000	400 000

Stretch zone: Difference is 234 594

Nearest 100 000 is 200 000

Nearest 10 000 is 230 000

Difference is 30 000

1C Ordering and comparing

Discover Student Book pages 12–13 • Practice Book page 18

Specific learning focus
- Order and compare numbers up to a million using the > and < signs.

Global skills
- **Creative skills:** problem solving/investigating
- **Interpersonal skills:** communication/teamwork

Key vocabulary
- hundred thousand, ten thousand, greater than, >, less than, <

Resources
- mini whiteboards and markers
- 0–9 digit cards
- paper
- scissors

Language support
- Add more larger numbers to the wall display, showing how to read them out loud, for example 27 858 *twenty-seven thousand eight-hundred and fifty eight*. Include guidance on how to read larger numbers, such as grouping the digits in threes from the right-hand end to separate the millions and the thousands.

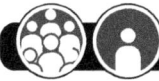 **Introductory activity**

Select four digit cards. Ask students to make two different 4-digit numbers using these digits and place them in order on their whiteboards, smallest first. For example, if you pick 2, 4, 5, 9 they can make: 2495 and 4925.

Unit 1 Number and place value

Now ask students to use either the < or the > sign between the numbers. Ask students which is correct and why. (< which means 'is **less than**')

2495   <   4925

Repeat with further 4-digit numbers but sometimes place the largest number first (> means 'is **greater than**'). For example:

6381   >   3861

Ask students to recap what the symbols > and < mean and to give examples of both.

Main activity

Tell students that they are going to work in pairs ordering and comparing numbers. With the class, read through the instructions on page 12 of the Student Book. Discuss what negative numbers are (numbers less than zero) and how they are written (with a negative sign to show that they are different from numbers greater than zero).

They should work together to choose and order their numbers and share their thinking, but record their answers individually. Encourage them to say each number aloud correctly throughout the activity. You can also vary the number of digits that students use as well as the quantity of numbers they work with.

For page 13 of the Student Book, students will need to identify the relative size of numbers in word problems. Ask students what sort of calculation is needed when the question asks 'by how many?' or 'by how much?' and agree that they are finding a difference, which is done by using subtraction.

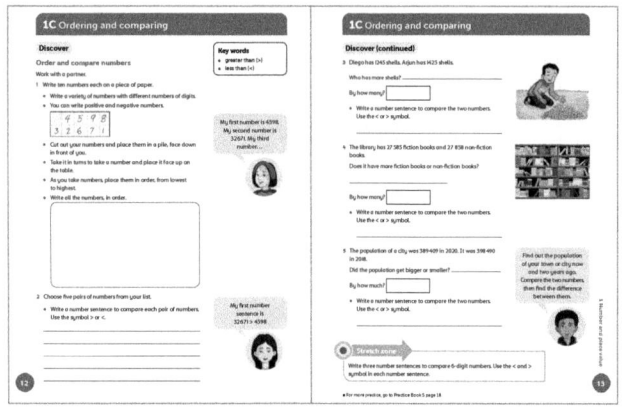

Differentiation

Supporting: Encourage students to use a place-value table and counters for support.

Consolidating: Encourage students to include 6-digit numbers.

Extending: Challenge students to work with numbers of 6 or more digits.

Stretch zone: *Write three number sentences to compare 6-digit numbers. Use the < and > symbol in each number sentence.*

For example, students could write 472 481 < 472 479 or 713 932 > 713 928. Ask them to read each of the numbers aloud and explain how they know that one number is greater than/less than the other.

Reflection time

Ask individual students to share their number sentences. Ask a question that prompts students to reflect on what they have learned in the lesson, for example: *The greater the first digit in a number, the greater the number. Is this always, sometimes or never true?*

Ask students to share their strategies and solutions for question 5 about population. Some students may have looked at the change in the population figures in the order they are written, but missed that the years were given in reverse order.

Practice Book: Students complete Practice Book page 18. They can do this directly after the Main activity, as homework, or as the focus of a separate mathematics session to help students consolidate their learning and build fluency.

Read through the instructions together and model how to complete each section as necessary. Students will need a set of digit cards or alternatively could close their eyes to point at random to the digit cards on page 18.

Differentiated outcomes	
All students	should order 5-digit numbers with place-value tables and counters for support.
Most students	will order 6-digit numbers, using < and > symbols.
Some students	may order positive and negative numbers up to 6 digits, using < and > symbols.

Answers

Student Book pages 12–13

1 and 2 Answers will vary because students make their own numbers. Check that they have been written in order from lowest to highest correctly.

Where students have used the > and < symbols to compare numbers, check that their comparisons are correct.

3 Arjun has 180 more shells. 1245 < 1425 or 1425 > 1245

4 The library has 273 more non-fiction books. 27 585 < 27 858 or 27 858 > 27 585

5 The population got smaller by 9081. 389 409 < 398 490 or 398 490 > 389 409

Practice Book page 18

Answers will vary because students use digit cards to make their numbers.

Check that the number sentences they have written using the < and > symbols are correct.

Check that students have followed the instructions and made the numbers that are closest to the specified number.

Stretch zone: 817 503

1C Ordering and comparing

Explore Student Book page 14 • Practice Book page 19

Specific learning focus
- Order and compare negative and positive numbers on a number line and temperature scale.
- Calculate a rise or fall in temperature.

Global skills
- **Creative skills:** exploring
- **Real-world skills** interpreting information
- **Interpersonal skills:** communication/teamwork

Key vocabulary
- ascending order, descending order, positive numbers, negative number, below freezing

Resources
- mini whiteboards and markers
- 0–9 digit cards

Language support

Explain that negative temperatures are described in different ways. English speakers may say, for -20 °C, 'twenty degrees below freezing', 'minus twenty degrees below freezing' (often leaving out the the word 'degrees') or 'negative 20 degrees'. 'Temperature' and 'thermometer' can be challenging to pronounce accurately. Model how to say both, clearly and precisely or play an aural example from the internet. Support students with the specific language of ordering and comparison by using it in different questions and statements. Say, for example, ... °C is colder/hotter than .... °C. 35 °C was the hottest temperature reached this summer.

Introductory activity

Ask a student to select five digit cards. Write the digits on the board. Ask students, working in pairs and using their whiteboards, to make the largest number they can, the smallest number they can and three numbers in between. Ask students to order the numbers in **ascending order**. Then they should order the numbers in **descending order**.

Come together as a class to share their answers and order the numbers in both ascending and descending order. Ask questions to encourage students to explain their thinking, for example: *How did you know which of these two numbers was larger? Did you need to look at all the digits to know which number was smaller? What does the 0 in this number tell us?*

Main activity

Ask students to look at the thermometers on page 14 of the Student Book. Ask them to work in pairs to make five statements about the thermometers, for example 'The hottest temperature the scale on these thermometers can record is 30 °C, and the coldest is -30 °C.' Share the statements as a class and discuss typical temperatures for where they live and what time of year is the coldest and hottest.

Write the following table on the board:

Place	Sahara Desert	North Pole	Bangkok	Moscow	Dubai
Average temperature in December	35 °C	-20 °C	24 °C	-8 °C	26 °C

Ask pairs to order the temperatures from coldest to warmest by placing them on a number line. When you take feedback, model the different ways to say the temperatures (for example either 'twenty degrees **below freezing**' or 'minus twenty degrees'). Ask questions so that students have a chance to talk about how they know -8 is warmer than -20, for example. Agree that negative temperatures are colder than positive ones and that the closer a negative temperature is to 0 °C, the warmer it is, so -8 is warmer than -20.

Explain the activity on page 14 of the Student Book. Ask students to look at the speech bubbles with examples of negative numbers being compared using < and >. Ask extension questions, for example: *Can you tell me another temperature that would be less than 30 degrees Celsius? Can you tell me another? Can you tell me one greater than -5 degrees Celsius but less than 5 degrees Celsius?*

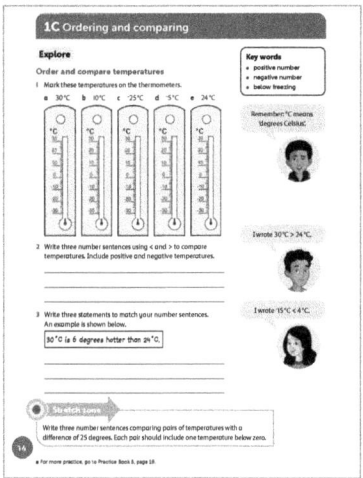

Differentiation

Supporting: Help students to order positive and negative numbers, using < and > symbols and a number line, using the thermometer as a vertical number line.

Consolidating: Ask students to explain how to order positive and negative numbers, using < and > symbols and a number line. Ask, *Which is warmer, -10 or -4? Why?*

Unit 1 Number and place value

Extending: Ask students to order positive and negative numbers, using < and > symbols, with up to 5 digits.

Stretch zone: *Write three number sentences comparing pairs of temperatures with a difference of 25 degrees. Each pair should include one temperature below zero.*

You could ask students to draw thermometers and place their temperatures onto them. They could then find the differences between pairs of temperatures. They could also calculate rises and falls in temperature. For example, you could ask: *It was ⁻4 °C this morning. The temperature has risen by 12 degrees. What is the temperature now?*

Reflection time

Invite pairs to share their number sentences. Ask other students to check to make sure that the number sentences make sense. Review the meanings of ascending and descending orders. To finish the lesson, write sets of numbers (including negative numbers) on the board for students to order in both ways.

Practice Book: Students complete Practice Book page 19. They can do this directly after the Main activity, as homework, or as the focus of a separate mathematics session to help students consolidate their learning and build fluency.

Point out that in question 1, average temperature in January means a typical January temperature.

Differentiated outcomes	
All students	should order positive and negative numbers on a number line with support.
Most students	will write a range of number sentences using positive and negative numbers.
Some students	will find temperature differences across zero.

Answers

Student Book page 14

1. Check that students have shown the given temperatures on the correct divisions on the thermometers.
2. Check that students' number sentences using < and > are correct.
3. Check that the statements match the number sentences.

Practice Book page 19

1. Check that the temperatures have been shown correctly on the thermometers. While they are working, ask students to explain how they know where to draw the mercury line.
2.
   a It is 29 degrees colder in Berlin than Bangkok.
   b It is 20 degrees warmer in Cairo than Helsinki.
   c It is 9 degrees colder in Cairo than Nairobi.
   d It is 8 degrees warmer in Berlin than Moscow.
   e It is 5 degrees colder in Nairobi than Bangkok.
3. Check that students' facts are correct.

Stretch zone: 4.5 °C.

1D Number sequences

Discover 1 Student Book page 15 • Practice Book page 20

Specific learning focus
- Recognise and extend number sequences.

Global skills
- **Creative skills:** investigating
- **Self-development skills:** reflecting on learning

Key vocabulary
- number sequence, count on/back in tens, hundreds, thousands, rule of the sequence, pattern

Resources
- 0–9 digit cards

Language support
Create a display with examples of sequences and labelled with key vocabulary such as 'difference', 'starting number' and 'linear'.

 **Introductory activity**

Write the following **number sequence** on the board: 4, 6, 9, … . Ask pairs to predict the next number. Then write 13 as the next number in the sequence. Ask pairs whether they were correct. If so, how did they know? Agree that there was a **pattern** to the sequence of numbers. Explain that they can describe this as the **rule of the sequence**. Ask them to tell you the rule (+ 2, + 3, + 4).

Repeat for the following sequence: double then subtract 1, double then subtract 2, double then subtract 3.

 Main activity

Ask students to look at the activities on page 15 of the Student Book. First, they will make two 2-digit numbers from a pair of digit cards and use these as the start numbers for each line of sequences. Each sequence is made by using the same rule on each number to get the next number. The rule in these cases will always be adding the same amount. Carefully go through the example, then ask students to complete the sequences in the tables.

In question 2, they start with two 3-digit numbers. Look together at how the sequences are the same and different for a number as well as how to predict and possibly calculate new numbers in the sequences.

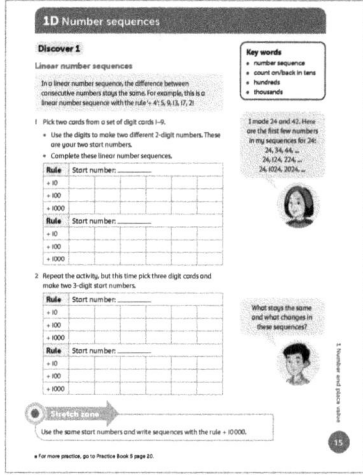

Differentiation
Supporting: Give students a place-value table to help continue each sequence.

Consolidating: Ask students to explain the pattern of change in their sequences.

Extending: Ask students to create sequences using two operations, for example 'multiply by 3 and add 2'.

Stretch zone: *Use the same start numbers and write sequences with the rule + 10 000.*

Check that students have completed the sequences correctly. Ask them to describe what stays the same and what changes in their sequence and predict what number will be in the *n*th position of their sequence.

 Reflection time

Invite pairs of students to share their sequences. Ask other students to identify the rule of the sequence and extend the sequences. To finish the lesson, write the sequence 18, 27, 36 on the board. Discuss with students how they can predict the next number and then the next and so on. Ask pairs to work out the tenth number in the sequence. They may realise that the sequence begins on the second multiple of 9, so the tenth number is the eleventh multiple of 9. Repeat for other similar sequences.

Practice Book: Students complete Practice Book page 20. They can do this directly after the Main activity, as homework, or as the focus of a separate mathematics session to help students consolidate their learning and build fluency. Explain to students that they are counting back as well as on in some sequences.

Differentiated outcomes	
All students	should create simple sequences, using a 2-digit starting number and an addition rule with support.
Most students	will create simple sequences, using a 2- or 3-digit starting number and an addition rule.
Some students	will create simple sequences using an addition rule and predict further numbers in the sequences.

Unit 1 Number and place value

Answers

Student Book page 15

1 Answers will vary because students use digit cards to pick the start number in their sequences.

For example, if they choose 4 and 7, they make 47 and 74 and the sequences for question 1 will be:

57, 67, 77, 87, 97, 107

147, 247, 347, 447, 547, 647

1047, 2047, 3047, 4047, 5047, 6047

84, 94, 104, 114, 124, 134

174, 274, 374, 474, 574, 674

1074, 2074, 3074, 4074, 5074, 6074

Similarly, they will record a sequence for 3-digit numbers in question 2.

Practice Book page 20

	Rule	Start number	Sequence			
1	count on in 150s	150	300	450	600	750
2	count on in 15s	150	165	180	195	110
3	count back in 15s	250	235	220	205	190
4	count on in 70s	250	320	390	460	530
5	count on in 110s	95	205	315	425	535
6	count back in 15s	95	80	65	50	35
7	count back in 15s	75	60	45	30	15
8	count on in 110s	75	185	295	405	515
9	count on in 400s	40	440	840	1240	1640
10	count on in 450s	40	490	940	1390	1840

Stretch zone: Check that students' descriptions are accurate. Students should notice, for example, that they are always counting on and back from a multiple of 5 and that each number in the sequence is also a multiple of 5.

1D Number sequences

Discover 2 Student Book page 16 • Practice Book page 21

Specific learning focus

Global skills

- **Creative skills:** investigating
- **Interpersonal skills:** communication
- **Self-development skills:** reflecting on learning

Key vocabulary

- number sequence, rule of sequence

Resources

- 0–9 digit cards (one set per pair)

Language support

Refer students to the display created in the previous lesson and use this to support their use of vocabulary when forming sequences.

 **Introductory activity**

Write on the board the sequence: 0, 4, 8, 12, …

Ask students to discuss in pairs what they think the rule could be for continuing the sequence. Ask students to share their suggestions and give reasons. They are very likely to say that the rule is to add 4.

Now, add the next number in your sequence: 0, 4, 8, 12, 24, …

Ask students whether this fits their rule. If not, can they change the rule so the sequence fits? You can give them a hint and say that the rule is made up of two operations that are repeated in turn. Give students some time in pairs to look at the sequence again and try to work out the rule. Agree that it is the following: add 4, then double, then add 4, then double and so on. Ask them to continue the sequence following this rule.

 **Main activity**

Ask students to look together at question 1 on page 16 of the Student Book. Talk through the instructions. You may prefer to adapt them to include some pair work. Suggest that students can use any of the four operations in their sequence rules. Direct students' attention to the

Unit 1 Number and place value

lower speech bubble and remind them that they can use decreasing sequences and sequences using negative numbers. Students then complete the activities on page 16 of the Student Book. As they work, circulate around the classroom asking questions such as:

- Can you give me a sequence with two rules?
- What is the rule for this sequence?
- What is the nth number in this sequence? How do you know?

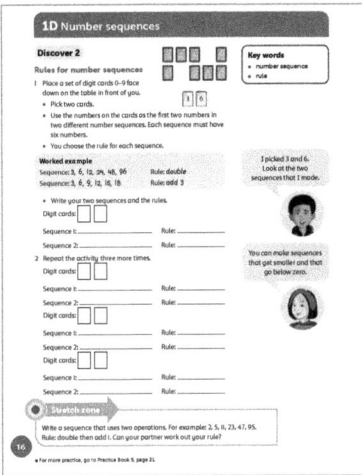

Differentiation

Supporting: Help students to describe ways of getting from their first number to the second as a way into thinking of rules for their two sequences.

Consolidating: Ask students to explain their thinking as they identify the rules for their partner's sequence.

Extending: Challenge students to use rules that include two operations.

Stretch zone: *Write a sequence that uses two operations. For example: 2, 5, 11, 23, 47, 95. Rule: double then add 1. Can your partner work out your rule?*

Encourage students to make sequences that they think are more difficult, then swap with a partner to check and extend the sequences.

 Reflection time

Ask several students to share their starting number and rules with the class. They can give the starting number and rule, then let the class find the first few numbers in the sequence, or they can give several numbers and ask the class to guess the rule.

Practice Book: Students complete Practice Book page 21. They can do this directly after the Main activity, as homework, or as the focus of a separate mathematics session to help students consolidate their learning and build fluency. Explain that the rule may use either addition, subtraction, multiplication or division.

Differentiated outcomes	
All students	should create two sequences using a starting number and one operation with support.
Most students	will create sequences, using two starting numbers and a choice of rules.
Some students	may create sequences using two or more starting numbers and more than one operation and predict further numbers in the sequences.

Answers

Student Book page 16

Answers will vary depending on the digit cards and sequence rules students choose.

Practice Book page 21

	Rule	Start number	Sequence					
	× 2	3	6	12	24	48	96	192
1	− 4	34	30	26	22	18	14	10
2	+ 3	14	17	20	23	26	29	32
3	× 3	1	3	9	27	81	243	729
4	+ 7	5	12	19	26	33	40	47
5	+ 9	2	11	20	29	38	47	56
6	− 9	54	45	36	27	18	9	0
7	+ 10	15	25	35	45	55	65	75
8	× 10	0.25	2.5	25	250	2500	25 000	250 000
9	− 3	6	3	0	-3	-6	-9	-12

Check that students' own sequences are correct.

Stretch zone: Check that students have created a suitable sequence, for example: 50, 160, 270, 380, 490, 600. (The rule is + 110.)

1D Number sequences

Explore 1 Student Book page 17 • Practice Book page 22

Specific learning focus
- Apply a rule to create a number sequence with a 5-digit start number.

Global skills
- **Creative skills:** investigating
- **Interpersonal skills:** communication

Key vocabulary
- count on/back in tens, hundreds, thousands; negative

Resources
- mini whiteboards and markers
- 0–9 digit cards

Language support
Discuss negative numbers and how we say them. For example, ⁻11 could be 'negative 11' or 'minus 11'. In the context of numbers in a sequence, encourage students to use **negative 11**, which distinguishes it from positive 11, whereas minus is the operation of subtraction.

Introductory activity

Write the following sequence on the board: 44, 42, 39, 35, … . Ask pairs to predict the next number. Then write 30 as the next number in the sequence. Ask pairs whether they were correct. They should tell you the rule (– 2, – 3, – 4, …).

Repeat for the sequence: 'If it's odd, add 3, if it's even, halve it', starting from 50.

Main activity

Draw a number line on the board and label it with this sequence: 10, 15, 20, 25.

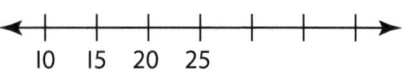

Say that the start number is 10. Can students work out the rule? (add 5) Ask them to extend the sequence to 50 on their whiteboards. Ask, *How can you find the tenth number in the sequence?* Give students some time to consider this with their partner and agree that they could either count on ten times or multiply 5 by 10. Ask, *How can you find the number that is in any position in the sequence?* Ask students to work with a partner to work out numbers of their choice within this sequence. Agree that they multiply the number position of their choice by 5, for example to find the number in the eighth position they multiply 8 by 5.

Next, say that that the start number is still 10 but that the rule is subtract 5. *What will the tenth number in the sequence be?* Before they begin, ask them to tell you what type of number it might be. Take feedback. They should realise that it will end in either a 0 or a 5 as they are counting back in fives. They should also say that the number will be negative.

Refer students to page 17 of the Student Book. They make two 5-digit numbers from a pair of digit cards and use these as the start numbers for each line of sequences that involve subtracting powers of 10. Ask, *Can you make any other sequences?* Encourage them to share their sequences with a partner. Then ask the partner to work out the sequences and to continue them for another five numbers.

Differentiation
Supporting: Help students to count back to find the terms of their sequences.

Consolidating: Ask students to explain their thinking while creating a decreasing sequence.

Extending: Ask students to create decreasing sequences using two operations. For example, the rule could be 'divide by 2 and add 3'.

Stretch zone: *Use the same start numbers and write sequences with the rule – 10 000.*

Check that students have completed the sequences correctly.

Reflection time

Invite pairs of students to share their sequences. Ask other students to identify the rule of the sequence and extend the sequences. To finish the lesson, write the sequence 72, 64, 56 on the board. Ask pairs to work out the sixth number in the sequence.

They may realise that the sequence begins on the ninth multiple of 8, the second number is the eighth multiple of 8, the third number is the seventh multiple of 8 and so on, so the sixth number is the fourth multiple of 8 because the multiple and position total 10 each time. Repeat for other similar sequences.

Practice Book: Students complete Practice Book page 22. They can do this directly after the Main activity, as homework, or as the focus of a separate mathematics session to help students consolidate their learning and build fluency.

Remind students to look at the numbers and decide whether they are getting bigger or smaller to help them work out which operation may be part of the rule.

Differentiated outcomes	
All students	should create simple sequences, using two starting numbers and a given subtraction rule with support.
Most students	will create simple decreasing sequences, using two starting numbers and a given subtraction rule.
Some students	may create decreasing sequences and two operations, and predict further numbers in the sequences.

Answers

Student Book page 17

1 Answers will vary because students use digit cards to pick the start numbers in their sequences.

For example, if they choose 1, 3, 4, 5 and 7, they could make 35 147 and 71 534 and the sequences for question 1 would be:

35 137, 35 127, 35 117, 35 107, 35 097

35 047, 34 947, 34 847, 34 747, 34 647

34 147, 33 147, 32 147, 31 147, 30 147

71 524, 71 514, 71 504, 71 494, 71 484

71 434, 71 334, 71 234, 71 134, 71 034

70 534, 69 534, 68 534, 67 534, 66 534

Similarly, they will record a sequence for 6-digit numbers in question 2.

Practice Book page 22

	Rule	Start number	Sequence				
1	+ 40	260	300	340	380	420	460
2	− 60	1050	990	930	870	810	750
3	− 40	670	630	590	550	510	470
4	+ 50 000	750 000	800 000	850 000	900 000	950 000	1 000 000
5	+ 10	⁻30	⁻20	⁻10	0	10	20
6	+ 1000	1500	2500	3500	4500	5500	6500
7	+ 1100	1400	2500	3600	4700	5800	6900
8	+ 50 000	0	50 000	100 000	150 000	200 000	250 000
9	+ 50 000	10 000	60 000	110 000	160 000	210 000	260 000
10	+ 75	⁻75	0	75	150	225	300
11	+ 10	980	990	1000	1010	1020	1030
12	+ 17	⁻5	12	29	46	63	80

Stretch zone: Subtract 160: 150, ⁻10, ⁻170, ⁻330, ⁻490, ⁻650

Unit 1 Number and place value

1D Number sequences

Explore 2 Student Book page 18 • Practice Book page 23

Specific learning focus
- Explore Fibonacci sequences.

Global skills
- **Creative skills:** exploring
- **Self-development skills:** reflecting on learning

Key vocabulary
- sequence, rule of the sequence, pattern

Resources
- mini whiteboards and markers

Language support
Add the Fibonacci sequence to the sequences display once students have discovered the rule. Support students in being able to describe the rule clearly.

Introductory activity

Ask students to sit in pairs or groups of three. Each group should contain a student who can spot patterns confidently. Write this number sequence on the board: 15, 27, 39, 51. Ask students to extend the sequence down to ⁻33 and up to 99. Discuss the sequence. Establish that all the answers are odd numbers and there is a pattern (add/subtract 1 to the tens and add/subtract 2 to the ones). Agree that the rule for this sequence is add 12. Ask students to make up their own number sequence. They should share their sequence with their partner or group. The partner or group then extends the sequence down and up again.

Invite students to share their sequences with the class. The class works out the rule for the sequence. Write on the board: 1, 4, 8 …. Ask students to extend this sequence, making up their own rule, for example: 1, 4, 8, 13, 19 (+ 3, + 4, + 5 …) or 1, 4, 8, 18, 36 (double then add 2, double …). Agree that number sequences like these can be made in different ways.

Main activity

Introduce the specific sequence known as the Fibonacci sequence, which starts 1, 1, 2, 3, 5, 8, 13. Ask students to extend the sequence for another ten numbers and to write the sequence on their whiteboards. Ask, *Can you discover what the rule is for the Fibonacci sequence?*

Explain that the Fibonacci sequence appears in many natural contexts, for example in the numbers of spirals on the pattern of a pineapple, the seed head spirals on a sunflower and the number of petals on different flowers. Find real examples or images to demonstrate these to students.

Refer students to page 18 of the Student Book and look at the photograph and drawings and read about Fibonacci. Students should then complete the activities, remembering to use the same rule for each sequence, adding two previous numbers to get the next.

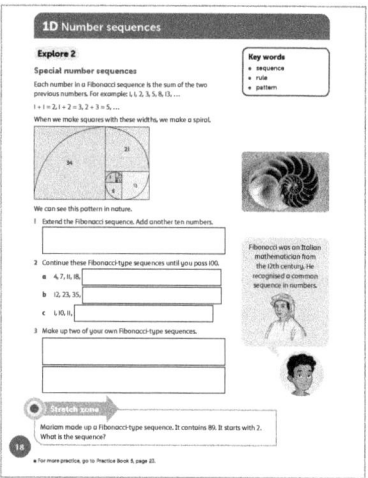

Differentiation

Supporting: Help students to make their own Fibonacci sequences with different starting numbers.

Consolidating: Ask students to explain the rule for Fibonacci-type sequences as they discover them.

Extending: Ask students to divide successive terms in the Fibonacci sequence and describe what happens.

Stretch zone: *Mariam made up a Fibonacci-type sequence. It contains 89. It starts with 2. What is the sequence?*

Students should discover that the sequence is: 2, 3, 5, 8, 13, 21, 34, 55, 89. They should also explain the rule of adding the two previous numbers to produce the next.

 Reflection time

Invite students to share the sequences they made up. Ask the rest of the class to extend them further. Write on the board: 2, 5, 7, 12, 19. Ask students to work out the rule for your sequence (add two numbers to get the next). Ask them to find the next five numbers in the sequence.

Practice Book: Students complete Practice Book page 23. They can do this directly after the Main activity, as homework, or as the focus of a separate mathematics session to help students consolidate their learning and build fluency. Some students may struggle to write a description of their sequence. You may choose to have students work in pairs or have adult support with their writing.

Unit 1 Number and place value

Differentiated outcomes	
All students	should create simple Fibonacci-type sequences.
Most students	will create Fibonacci-type sequences.
Some students	may explore the ratio of consecutive terms in Fibonacci-type sequences.

Answers

Student Book page 18

1 21, 34, 55, 89, 144, 233, 377, 610, 987, 1597

2 a 29, 47, 76, 123

   b 58, 93, 151

   c 21, 32, 53, 85, 138

Check that students' Fibonacci-type sequences follow the rules set.

Practice Book page 23

Check that students have created a number sequence using a starting number and a rule, and described features of their sequence appropriately. Read the instructions and discuss the worked example before they begin the activity.

Stretch zone: Check that students have formed sequences according to the given information. For example, 'subtract 1 and double it' gives 5, 8, 14, 26, 50.

1E Odd and even numbers

Discover Student Book pages 19–20 • Practice Book page 24

Specific learning focus

- Make general statements about sums and differences of odd and even numbers.

Global skills

- **Creative skills:** exploring
- **Interpersonal skills:** communication/teamwork
- **Self-development skills:** reflecting on learning

Key vocabulary

- odd number, even number, generalisation

Resources

- interlocking cubes for pairs of students
- digit cards 0–9 for each pair
- large sheets of paper

Language support

Ask students to create a poster that illustrates odd and even numbers using towers as an image and includes the general statements they have discovered. Display the poster for students to refer to.

 Introductory activity

Give each pair of students a set of digit cards. Play a game where they each make a set of three 5-digit numbers by randomly choosing digit cards (replacing the card after each pick). They then check to see whether they have formed **odd numbers** or **even numbers**. They can agree beforehand what decides the winner – perhaps it is the person with most odd numbers. Ask them to explain how they know whether their numbers are odd or even.

 Main activity

Ask students to look at the Think back statement on page 19 of the Student Book. Discuss how to check whether a number is even or odd and how the ones digit is significant.

Ask students to work on the activities on pages 19–20, using interlocking cubes to explore questions about odd and even numbers. As students are working, circulate and ask them to explain their answers.

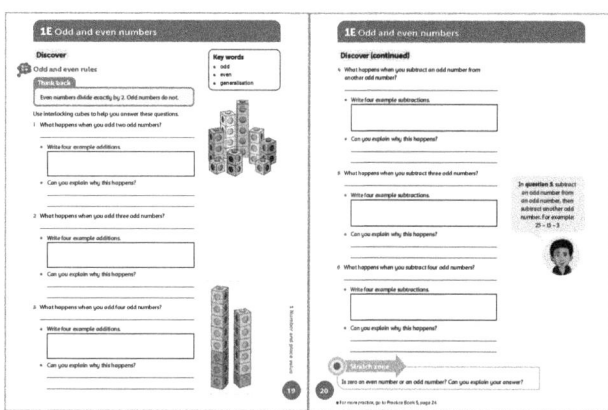

Unit 1 Number and place value

Differentiation

Supporting: Provide students with a list of larger odd and even numbers to refer to during the activities as well as working with interlocking cubes for smaller numbers.

Consolidating: Ask students to explain general statements about odd and even numbers and say how they know they are true, providing you with examples.

Extending: Ask students to demonstrate their findings using larger odd and even numbers.

Stretch zone: *Is zero an even number or an odd number? Can you explain your answer?*

It is an even number. Students may explain their thinking by saying that zero is even because it divides by 2 with no remainder, and is part of the sequence of adding (or subtracting) twos: ⁻6, ⁻4, ⁻2, 0, 2, 4, 6, …

Reflection time

Take feedback from the activity. Ask different pairs to share their findings. Establish that you can make **general** statements, for example: 'Adding/subtracting an even number of odd numbers gives an even answer'; 'Adding/subtracting an odd number of odd numbers gives an odd answer.'

Practice Book: Students complete Practice Book page 24. They can do this directly after the Main activity, as homework, or as the focus of a separate mathematics session to help students consolidate their learning and build fluency.

This activity is a good review of this lesson's learning; you may prefer to do this activity at a later date to revisit this work.

Differentiated outcomes	
All students	should explore sums and differences of odd and even numbers using cubes with support.
Most students	will explore sums and differences of odd and even numbers using interlocking cubes.
Some students	may be able to prove their general statements about odd and even numbers.

Answers

Student Book pages 19–20

Check that students' examples and explanations match the investigations.

1 Adding two odd numbers gives an even answer.
2 Adding three odd numbers gives an odd answer.
3 Adding four odd numbers gives an even answer.
4 Subtraction with two odd numbers gives an even answer.
5 Subtraction with three odd numbers gives an odd answer.
6 Subtraction with four odd numbers gives an even answer.

Practice Book page 24

	Statement	True?	Examples
1	Even numbers are divisible by 2	Always	$2 \div 2 = 1$ $34 \div 2 = 17$
2	Odd numbers are divisible by 3	Sometimes	$15 \div 3 = 5$ $17 \div 3 = 5r2$
3	The sum of two even numbers is even.	Always	$4 + 6 = 10$ $26 + 52 = 78$
4	The sum of two odd numbers is odd.	Never	$3 + 5 = 8$ $27 + 41 = 68$
5	If I subtract an odd number from an even number, the answer is odd.	Always	$42 - 17 = 25$ $86 - 19 = 67$
6	If I subtract an odd number from an odd number, the answer is odd.	Never	$37 - 23 = 14$ $77 - 35 = 42$
7	The difference between two even numbers is even.	Always	$40 - 16 = 24$ $106 - 64 = 42$
8	Even numbers are divisible by 4.	Sometimes	$28 \div 4 = 7$ $34 \div 4 = 8r2$

Stretch zone: Check that students' three statements about odd and even numbers are correct.

1E Odd and even numbers

Explore Student Book page 21 • Practice Book page 25

Specific learning focus
- Make general statements about multiplying odd and even numbers by 4 and 5.

Global skills
- **Creative skills:** investigating
- **Interpersonal skills:** communication/teamwork
- **Self-development skills:** reflecting on learning

Key vocabulary
- odd number, even number, generalisation, general statements

Resources
- mini whiteboards and markers
- interlocking cubes
- base-10 equipment

Language support
Make sure that students are clear about what a prediction and a generalisation are. Use both mathematical and non-mathematical examples to support your definitions. Examples could be: *'It has been raining all night so I predict it will be rainy this morning.' 'March is usually a rainy month in England.'*

Introductory activity
Review the work done in 1E Discover. Ask students to tell you the **general statements** they have made about adding and subtracting odd numbers and even numbers (adding even numbers gives an even answer; adding an even number of odd numbers gives an even answer; adding an odd number of odd numbers gives an odd answer). Remind students how they came up with these conclusions, using different numbers of interlocking cubes.

Ask, *Do your generalisations always work?* Ask students to add pairs of 3-digit and 4-digit even numbers and then odd numbers on their whiteboards. Repeat for three and four 3-digit and 4-digit numbers. Encourage students to explain their findings to the class and see whether they can justify the general statements.

 **Main activity**

Explain the activities on page 21 of the Student Book. Discuss how students can make a sensible prediction about what they think will happen when they multiply by 4 and 5, then use this to generalise about the result of multiplying by odd or even numbers. They can test their predictions, using interlocking cubes, and make general statements based on their test results. Students work in pairs. Encourage them to discuss and explain their predictions to their partner and then test their predictions with interlocking cubes. They should aim to recall all multiplications mentally and check them with cubes rather than using the cubes to calculate.

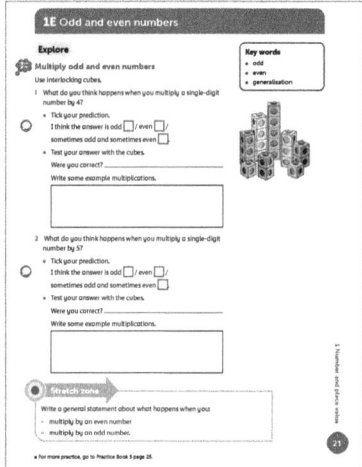

Differentiation
Supporting: Support students to find products using interlocking cubes, or base-10 equipment for larger numbers.

Consolidating: Ask students to make and explain general statements about the products of odd and even numbers.

Extending: Ask students to explain why the general statements are true.

Stretch zone: *Write a general statement about what happens when you:*
- *multiply by an even number*
- *multiply by an odd number.*

Students should find that multiplying by an even number always gives an even answer, but that multiplying by an odd number can give an odd or even answer.

 Reflection time

Invite individual students to share their predictions and their general statements. Ask them to demonstrate how they tested their predictions using the interlocking cubes. Ask the rest of the class: *Do you think they are correct?* Agree that: multiplying by even numbers gives an even answer; multiplying an odd number by an even number gives an even answer; multiplying two odd numbers gives an odd answer.

Unit 1 Number and place value

Practice Book: Students complete Practice Book page 25. They can do this directly after the Main activity, as homework, or as the focus of a separate mathematics session to help students consolidate their learning and build fluency.

For each question, encourage students to test a range of numbers, both small and large, and convince themselves of the result before writing their general statement.

Differentiated outcomes	
All students	should find products and check that they are odd or even using interlocking cubes.
Most students	will make a general statement about the products of odd and even numbers.
Some students	may write and prove a general statement about the products of odd and even numbers.

Answers

Student Book page 21

Students' examples will vary.

1. All answers will be even.
2. Some answers will be even and some answers will be odd.

Practice Book page 25

Check that students' statements about multiplying odd and even numbers are correct and that they have included appropriate examples.

1F Roman numerals

Discover   Student Book page 22 • Practice Book page 26

Specific learning focus
- Write year dates in Roman numerals.

Global skills
- **Creative skills:** exploring
- **Real-world skills:** interpreting information

Key vocabulary
- Roman numerals, Hindu-Arabic numerals

Resources
- none needed

Language support

Focus on how to say different dates, for example 2023 as twenty twenty-three, and any variations, for example 1901 as nineteen oh-one, and 1600 as sixteen hundred, as conventions will vary from language to language.

 Introductory activity

Ask students to look at the chart with **Roman numerals** and **Hindu-Arabic numerals** on page 22 of the Student Book. Give them two minutes to discuss with a partner what they can see on the chart. *What do you notice? Where do you see Roman numerals today?* (Old buildings, statues, clocks and at the end of films and TV shows for the year they were made)

Ask students to share what they noticed. Add some other number equivalences, for example I–XXI (1–21), on the board and ask students to see what patterns they can notice there as well.

Summarise the general pattern for smaller Roman numerals, as follows.

If a smaller numeral comes after a larger numeral, add the smaller numeral to the larger numeral. For example, XI: X + I is 11.

If a smaller numeral comes before a larger numeral, subtract the smaller numeral from the larger numeral. For example, IX: X – I is 9.

Discuss with students the names of the number words in the chart. Can they see parts of words that they recognise? For example, they might mention decimal, decimetre and December. Explain that X is *decum*, which is linked to the Latin root word of *Decem*, meaning ten. December got its name because it was the tenth month in the Roman calendar, which began in March. *Deci* is latin for tenth.

 Main activity

Look together at the rules about how to make Roman numbers and the example of the plaque showing when the school was built on page 22 of the Student Book. Check that the number shown follows the rules for Roman numbers and then ask pairs to work out what year this would be in Hindu-Arabic numbers (1877) and read it aloud.

Students should then complete the table on page 22. You may want to suggest they fill in the column of Hindu-Arabic numerals first, before working in pairs to convert them to Roman numerals.

Unit 1 Number and place value

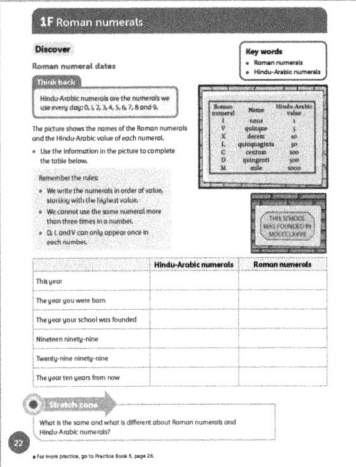

Differentiation

Supporting: Help students to identify the value of each symbol by using the chart and the rules on the board.

Consolidating: Ask students to write the nine times table in Roman numerals.

Extending: Challenge students to write the 12 times table, or the first 10 numbers of the Fibonacci sequence in Roman numerals.

Stretch zone: *What is the same and what is different about Roman numerals and Hindu-Arabic numerals?*

Students might describe the use of letters in Roman numerals, and the fact that larger amounts appear first in numbers.

Reflection time

Work through the table of years as a class. Ask pairs of students to say what they put for each year in turn and explain why they decided it was correct. For each year, ask questions such as *What was the previous year? What is the following year? Which year was 20 years earlier?*

Practice Book: Students complete Practice Book page 26. They can do this directly after the Main activity, as homework, or as the focus of a separate mathematics session to help students consolidate their learning and build fluency.

If appropriate, assign each student one event to research, and then they can share what they learned with the rest of the class.

Differentiated outcomes	
All students	should write the years in Roman numerals with support.
Most students	will write the years using Roman numerals.
Some students	may calculate and write years before and after a given year in Roman numerals.

Answers

Student Book page 22

	Hindu-Arabic numerals	Roman numerals
This year	e.g. 2024	MMXXIV
The year you were born	e.g. 2014	MMXIV
The year your school was founded	e.g. 1991	MCMXCI
Nineteen ninety-nine	1999	MCMXCIX
Twenty-nine ninety-nine	2999	MMCMXCIX
The year ten years from now	e.g. 2034	MMXXXIV

Practice Book page 26

	Event	Year (Hindu-Arabic numerals)	Year (Roman numerals)
1	fall of the Roman Empire in the West	476	CDLXXVI
2	Muhammad ibn Musa al-Khwarizmi wrote the first book of algebra	820	DCCCXX
3	start of the Hundred Years' War	1337	MCCCXXXVII
4	sea route from Europe to India discovered	1498	MCDXCVIII
5	outbreak of the Great Plague in England	1665	MDCLXV
6	Emancipation Proclamation signed in America	1863	MDCCCLXIII
7	Republic of China established	1912	MCMXII
8	summit of Mount Everest reached for the first time	1953	MCMLIII

Stretch zone: Check that the rules are correct, as shown on page 22 of the Student Book.

1F Roman numerals

Explore Student Book pages 23–24 • Practice Book page 27

Specific learning focus
- Write dates in Hindu-Arabic and Roman numerals and read them aloud.

Global skills
- **Creative skills:** investigating
- **Real-world skills:** research/presenting information/interpreting information
- **Interpersonal skills:** communication/teamwork

Key vocabulary
- Roman numerals, Hindu-Arabic numerals

Resources
- books or the internet
- short list of key historical events

Language support
Display the rules for smaller and larger Roman numbers. As larger numbers are written in Roman numerals, add these to the display with their equivalents and ask students to say the numbers aloud.

Introductory activity

Write on the board some Roman numbers that use three different symbols, for example:

MCXVI   DCLXI   MDLVI   MDCLXV

Ask students to work in pairs to see whether they can write the equivalents using the examples and clues they have used previously, referring to the class display or Student Book as necessary.

Talk through each one, asking pairs to describe how they found the equivalent number in Hindu-Arabic numerals. They should have worked out that the numbers are 1116, 661, 1556 and 1665. Remind students that the numbers we normally use are called Hindu-Arabic numerals because of the area of the world where they started. Students may recall that Roman numbers came from the Roman Empire.

Main activity

Before completing page 23 of the Student Book, students will need to research key events in their country's history using books, the internet, or by conducting interviews. Discuss what a key event could be, perhaps starting with key events in their own lives.

Ask students to look at the examples in the Student Book that appear in the speech bubbles. Explain the abbreviations CE (common era) and BCE (before common era), the common era beginning in year 1 of the Gregorian calendar, the most commonly used calendar in the world.

Students will continue to convert date years from Hindu-Arabic numerals to Roman numerals and vice versa. They should use the chart given on page 22 of the Student Book as well as their knowledge from previous examples as they write the years of historical events and film releases in both types of numerals. As they work, ask them to explain how they knew to write the symbols as they did, reminding them of the rules for writing Roman numerals.

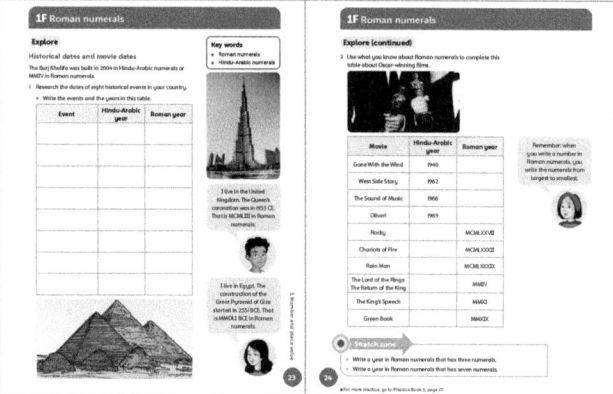

Differentiation

Supporting: Help students to identify the value of each symbol for the dates by using the chart on page 22 of the Student Book.

Consolidating: Ask students to explain how they use the chart to write the years in Roman numerals.

Extending: Challenge students to start a sequence in Roman numerals for a partner to continue.

Stretch zone: *Write a year in Roman numerals that has three numerals.*

Write a year in Roman numerals that has seven numerals.

Students may provide various answers from III to MMM for three numerals. There are also various possibilities with seven numerals, for example MCMLXXI = 1971.

 Reflection time

Draw a timeline on the board. Ask pairs to share some of their key events and what year they took place. For each year, ask, *How do we write that year? How is it written in Roman numerals? How do we say it?* After a few examples ask, *Which was the earliest of these dates?* Then ask the class whether anyone had an earlier date. Begin to add dates to a timeline together, asking students to suggest where they should go on the timeline. As they do, encourage students to say the dates and describe their position in relation to other dates. For example, the first moon landing was in 1969 (MCMLXIX) so on the timeline we need to put it to the

Unit 1 Number and place value

left of the London Olympic Games, which took place in 2012 (MMXII). Share some key dates from your own list, explaining why you chose them, for students to record in Roman numerals and add them to the timeline together.

Practice Book: Students complete Practice Book page 27. They can do this directly after the Main activity, as homework, or as the focus of a separate mathematics session to help students consolidate their learning and build fluency.

Students will need resources or access to the internet to conduct research on important historical events from around the world.

Differentiated outcomes	
All students	should write the years in Roman numerals with support.
Most students	will write the years using Roman numerals.
Some students	may calculate and write years in Roman numerals before and after a given year.

Answers

Student Book pages 23–24

1. Answers will vary depending on which events have been chosen.

2.

Movie	Hindu-Arabic Year	Roman year
Gone with the Wind	1940	MCMXL
West Side Story	1962	MCMLXII
The Sound of Music	1966	MCMLXVI
Oliver!	1969	MCMLXIX
Rocky	1977	MCMLXXVII
Chariots of Fire	1982	MCMLXXXII
Rain Man	1989	MCMLXXXIX
The Lord of the Rings	2004	MMIV
The King's Speech	2011	MMXI
Green Book	2019	MMXIX

Practice Book page 27

Answers will vary according to which events and years students choose.

Stretch zone: Check that the difference in years has been calculated correctly from those in the table.

1G Number problems

Discover Student Book page 25 • Practice Book page 28

Specific learning focus
- Ordering, comparing and rounding to the nearest 1000, 10 000 and 100 000.

Global skills
- **Creative skills:** problem solving
- **Real-world skills:** interpreting information
- **Interpersonal skills:** communication/teamwork

Key vocabulary
- round, order, nearest 1000/10 000/100 000

Resources
- mini whiteboards
- place-value cards
- sheets of paper

Language support

Continue to model saying 5-digit and 6-digit numbers in full (for example 732 246 as *seven hundred and thirty-two thousand two hundred and forty-six*, 99 206 as *ninety-nine thousand two hundred and six*) and to check that students don't say each digit separately (for example *seven three two two four six*).

 **Introductory activity**

Ask pairs of students to write on a whiteboard their answers to puzzles you set, for example: *Can you write me a 6-digit number with:*
- *all odd digits*
- *the thousands digit double the hundreds digit*
- *a digit total of 17?*

Pairs should take it in turns to give their answers. Repeat each answer as a whole class, making sure that you say the number in full.

 **Main activity**

Display a number line marked in 10 000s with ends from 70 000 to 120 000 and write a number in this range on the board, for example 98 371. Ask, *What is value of the 9?* (90 000) *How many ten-thousands is this?* (9) Repeat the question for each digit.

Now ask, *Which digit will you look at first to help you place the number on the number line?* Students should see that the 9 ten-thousands are needed to locate the position on the line. *What digit next?* (8) They might need prompting to find the midpoints on the number line to help them place it. Ask, for example, *Will the number be closer to 90 000 or closer to 100 000? How do you know? Is it closer to 98 000 or 99 000? How do you know?* Remind them of their earlier work on rounding, as necessary.

Unit 1 Number and place value

Repeat with another number, say 105 739, or make one from choosing digit cards.

Ask students to work with a partner to complete the activities in the Student Book on page 25. One of the pair could make the stadium capacity using place-value cards. Then the other should say the number aloud before they answer the question. Listen to the pairs talking when they work on the questions and ask them to say the value of different digits and justify their answers. Encourage students to continue to draw simple number lines to help them round to the nearest 1000, as necessary. For example:

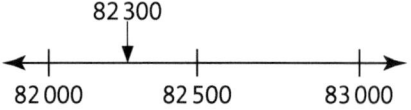

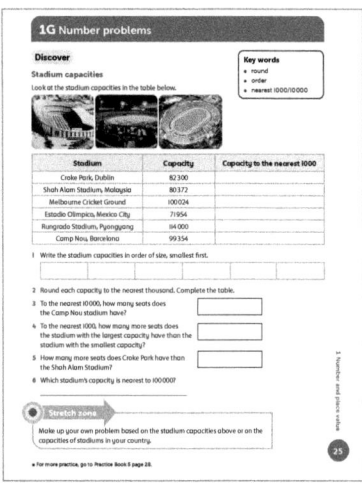

Differentiation

Supporting: Ask students to repeat saying 5- and 6-digit numbers after you have said them. Model how to write the numbers.

Consolidating: Ask students to explain their strategies for ordering capacities and finding the difference between them, for example.

Extending: Ask students to write and say 6-digit numbers to you and round to the nearest hundred thousand.

Stretch zone: *Make up your own problem based on the stadium capacities above or on the capacities of stadiums in your country.*

Students should use realistic data and make up appropriate questions that involve rounding. Provide access to the internet or collect some relevant data prior to the lesson.

 Reflection time

On the board, draw a 0–100 000 number line marked off in multiples of 10 000. Pick five students and ask them to call out 5 digits which you write on the board. In pairs, students write as many different 5-digit numbers as they can using these digits. Pick a pair to say one of their numbers and then come to the board and write it on the number line. Discuss any difference between how the numbers are said and written, tackling any misconceptions that may have caused differences to arise and then rewrite the number or say it correctly, as appropriate.

Practice Book: Students complete Practice Book page 28. They can do this directly after the Main activity, as homework, or as the focus of a separate mathematics session to help students consolidate their learning and build fluency.

Look at the map together and discuss what it shows. Say that it is likely they will need to do written calculations to answer the questions so they should record their workings on a separate sheet of paper.

Differentiated outcomes	
All students	should round and order 5-digit numbers using place-value cards and number lines with support.
Most students	will order and round 5-digit and 6-digit numbers using place-value cards and number lines.
Some students	may create number lines and order their own 6-digit numbers accurately by rounding.

Answers

Student Book page 25

Stadium	Capacity	Capacity to the nearest 1000
Croke Park, Dublin	82 300	82 000
Shah Alam Stadium, Malaysia	80 372	80 000
Melbourne Cricket Ground	100 024	100 000
Estadio Olimpico, Mexico City	71 954	72 000
Rungrado Stadium, Pyongyang	114 000	114 000
Camp Nou, Barcelona	99 354	99 000

1 71 954, 80 372, 82 300, 99 354, 100 024, 114 000

2 See table.

3 100 000

4 42 000

5 1928

6 Melbourne Cricket Ground

Practice Book page 28

1 52 000 m

2 38 169 m – shorter than question 1.

3 51 925 m

4 50 971 m

5 33 717 − 26 652 = 7065 m

Stretch zone: BCACED = 63 400 m

1G Number problems

Explore Student Book page 26 • Practice Book page 29

Specific learning focus
- Ordering, comparing and rounding to the nearest 1000, 10 000, 100 000 and 1 000 000.

Global skills
- **Creative skills:** problem solving
- **Real-world skills** interpreting information
- **Interpersonal skills:** communication/teamwork

Key vocabulary
- round, order, nearest 100/1000/10 000

Resources
- place-value cards
- 0–9 digit cards
- mini whiteboards and markers

Language support
Continue to model for students the language of 'rounding to the nearest' when working on the problems. For example, say, *896 445 rounded to the nearest thousand is 896 thousand*.

 **Introductory activity**

Ask a student to choose six cards from digit cards, say 2, 5, 6, 8, 9, 0. The class should work in pairs to form six different 6-digit numbers that all start with the same digit, for example 5. They might make numbers such as 562 980, 586 029, 590 268, 528 960, 509 628 and 596 208.

They should now list their numbers in **order** of size from largest to smallest on their whiteboards. As they work, ask them, for example, *How you know which number is largest? How do you decide the second largest?*

Ask pairs to draw a number line marked in 10 000s from 500 000 to 600 000. They should work together to place their six numbers on the number line as accurately as possible. Circulate around the classroom and ask pairs to explain as they work how they decide on each position. Ask, for example, *Does the ones digit help you place the numbers? Why?*

 **Main activity**

Refer students to the table on page 26 of the Student Book. Show students a world map, if possible, and point out each island in turn.

Discuss the populations and make sure that students can say each correctly. Ask questions such as *What do you notice about the populations? Which has the highest/lowest population?*

Point out that, as in the last activity, they will need to round larger numbers but this time they will need to round them to the nearest 100 and 10 000 as well as the nearest 1000. Students should complete the activities individually and then compare answers with a partner.

Students can continue to use place-value cards and numbers lines to help them round the populations.

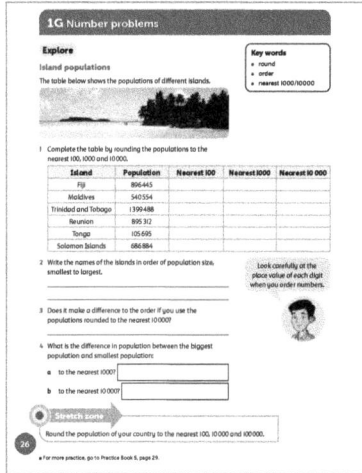

Differentiation

Supporting: Encourage students to continue to use number lines to help them round numbers. They may also find them useful for finding differences between populations.

Consolidating: Ask students to explain their strategies for rounding each population to the nearest 100, 1000 or 10 000.

Extending: Ask students to make up their own questions that involve rounding 6-digit numbers.

Stretch zone: *Round the population of your country to the nearest 100, 10 000 and 100 000.*

Answers will vary. Ask students to explain their strategy.

 Reflection time

Look together at the answers to questions 2 and 3. Discuss why the order can sometimes change if the populations are rounded. Students should see that, if you look only at the populations rounded to 10 000, you can no longer tell that the population of Fiji is greater than the population of Reunion Island. However, when the populations are rounded to the nearest 100 and 1000 you still can.

Ask students to describe their strategies for finding the differences between rounded populations.

Practice Book: Students complete Practice Book page 29. They can do this directly after the Main activity, as homework, or as the focus of a separate mathematics session to help students consolidate their learning and build fluency.

Unit 1 Number and place value

Check that students have a clear understanding of the real-life context of this activity.

Differentiated outcomes	
All students	should round 6-digit numbers to the nearest 100, 1000 and 10 000 using number lines and with support.
Most students	should round 6-digit numbers to the nearest 100, 1000 and 10 000 using number lines and place-value cards.
Some students	may begin to round 6-digit numbers to the nearest 100, 1000 and 10 000 confidently using mental methods.

Answers

Student Book page 26

1

Island	Population	Nearest 100	Nearest 1000	Nearest 10 000
Fiji	896 445	896 400	896 000	900 000
Maldives	540 554	540 600	541 000	540 000
Trinidad and Tobago	1 399 488	1 399 500	1 400 000	1 400 000
Reunion	895 312	895 300	895 000	900 000
Tonga	105 695	105 700	106 000	110 000
Solomon Islands	686 884	686 900	687 000	690 000

2 Tonga, Maldives, Solomon Islands, Reunion, Fiji, Trinidad and Tobago

3 Fiji and Reunion are the same when rounded to nearest 10 000, so listing them in order from the table puts Fiji ahead of Reunion.

4 a 1 294 000

b 1 290 000

Practice Book page 29

Ship	Capacity (in TEU)	1 Rounded to the nearest 10	2 Rounded to the nearest 1000
HMM Algeciras	23 964	23 960	24 000
HMM Oslo	23 820	23 820	24 000
MSC Leni	23 756	23 760	24 000
OOCL Indonesia	21 413	21 410	21 000
MCS Arina	23 656	23 660	24 000

3 OOCL Indonesia, MSC Arina, MSC Leni, HMM Oslo, HMM Algeciras

4 2551

5 47 784

6 308

Stretch zone: Exact total = 116 609

Rounded total = 117 000

Difference = 391

---

1 Number and place value

Connect Student Book page 27

Big idea

- I can read and write numbers up to one million. I can order them and use them to solve problems. I can read and write Roman numerals.

Global skills

- **Creative skills:** problem solving
- **Real-world skills** interpreting information
- **Self-development skills:** reflecting on learning

Key vocabulary

- hundred thousand, ten thousand, rounding to the nearest …

Resources

- place-value cards
- place-value tables and counters

Language support

Remind students of the language of comparison – small, smaller, smallest, for example, and ordinal numbers, such as *It has the second smallest population*.

 Introductory activity

Ask students in pairs to write down a 5-digit number. Then they should then say facts about their number. Encourage them to include, for example: the number of digits, the positions of the digits, if it is odd or even, the sum of the digits, what the number is when rounded to the nearest 1000 and nearest 10 000. Start with 5-digit numbers and then move to 6-digit numbers.

 **Main activity**

Set students the following task to complete in groups. Research the populations of six of the larger cities in your country. Record their populations as exactly as you can find out, then round the populations to the nearest 10, 100, 1000, 10 000 and 100 000. List the cities in order from smallest to largest and write some sentences about the populations, comparing the numbers, for example 'City A has 23 476 more people than city B'. Students could also research to try to find the five countries that have the closest population numbers to the population of their country.

Refer students to the activity on page 27 of the Student Book. They can work in pairs to complete the table and questions on populations in US cities. Students can use number lines to help them. Encourage them to use all the facts from the unit. Ask questions to prompt them, for example:

- Is your number greater than … ?
- Is your number less than … ?
- Does your number have a digit in the tens of thousands?
- Is your number between … and … ?

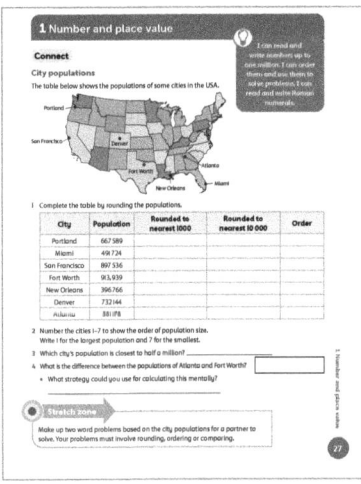

Stretch zone: *Make up two word problems based on the city populations for a partner to solve. Your problems must involve rounding, ordering or comparing.*

Students should make up questions that their partner can solve using data from the table.

 Reflection time

Ask pairs of students to share their questions on the populations in the table. Ask them to explain how they solved the problems and whether they estimated the answers first. Did they plot their numbers on a number line to help compare and round them?

Differentiated outcomes	
All students	should solve problems that involve 6-digit numbers using number lines or place-value tables and counters, with support.
Most students	will solve problems that involve 6-digit numbers using number lines, place-value tables and counters, or place-value cards.
Some students	may create questions that involve comparing and ordering 6-digit numbers.

Answers

Student Book page 27

1

City	Population	Rounded to nearest 1000	Rounded to nearest 10 000	Order
Portland	667 589	668 000	670 000	4
Miami	491 724	492 000	490 000	6
San Francisco	897 536	898 000	900 000	2
Fort Worth	913 939	914 000	910 000	1
New Orleans	396 766	397 000	400 000	7
Denver	732 144	732 000	730 000	3
Atlanta	501 178	501 000	500 000	5

2 See table.

3 Atlanta

4 412 761

Check that students have used an appropriate strategy to reflect the numbers involved, for example a written method.

Unit 1 Number and place value

1 Number and place value

Review Student Book page 28 • Practice Book page 30

Global skills

- **Creative skills:** problem solving
- **Real-world skills:** interpreting information
- **Interpersonal skills:** communication/teamwork
- **Self-development skills:** reflecting on learning

Student Book

With young students, assessment activities are most effective when carried out as an everyday classroom activity.

The first part of the activity is done in groups of six students. They can complete questions 3–6 independently.

Students should have access to number lines and place-value cards, place-value tables and counters to support them. Watch as students round and order large numbers. Listen as they say the value of each digit and say aloud the whole numbers. Listen to check that they understand how the place of a digit affects its value. If there are errors, help them partition so they can understand the value of different digits. It may help to model how to place numbers on number lines, using the most significant digits, and to model the partitioning of large numbers using place-value cards.

Some students may struggle to structure their questions clearly for others in their group. Support them as necessary so that all students have the opportunity to ask as well as answer questions.

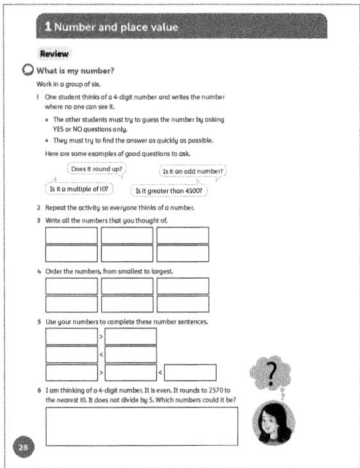

Answers

Student Book page 28

1. Students choose their own numbers, so answers will vary. Check that suitable questions are being asked to help find out the number.
2. As for 1.
3. Check that six 4-digit numbers have been used.
4. Check that the numbers have been ordered correctly.
5. Check that the number sentences have been completed accurately using < and >.
6. 2566, 2568, 2572, 2574.

Practice Book

With students in the upper primary years, it is appropriate to complete this as a whole-class discussion. You may choose to keep a record of the class discussion or a copy of the Review page for your own records. Use the Student Book to briefly remind students of the areas of mathematics that they have worked on in this unit.

Ensure that students have a copy of the Student Book to support them as they discuss and answer the questions in the Practice Book.

Allow students plenty of time for discussion before asking them to share their responses with the rest of the class. If students complete this assessment at home, encourage them to discuss this with adults.

Make a note of areas that students still feel unsure about. As number is at the heart of many of the other units, you can revisit this regularly. You can also build number into everyday practice. For example, you could look for large numbers in the media to use with students.

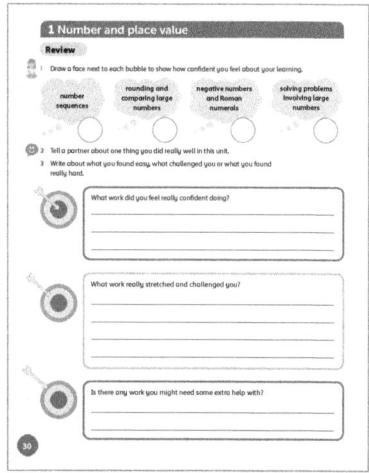

Additional material

Additional end-of-unit assessment are available on the *Oxford Owl for School* website.

Unit 1 Number and place value

2 Addition and subtraction

Overview

Big idea

The Big idea for this unit is that we can draw upon various strategies when adding or subtracting, depending on the context and the numbers within a problem.

The unit develops mental and written strategies. Students decide the appropriate method to answer questions, in particular whether to use a mental or a written method. They learn to add and subtract 4-digit and 5-digit numbers, including using written algorithms involving carrying figures and decomposition, and applying the strategies to solve problems. Strategies that students know from earlier stages are extended in this unit, for example:

- extending partitioning to 5- and 6-digit numbers
- adjusting from near multiples, such as adding 3999 by adding 4000 and subtracting 1
- estimating by rounding to the nearest 100 000, 10 000 and 1000.

Encourage students to use what they already know to extend familiar strategies and develop new ones.

Look out for

- **Students who use the formal written methods (algorithms) they know for all calculations, even when they are not appropriate.** Help students to build a repertoire of mental and written strategies. Encourage them to look at the numbers in a calculation and decide how to approach it.
- **Students who are not able to make the decision that they can, for example, work out the calculation in their head or use jottings or an algorithm.** Give them opportunities to choose which is more suitable. For problem solving with real-life data, students also need to decide when it is appropriate to use a calculator.
- **Students who do not apply the correct operation to word problems.** Provide a range of contexts so that they are aware of when to add or subtract.

Possible misconceptions

- **Students may use the wrong digits to round numbers.** Model the use of leading digits and suitable place value. For example, if trying to round 28 496 to the nearest ten thousand, a student might use the 9 tens to round up the 4 hundreds to a 5, then think that the 5 hundreds rounds up the 8 thousands to a 9.
- **Students may think that all calculations involving larger numbers are best solved using a written method.** Use calculations that suit mental calculation rather than a written method, for example 25 000 − 24 999.

Key vocabulary

- mental strategy, calculation strategy, efficient strategy, mental methods, written methods
- doubles, doubling, near-doubling, halving, rounding
- number bonds, partitioning, count on, count back, complementary addition, sequencing
- multiple, adjust, compensation, compensating
- total, difference, estimate, word problem
- inverse

Coverage in lessons

Learning objective	E	2A	2B	2C	2D	2E	C	R
Add and subtract whole numbers with more than four digits, including using formal written methods (columnar addition and subtraction).	✓	✓	✓	✓	✓	✓	✓	✓
Add and subtract numbers mentally with increasingly large numbers.	✓	✓	✓	✓	✓	✓	✓	✓
Use rounding to check answers to calculations and determine, in the context of a problem, levels of accuracy.			✓		✓	✓	✓	✓
Solve addition and subtraction multi-step problems in contexts, deciding which operations and methods to use and why.	✓	✓		✓	✓	✓	✓	✓

Unit 2 Addition and subtraction

2 Addition and subtraction

Engage — Student Book page 29

Big question
- What strategies can I use for adding and subtracting large numbers?

Global skills
- **Creative skills:** exploring
- **Interpersonal skills:** communication
- **Self-development skills:** reflecting on learning

Key vocabulary
- mental calculation strategies, rounding, doubles, near doubles, halving, number bonds, complementary addition, sequencing

Resources
- mini whiteboards and markers

Language support
Make notes on the 'poster' that has been produced during Reflection time. Include it as part of a classroom display or working wall that you build up across the unit.

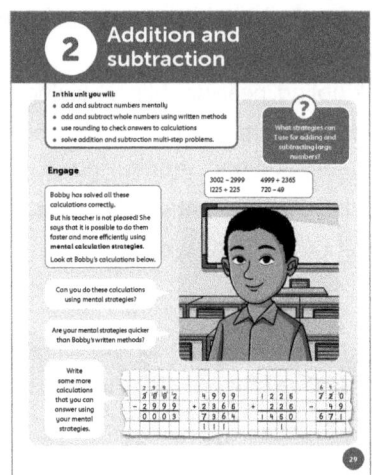

Introductory activity

Write 23 + 22 on the board. Ask pairs of students to work out the answer in several different ways. Tell them to show their methods on their whiteboards. Invite students to share their methods. They may have used, for example:

- rounding and adjusting (20 + 20 + 5)
- near doubles (double 22 + 1, double 23 − 1)
- partitioning (23 + 20 + 2).

Demonstrate any strategies that students do not use on the board.

Ask, *Which method do you think is the most efficient? Can you explain your thinking?* Students may suggest factors such as speed, how easy it was to do any calculations mentally, how many steps were involved, for example.

Write the following calculations on the board.

99 + 17   56 − 14   105 − 98   502 − 408

Ask students to work in pairs to decide which calculation should be tackled in which way, drawing upon the strategies just discussed, and to give a reason.

Discuss their answers as a class.

Main activity

Look together at the calculations on page 29 of the Student Book. If you have access to an IWB, you could use this.

Students need to work out the answers to those calculations using appropriate **mental calculation strategies**. They should work in pairs and try several different strategies for each calculation. Each pair should create at least two mental calculations for others to try. They must be able to explain how to calculate the answer mentally themselves.

As pairs work, circulate, asking them questions to assess their understanding, for example:

- *What do you notice about the numbers in this calculation?*
- *Which strategy did you use for this calculation? Why did you choose that one?*
- *Can you give me an example of two calculations with 4-digit numbers, one that is easy to solve mentally and one that is easier to solve using a written method?*

Differentiation

Supporting: Model mental calculation strategies for students.

Consolidating: Ask students to share their mental calculation strategies with you or a partner.

Extending: Ask students to explain why they are choosing a particular mental calculation strategy and compare its efficiency to another mental strategy.

Reflection time

Take feedback from the activity. Invite students to share their mental calculation strategies and demonstrate them on the board. Ask them to share the calculations that they made up. Ask students whether their mental strategies are quicker than Bobby's written methods. Other students in the class can answer these using one of the mental calculation strategies or another strategy of their choice.

Record the different strategies on a piece of flipchart paper and display this as part of a class display for the rest of the unit.

Unit 2 Addition and subtraction

2A Partitioning to add or subtract

Discover Student Book page 30 • Practice Book page 31

Specific learning focus
- Partition to add or subtract.

Global skills
- **Creative skills:** problem solving
- **Interpersonal skills:** communication/teamwork

Key vocabulary
- count on, count back, partitioning

Resources
- mini whiteboards and markers
- 0–9 digit cards

Language support
Add a pictorial example of adding and subtracting using partitioning, for example with base-10 equipment, alongside a step-by step written explanation.

 **Introductory activity**

Set the following problem for students to work on in pairs.

Cara counted the number of people going into a store. In the morning, she counted 746 people. In the afternoon, she counted 231. How many went into the store altogether?

Ask pairs to solve the problem on their whiteboards. Take feedback on all the different strategies that students used. If no one has used **partitioning**, model that strategy, for example:

746 + 200 + 30 + 1 = 946 + 30 + 1 = 976 + 1 = 977

 **Main activity**

Choose a student to come to the front of the class. Model the activity in the Student Book on page 30 with them. Say, *Each of you should explain very carefully the strategy you are using for each calculation.* Students work on the activity in pairs so that they can share strategies and check one another's answers. It may be appropriate for some students to begin with 2-digit numbers and some students to extend the activity to include some 4-digit numbers.

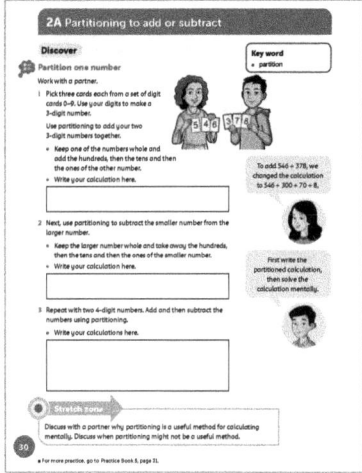

Differentiation

Supporting: Ask students to start by adding or subtracting 2-digit and 3-digit numbers, using base-10 equipment to support their calculations.

Consolidating: Suggest that students explain their thinking to other students.

Extending: Encourage students to extend the activity to 4-digit numbers.

Stretch zone: *Discuss with a partner why partitioning is a useful method for calculating mentally. Discuss when partitioning might not be a useful method.*

Students might say that it is easier to partition when the totals in each column remain below 10. You can also ask students to make up word problems to go with their calculations.

 Reflection time

Invite individuals to share the calculations they made up. Ask, *How did you work out your answers?* Discuss their opinion of this strategy. *Is this strategy helpful?* Ask students to give reasons for their opinions. Students might say that they can add or subtract more quickly when they use mental partitioning.

Practice Book: Students complete Practice Book page 31. They can do this directly after the Main activity, as homework, or as the focus of a separate mathematics session to help students consolidate their learning and build fluency.

Support some students to make and record their six numbers if you think they will struggle to work systematically to make their numbers. Students may prefer to complete the Stretch zone question orally.

Differentiated outcomes	
All students	should use partitioning strategies with 2-digit numbers with support.
Most students	will use partitioning strategies with 3-digit numbers.
Some students	may use partitioning strategies with 3- and 4-digit numbers to add and subtract mentally.

Unit 2 Addition and subtraction

Answers

Student Book page 30

Answers will vary because students make their own numbers. While students are working, listen to their explanations of their methods and strategies for adding and subtracting their numbers. Note who has chosen appropriate strategies and can explain their thinking clearly.

Practice Book page 31

Students make their own 4-digit numbers to work with, so answers will vary. Check that they have written two subtraction and two addition sentences using their chosen numbers and have used partitioning to solve them.

Stretch zone: Answers will vary. Check that students have explained their thinking clearly and have supported it with examples.

2A Partitioning to add or subtract

Explore Student Book page 31 • Practice Book page 32

Specific learning focus
- Partition one or both numbers to add or subtract.

Global skills
- **Creative skills:** problem solving/exploring
- **Interpersonal skills:** communication/teamwork

Key vocabulary
- mental method, written method, efficient strategy, count on, count back

Resources
- mini-whiteboards and markers
- 0–9 digit cards

Language support

Using the same calculation you presented with base-10 equipment as part of your classroom display, model partitioning using place-value cards, so that students have an additional visual representation to refer to.

 **Introductory activity**

Use digit cards to create a 4-digit number. Practise **counting on** and **counting back** in thousands, hundreds and tens, for example:

4572, 5572, 6572, 7572, 8572

4572, 4672, 4772, 4872, 4972

4572, 4582, 4592, 4602, 4612

Use the digit cards to create another 4-digit number. Pairs of students should list the first six numbers counting on in thousands, hundreds and tens and then count back in thousands, hundreds and tens.

 **Main activity**

Set the following problem for students to work on in pairs using their whiteboards.

Labeeb counted 2876 home team supporters at the football match. Kishwar counted 1453 away team supporters. How many supporters were there altogether?

Write the two numbers on the board. Invite a student to demonstrate on the board how they found their answer. Next, ask students to subtract in the same way to find out how many more home than away supporters there were. Say that you can use **partitioning** to solve this and model finding the answer by partitioning, for example:

2876 – 1000 – 400 – 50 – 3

Ask students to look at the Think back information on page 31 of the Student Book and discuss partitioning with the addition examples. Explain that as students work through the activities in pairs, they should explain very carefully the strategy they are using for each calculation. Students can share strategies and check one another's answers. It may be appropriate for some students to begin with 3-digit numbers and some students to extend the activity to include some 5-digit numbers.

Ask students questions to assess their understanding of the strategy, for example:
- *What is partitioning?*
- *How can you use partitioning to subtract 354 from 876?*
- *Can you explain your partitioning strategy for these two numbers?*

Unit 2 Addition and subtraction

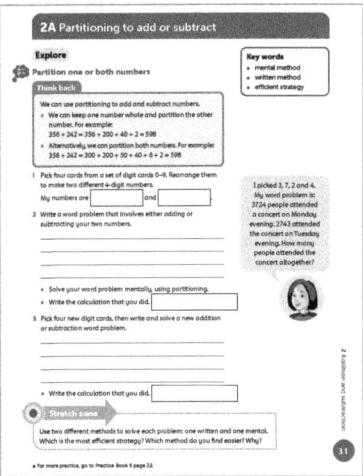

Practice Book: Students complete Practice Book page 32. They can do this directly after the Main activity, as homework, or as the focus of a separate mathematics session to help students consolidate their learning and build fluency.

This activity can be done in pairs. Look at the instructions together and discuss strategies for adding and subtracting mentally. Explain that for their different strategy they could use a **mental method** or formal **written method**. Encourage them to record their thinking in jottings to explain their mental method.

Differentiated outcomes	
All students	should use partitioning strategies with 3-digit numbers.
Most students	will use partitioning strategies with 4-digit numbers.
Some students	may judge whether, for example, partitioning both numbers or just one number was more efficient or partition in a way other than by hundreds, tens and ones.

Answers

Student Book page 31

Answers will vary because students make their own numbers. While students are working, ask them to explain their methods and strategies for adding or subtracting their numbers. Why did they choose a particular method for a particular pair of numbers?

Practice Book page 32

For each calculation, check that students' strategies are appropriate and efficient. They should be able to offer two different strategies for each calculation, either mental or written.

Differentiation

Supporting: Suggest that students start the activity by using base-10 equipment and a place-value table to add or subtract 3-digit numbers.

Consolidating: Ask students to explain their choice of strategy.

Extending: Challenge students to make up their own problem using 5-digit numbers.

Stretch zone: *Use two different methods to solve each problem: one written and one mental. Which is the most efficient strategy? Which method do you find easier? Why?*

You could ask students to compare strategies to go with their explanations.

 Reflection time

Invite individuals to share the word problems they made up. Ask, *How did you work out your answers?* Discuss their opinion of using a partitioning strategy to add and subtract. Ask, *Is this strategy efficient? Is there another strategy you think could be more helpful? Why? Do the numbers in your calculations make a difference?* Ask students to give reasons for their opinions.

2B Adding and subtracting near multiples

Discover Student Book page 32 • Practice Book page 33

Specific learning focus
- Add or subtract near multiples of 10 or 100.

Global skills
- **Creative skills:** problem solving
- **Interpersonal skills:** communication

Key vocabulary
- multiple, near multiple, calculation strategy, adjust

Resources
- mini whiteboards and markers
- blank number lines (optional)

Language support

Include a worked example on your working wall to model adding by rounding and adjusting **near multiples** of 10 for students to refer to. Include speech bubbles that use key words to support students when explaining their thinking.

 **Introductory activity**

Write on the board:

27   54   88   143   175   192

Unit 2 Addition and subtraction

Ask students to write the numbers on their whiteboards and, in discussion with a partner, round each number to the nearest 10. Choose students to share their answers and their strategy.

 **Main activity**

Students should work in pairs on the activities on page 32 of the Student Book so they can support each other. It may be appropriate to simplify or extend the first activity for some students. For example, they can choose to add each number to 65 or 1234 rather than 157.

Students should then work on the word problems in pairs, or groups of four, if students need additional support. As students work, encourage them to consider whether rounding both numbers to the nearest **multiple** of 10 would be helpful. For example, in question 3b, they could round $249 to $250 as well as $49 to $50, add $250 and $50 and then **adjust** by subtracting 2 to find the total amount of money Amira had.

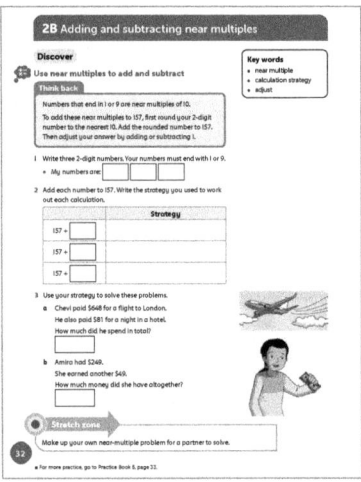

Differentiation

Supporting: Students could use number lines to help them recognise which is the nearest multiple of 10.

Consolidating: Ask students to share their strategies for solving the word problems with other students.

Extending: Challenge students to create 2-step word problems using near multiples.

Stretch zone: *Make up your own near-multiple problem for a partner to solve.*

Students should make up calculations that can be answered using the near-multiples strategy.

 Reflection time

Invite students to share their strategies from the activities in the Student Book. Together, work through the problems. When someone shares their strategy, other students can offer different answers or strategies, which can lead to fruitful discussion.

Practice Book: Students complete Practice Book page 33. They can do this directly after the Main activity, as homework, or as the focus of a separate mathematics session to help students consolidate their learning and build fluency.

Look at the worked example together. Ask, *What digit does the first number end in?* (8) Point out that in the example they have rounded 178 to 180. Can they explain why? (78 is between 70 and 80, but closest to 80 because when the one digit is 5 or greater you round up) *What is the difference between 178 and 180?* (2) Remind them that they will always need to also subtract or add, depending on the calculation, the difference to find the answer. Reinforce this with the worked example. Tell students that they can round either or both of the numbers, depending on what makes the calculation easier for them. They just need to remember to adjust the calculation accordingly. For example: 261 − 58 = 260 − 60 + 1 − 2 = 200 − 1 = 199.

Differentiated outcomes	
All students	should use near multiples to solve problems bridging 100 with support.
Most students	will solve the word problems by rounding near multiples of 10 and adjusting.
Some students	may create their own word problems using near multiples of 10.

Answers

Student Book page 32

Answers to the first activity will vary because students choose their own numbers to work with. Check that students used the 'near multiple' strategy correctly.

3 **a** Chevi spent $729.

 b Amira had $298.

Practice Book page 33

Check that students have used efficient strategies. Answers are:

1 252

2 262

3 261

4 263

5 203

6 193

7 183

8 193

9 49

Stretch zone: Check students' methods for accuracy and efficiency. For example, having solved 168 + 84 = 252, did students then calculate 168 + 94, or did they recognise that the answer will be 10 more because 94 is 10 more than 84?

2B Adding and subtracting near multiples

Explore Student Book page 33 • Practice Book page 34

Specific learning focus
- Add or subtract near multiples of 100 or 1000.

Global skills
- **Creative skills:** problem solving/exploring
- **Interpersonal skills:** communication/teamwork

Key vocabulary
- multiple, near multiple, calculation strategy

Resources
- mini whiteboards and markers
- 1–9 digit cards

Language support
Look at the word problems with students, explaining any unfamiliar vocabulary. Encourage them to identify key words or phrases that tell them what calculation they will need to do, for example 'How many?' and 'How much more?'

Introductory activity

Set the following problem for students to work on in pairs.

Vishwa and Kelvin played a game. Vishwa scored 2003 points, Kelvin scored 1996 points. How many more points did Vishwa score?

Ask pairs, *What mental calculation strategy could you use to solve this problem?* Take feedback. Accept any strategies that give a correct answer, focusing on using near multiples. Remind students of what they learned about in 2B Discover. Model or invite a student to show and explain how they could round to the nearest multiple of 1000 and then adjust. For example, use 2003 to 2000: 2000 − 1996 + 3 = 7.

Next, draw a number line and place the two numbers on it. Model how to count on 4 to 2000 (the nearest multiple of 1000) and then count on 3 from 2000 to find the difference.

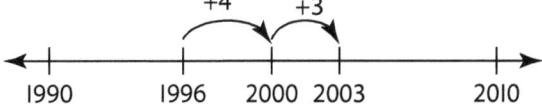

Agree that the difference between these two numbers is 7. You could also show students on the number line how they could 'slide the difference along' by adding 4 to both numbers to simplify the calculation, making 2007 − 2000.

Write on the board other numbers that are close together and that cross a thousand boundary. Ask students to work out the differences between them on their whiteboards.

Main activity

Ask students to work in pairs to calculate 5001 − 4988. They should share their strategy. Illustrate using a number line counting on from 4988 to 5000 (12) and then adding another 1.

Next, ask pairs to calculate 3001 − 2988. They should share their strategy. Illustrate using a number line counting on from 2988 to 3000 (12) and then adding another 1. Show how the strategy is the same whether using 3- or 4-digit numbers.

Students should then work on the activities on page 33 of the Student Book in pairs.

Ask students questions to encourage them to explain their thinking, for example:
- *What strategy would you use for subtracting 999 from 1012? Can you think of another strategy?*
- *Can you show me your strategy using a number line?*

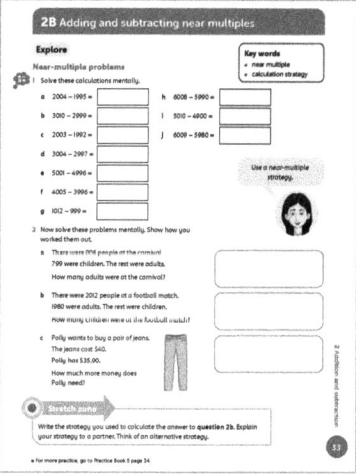

Differentiation

Supporting: Encourage students to draw number lines. It may be appropriate to simplify the first activity for some students and use 3-digit numbers so that the students are bridging hundreds rather than thousands.

Consolidating: Ask students to describe to you the difference between the strategies they have used in the previous two lessons. When is one strategy more appropriate than another?

Extending: Ask students to create subtraction word problems using near multiples.

Stretch zone: *Write the strategy you used to calculate the answer to question 2b. Explain your strategy to a partner. Think of an alternative strategy.*

Check that students have used an appropriate strategy to answer question 2b and that their explanation is clear. Encourage their partner to ask questions.

Reflection time

Invite students to share their explanations from the activity on page 33 of the Student Book. Choose three calculations from question 1 and ask students to draw them on a number line and describe their strategy. Then work through the word problems together to bring out the strategies that students used.

Practice Book: Students complete Practice Book page 34. They can do this directly after the Main activity, as homework, or as the focus of a separate mathematics session to help students consolidate their learning and build fluency.

Look at the worked example together. Tell students that they can use this near-multiples strategy or another, if they think it is more efficient. Look at the word problems and how rounding will change the prices. Agree that, as they are 1p less than a whole pound, the prices will round up to a whole-pound amount. Remind them they will then need to adjust their total by subtracting the correct number of pence.

Differentiated outcomes	
All students	should use near multiples to solve problems bridging hundreds with support.
Most students	will solve the word problems using near multiples of 100 or 1000.
Some students	may create their own word problems using near multiples of 100 or 1000.

Answers

Student Book page 33

1 a 2004 − 1995 = 9

   b 3010 − 2999 = 11

   c 2003 − 1992 = 11

   d 3004 − 2997 = 7

   e 5001 − 4996 = 5

   f 4005 − 3996 = 9

   g 1012 − 999 = 13

   h 6008 − 5990 = 18

   i 5010 − 4900 = 110

   j 6009 − 5980 = 29

2 a 107

   b 32

   c $4.10

Practice Book page 34

1 2965

2 3365

3 2764

4 2067

5 2268

6 7112

7 7989

8 8585

9 7783

10 6816

11 a $24.97

    b $18.97

    c $32.97

12 $2.02

Stretch zone: Check that the answers to students' own word problems are correct.

Unit 2 Addition and subtraction

2C Which strategy?

Discover Student Book page 34 • Practice Book page 35

Specific learning focus
- Choose and use appropriate mental strategies or written methods to add or subtract pairs of 5-digit numbers, including jottings where necessary.

Global skills
- **Creative skills:** exploring
- **Interpersonal skills:** teamwork
- **Self-development skills:** reflecting on learning

Key vocabulary
- mental strategy, written method

Resources
- mini whiteboards and markers
- digit cards 0-9

Language support
Display a worked example with steps of how to complete the calculation, labelled to support students in how to explain their written methods.

 Introductory activity

Set the following problem.

Ben and Bethany collected some coins. Ben collected 5206 coins. Bethany collected 5294 coins. How many coins did they have altogether?

Ask pairs, *How can you answer this problem?* (Use a mental strategy, then another and then another) Take feedback. Share ways of finding the total and discuss which strategy students think is the most efficient. As a class, look at these possible methods for addition:

- number bonds to 10 and 1000 and multiple of 100: 6 + 4 = 10, 90 + 10 = 100, 200 + 200 + 100 and 5000 + 5000 = 10 500
- partitioning : 5294 + 5000 + 200 + 6 = 10 494 + 6 = 10 500
- near multiples : 5206 + 5300 − 6 = 10 506 − 6 = 10 500.

Choose a student to use a written method and record it on the board.

Ask, *How many more coins did Bethany have than Ben?* Take feedback. Look at the strategies students used for finding the difference as you did for finding the total.

Ask, *Which strategy do you think is best for this problem?*

 Main activity

Ask students to work on the activities on page 34 of the Student Book. Go through the instructions together to ensure that students are clear on what the activity involves. Suggest that students estimate each total first, write the estimate on their whiteboards and then do their calculations.

While they are working, select three pairs to share problems with the rest of the class at the end of the lesson. As you circulate, ask pairs to explain how they decided between using a written or mental method. Encourage them to try another method to check its efficiency.

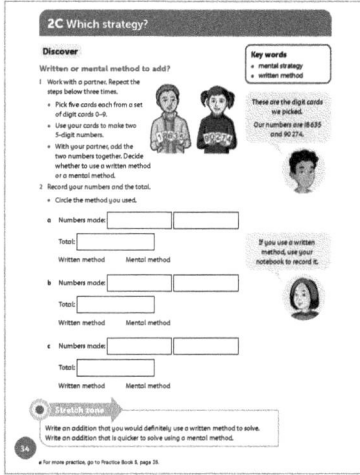

Differentiation
Supporting: Model possible strategies for students, then ask them to choose a strategy they can work with.

Consolidating: Ask students to explain their choices of strategies.

Extending: Challenge students to evaluate strategies and find more efficient alternatives.

Stretch zone: *Write an addition that you would definitely use a written method to solve. Write an addition that is quicker to solve using a mental method.*

Students could suggest 5879 + 7488 for a written method and 6134 + 2823 for a mental method, for example.

 Reflection time

Ask the pairs you selected earlier to share their calculations. Students who completed the Stretch zone could share the calculations they wrote and model their solutions, describing the strategies they used. Ask the class whether they agree with the choices and why.

Practice Book: Students complete Practice Book page 35. They can do this directly after the Main activity, as homework, or as the focus of a separate mathematics session to help students consolidate their learning and build fluency.

Unit 2 Addition and subtraction

Explain to students that, in the table, the total column is the result of adding the two numbers and the difference column is for the result of subtracting them. The strategy columns are for written workings.

Differentiated outcomes	
All students	should choose suitable addition and subtraction strategies with support.
Most students	will choose suitable addition and subtraction strategies.
Some students	may evaluate strategies and describe alternatives.

Answers

Student Book page 34

Answers will vary because students make their own numbers to use in the word problems they create. While students are working, listen to their explanations of their methods and strategies for adding their numbers. Note who has chosen appropriate strategies and can explain their thinking clearly.

Practice Book page 35

1 Total = 3790, Difference = 632
2 Total = 3755, Difference = 2819
3 Total = 4198, Difference = 3800
4 Total = 8350, Difference = 7800
5 Total = 11 549, Difference = 8449

Check that students have demonstrated their strategies clearly.

Stretch zone: Check that students have chosen suitable calculations and demonstrated their strategies clearly.

2C Which strategy?

Explore Student Book page 35 • Practice Book page 36

Specific learning focus

- Choose and use appropriate mental strategies or written methods to add or subtract pairs of 5-digit numbers, including jottings where necessary.

Global skills

- **Creative skills:** investigating
- **Self-development skills:** reflecting on learning

Key vocabulary

- mental strategy, written method

Resources

- mini whiteboards and markers
- digit cards

Language support

Continue to support students with reading larger numbers correctly by modelling how to say them.

 Introductory activity

Set the following problem.

A large number of people attended a baseball match. There were 10 567 people in zone A. There were 10 689 people in zone B. How many people were in zones A and B altogether? How many more were in zone B than zone A?

Ask pairs, *How can you answer this problem?* Ask them to use one strategy, then another and then another. Take feedback. Share ways to add to find the total and subtract to find how many more people were in zone B. Remind students of the methods for addition they discussed in 2C Discover. As a class, look at these possible methods for subtraction and relate each to the problem:

- near multiples: 10 689 − 10 570 + 3 or 10 690 − 10 570 + 4
- complementary addition: 10 567 + 3 + 30 + 89
- partitioning: 10 689 − 10 000 − 500 − 60 − 7

Discuss how students can check their answers: adding in a different order, using another strategy, using the inverse operation.

 Main activity

Point out the activity on page 35 of the Student Book. Read the instructions for question 1 and look at the worked example. Model the counting-on strategy along a blank number line.

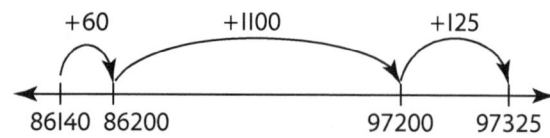

Ask, *Are there any other strategies you could use?* Take and discuss any suggestions, drawing upon strategies from the Introductory activity. Invite a student to model using a written method on the board.

$$\begin{array}{r} \overset{2\ 1}{97\overset{}{3}25} \\ -\,86140 \\ \hline 11185 \end{array}$$

Discuss which method students found most efficient. Can they explain why? Ask, *Do you think that method you find most efficient for this calculation will always be the most efficient?* Agree that it will depend on the numbers they make.

Students then work in pairs to complete the questions on page 35 of the Student Book.

Ask students questions to check that they understand how to use the strategies, for example:
- *How can you find the difference between 14 783 and 12 698? Is there another way?*
- *How can you check that your answer is correct?*
- *What do I mean when I say I will use complements to find the answer?*

Note who has chosen appropriate strategies and can explain their thinking clearly so you can ask them to share with the class in Reflection time.

If time allows, students can make pairs of a mixture of 2-digit to 5-digit numbers and find the difference between them.

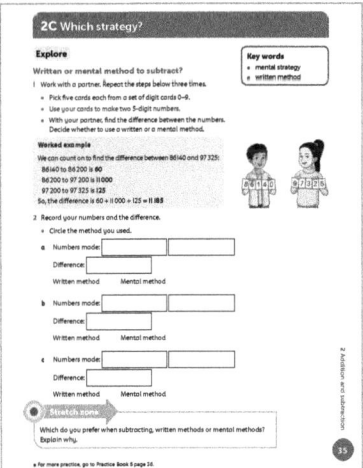

Differentiation

Supporting: Model possible strategies for students, then ask them to choose a strategy they can work with.

Consolidating: Ask students to explain their choices of strategies.

Extending: Challenge students to evaluate strategies and find more efficient alternatives.

Stretch zone: *Which do you prefer when subtracting, written methods or mental methods? Explain why.*

Listen for clear explanations, which include examples, to support their preference. Pair up a student with another with opposing preferences to discuss.

Reflection time

Choose pairs to share their calculations and model their solutions, describing the strategies they used. Ask the class whether they agree with the choices and why.

Practice Book: Students complete Practice Book page 36. They can do this directly after the Main activity, as homework, or as the focus of a separate mathematics session to help students consolidate their learning and build fluency. Point out to students that for this activity they will be thinking about written and mental methods for both addition and subtraction.

Differentiated outcomes	
All students	should choose suitable subtraction strategies with support.
Most students	will choose suitable subtraction strategies.
Some students	may evaluate strategies and describe alternatives.

Answers

Student Book page 35

Answers will vary because students make their own numbers to use in the problems they create.

Practice Book page 36

Students choose their own numbers by picking digit cards. Check that the calculations are correct for the numbers they made and look for efficient strategies.

Stretch zone: Check that students have identified the problems they would solve with a mental method. Ask them to explain why.

Unit 2 Addition and subtraction

2D Written methods of adding and subtracting

Discover 1 Student Book page 36 • Practice Book page 37

Specific learning focus
- Add or subtract any pair of 5-digit numbers using an efficient strategy.

Global skills
- **Creative skills:** exploring
- **Real-world skills:** presenting information
- **Interpersonal skills:** communication

Key vocabulary
- mental strategy, written methods, efficient strategy, column method, compensation

Resources
- mini whiteboards and markers
- digit cards (one set per pair)

Language support
Add to your working wall an explanation of the compensation strategy, with key language highlighted alongside visual support.

 **Introductory activity**

Write the following problem on the board and read it aloud.

There were 2567 people at one game and 2879 people at another. How many people attended the two baseball games altogether? How many more people were at the second game than the first?

Ask, *What do we need to do to answer these questions?* Take feedback, sharing possible strategies to add to find the total, and ways to subtract to find how many more people attended the second game. Remind students of any methods for addition and subtraction they have already looked at in this unit. Ask pairs to work on the problem, making jottings on their whiteboards. Tell them to estimate the answers first by rounding.

Ask students to show how to calculate the difference on their whiteboards using partitioning and **compensation**.

For partitioning, they should start with the larger number, 2879, and then subtract in turn the numbers for each digit in the smaller number:
2879 − 2000 − 500 − 60 − 7 = 312

For compensation, start with the smaller number and add to bring the total up to the larger number:
2567 + 3 = 2570; 2570 + 30 = 2600; 2600 + 200 = 2800 and 2800 + 79 = 2879. (The total amount added on = 200 + 79 + 30 + 3 = 312).

Finally, choose students to demonstrate how to solve each problem using column addition and subtraction. Take students through how to do this step by step, as necessary.

```
  2567        2879
+ 2879      − 2567
------      ------
  5446         312
  1 1 1
```

 **Main activity**

Ask students to look at page 36 of the Student Book. Using the two numbers given as an example, take students through how to complete the activity, explaining that they should estimate their answer first, and then choose a strategy to find the total and difference. Encourage them to try to find the most efficient strategy when choosing their favourite.

Ask pairs of students to find the total and difference of 20 581 and 18 502. You may wish to group students by prior experience. They should think about their preferred strategy and then discuss with their partner why they think it is efficient.

Share strategies as a class, explaining why they are efficient. Agree that the total is 39 083 and the difference is 2079.

Repeat by asking students to make the largest and the smallest possible 5-digit numbers they can with these digits. They then find the total and difference of these two numbers.

Ask questions to monitor students' learning of the vocabulary, for example:

- *Can you describe the **column method** for addition? How is it the same as partitioning? How is it different?*
- *How can we find the difference between two numbers?*

While they complete the activity, ask students to explain why they have chosen to use the strategies and methods that they have. Which operation and strategy do they use to check an answer?

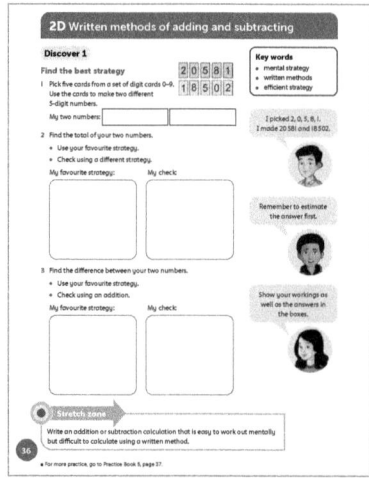

Differentiation

Supporting: Help students to find totals and differences of pairs of 3- or 4-digit numbers by modelling strategies for them, moving on to 5-digit numbers when they are comfortable working with the smaller numbers.

Consolidating: Ask students to find totals and differences of pairs of 5-digit numbers. Ask them to explain the strategies they used.

Extending: Challenge students to find totals and differences of pairs of 6-digit numbers using at least two strategies and comparing their efficiency.

Stretch zone: *Write an addition or subtraction calculation that is easy to work out mentally but difficult to calculate using a written method.*

Students may suggest 1005 − 997 as such a calculation. Ask them to explain their choice.

 Reflection time

Invite students to share examples of their addition and subtraction calculations and explain their strategies. *Which was more efficient, your favourite or the check? Why do you think that was? Can anyone else suggest a more efficient strategy?* Make sure that you include students from all of the different pairings.

Practice Book: Students complete Practice Book page 37. They can do this directly after the Main activity, as homework, or as the focus of a separate mathematics session to help students consolidate their learning and build fluency.

After students have completed the activity, ask them why they chose the strategies that they did for each calculation. Would a different strategy have been more efficient for a particular pair of numbers?

Differentiated outcomes	
All students	should find totals and differences of pairs of 4-digit numbers with support.
Most students	will find totals and differences of pairs of 5-digit numbers.
Some students	may find totals and differences of pairs of 6-digit numbers and explain their strategies.

Answers

Student Book page 36

Answers will vary because students make their own 5-digit numbers. Check that they have calculated the totals and differences for each pair of numbers correctly.

Practice Book page 37

Answers will vary because students make the 5-digit numbers that they will add and subtract. Check that the answers are correct and that students have set their work out clearly.

Stretch zone: Check that students have used appropriate strategies and found the same answer by both mental and written methods for each calculation.

2D Written methods of adding and subtracting

Discover 2 Student Book page 37 • Practice Book page 38

Specific learning focus
- Write problems that require addition or subtraction of a pair of 5-digit numbers.

Global skills
- **Creative skills:** problem solving/exploring
- **Interpersonal skills:** communication/teamwork

Key vocabulary
- partitioning, compensating, column method

Resources
- mini whiteboards and markers

Language support

Provide students with question frames to support them in writing their word problems, linking them to addition and subtraction. For example, use *How many _____ altogether? How many more _____ than _____?* Students may need additional support with appropriate contextual vocabulary for their word problems. Encourage them to use a bilingual dictionary if appropriate.

 **Introductory activity**

Write the following problem on the board and read it aloud. Ask students to work on it in pairs, using their whiteboards.

Nabila records the number of people who visit a superstore. In one week, 12 468 people visited. In the following week, 13 739 people visited. How many people visited altogether? How many more people visited in the second week than in the first?

Take feedback on the strategies they use. Agree that you need to add to find out how many people there are altogether and to subtract to find out how many more people there are in the second week than in the first. Review the column method for both addition and subtraction.

```
    12468           13739
  +13739          -12468
   ─────           ─────
    26207            1271
      111
```
(Note: subtraction shows borrowed digits 6 1 above the 7 3)

 **Main activity**

Ask students to read the instructions for the activity on page 37 of the Student Book. Remind them of the activity on page 35 of the Student Book, 2C Explore. Explain that this activity is similar to that one, although here they will do addition as well as subtraction and are creating word problems.

Before the activity, you may want to spend some time brainstorming to generate examples of real-life contexts of 5-digits numbers and any associated vocabulary. This will support students in coming up with a range of ideas for their word problems.

Encourage students to estimate first and then check their answers by swapping with a partner.

While they complete the activity, ask students to explain why they have chosen to use the strategies and methods that they have.

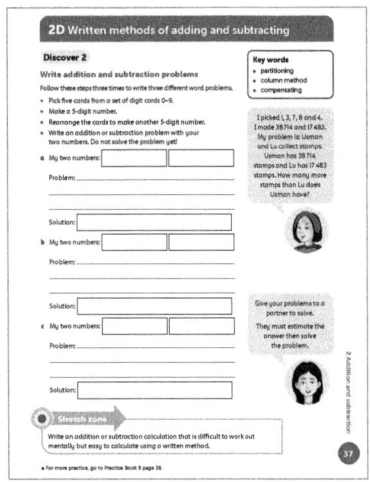

Differentiation

Supporting: Help students to solve problems by modelling the written methods.

Consolidating: Ask students to explain how they solved the problems using written methods.

Extending: Challenge students to make up two-step problems that involve addition and subtraction.

Stretch zone: *Write an addition or subtraction calculation that is difficult to work out mentally but easy to calculate using a written method.*

One example might be 5487 + 3976.

 Reflection time

Invite students to share examples of their word problems. Make sure that you include students from all of the different pairings. Ask the rest of the class to estimate the answer first and then solve the problem in their pairs. Share and compare strategies and agree on answers.

Practice Book: Students complete Practice Book page 38. They can do this directly after the Main activity, as homework, or as the focus of a separate mathematics session to help students consolidate their learning and build fluency.

If necessary, remind students that 'sum' is another way of saying the total found by adding. After students have completed the activity, ask them why they chose the strategies that they did for each calculation. Would a different strategy have been more efficient for a particular set of numbers?

Differentiated outcomes	
All students	should solve problems about totals or differences of pairs of 4-digit numbers with support.
Most students	will make and solve word problems about totals or differences of pairs of 5-digit numbers.
Some students	may make and solve word problems about totals or differences of pairs of 5-digit numbers, including two-step problems.

Answers

Student Book page 37

Answers will vary because students make their own 5-digit numbers. Check that their word problems are appropriate for the numbers they are using, and that they have calculated the totals and differences for each pair of numbers correctly.

Practice Book page 38

Check that the students have made reasonable estimates, answers are correct and students have set their work out clearly.

1   2878 + 3851 = 6729

2   79 280 − 8450 = 70 830

3   55 746 − 809 = 54 937

4   54 937 + 809 = 55 746

Stretch zone: Questions 3 and 4 are inverse calculations.

Unit 2 Addition and subtraction

2D Written methods of adding and subtracting

Explore 1 Student Book page 38 • Practice Book page 39

Specific learning focus
- Add and subtract 4-, 5- and 6-digit numbers in a money context.

Global skills
- **Creative skills:** problem solving
- **Real-world skills:** presenting information/financial literacy
- **Interpersonal skills:** communication

Key vocabulary
- partitioning, compensating, column method

Resources
- mini whiteboards and markers

Language support
Model questions to support the language of money, for example:
- *How much does this cost?*
- *What is the price of …?*

 **Introductory activity**

Write the following problem on the board and read it aloud.

Sam spent $2499 on a plane ticket and $1375 on a hotel booking. He started with $4000, which he had saved up for his holiday. How much did he have left?

Ask students to estimate the answer first by rounding. They should say that the total paid was approximately $2500 and $1400 for a total spend of $3900, $100 less than $4000 so he had which is approximately $100 left. Then ask them to work with a partner to solve the problem. Ask students to write the amounts on their whiteboards.

Take feedback. Discuss all methods used. Revise how they would record column methods if students are lacking confidence in this method. Agree that the answer is $126 and that their estimate was reasonable.

 **Main activity**

Show students examples of car prices locally. Ask students about the totals and differences between pairs of car prices. Ask them to work in pairs and describe the strategies they use to complete the calculations.

Move on to the car sales activity on page 38 of the Student Book. Discuss possible strategies for finding how much is left over from $250 000 for each car cost shown. Encourage students to use one strategy covered in this unit, such as compensation, counting on or partitioning and to check their answers using another. They might, for example, use a **compensating** strategy and then check using the written column method. If students are tending to use one strategy only, regardless of how efficient it may be, model another, explaining why you chose it. Remind students to round the amounts first to estimate their answers.

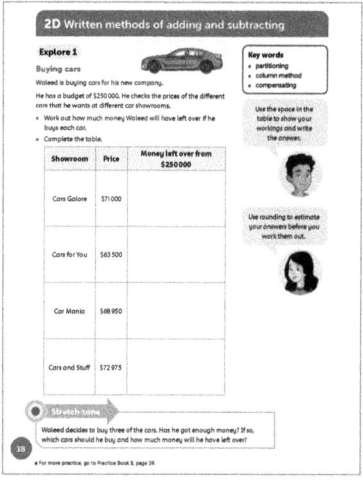

Differentiation

Supporting: Help students to round the car prices before estimating the difference.

Consolidating: Ask students to describe the different strategies they are using to solve and check each problem.

Extending: Ask students to create similar problems.

Stretch zone: *Waleed decides to buy three of the cars. Has he got enough money? If so, which cars should he buy and how much money will he have left over.*

He has enough money. Check that students have added three of the prices and have correctly calculated how much money Waleed would have left.

 Reflection time

Invite individual students who completed the Stretch zone question to share their methods for finding out whether Waleed can buy three cars and which showroom he should go to in the problem on page 38 of the Student Book. Ask students why they chose the strategy they used and compare it to other strategies used by other students.

Practice Book: Students complete Practice Book page 39. They can do this directly after the Main activity, as homework, or as the focus of a separate mathematics session to help students consolidate their learning and build fluency.

Remind students that they must make a reasonable estimate first and then calculate the answer. Say that they can also do their working on a separate sheet of paper if they find this easier.

Differentiated outcomes	
All students	should carry out the estimations and calculations with support.
Most students	will carry out the estimations and calculations independently.
Some students	may create their own similar problems involving estimating and calculating the amount of the budget left over.

Answers

Student Book page 38

Cars Galore $179 000

Cars for You $186 500

Car Mania $181 050

Cars and Stuff $177 025

Practice Book page 39

1 Estimate $170 000, total $167 965

2 Estimate $110 000, total $106 120

3 Estimate $50 000, difference $48 640

4 Estimate $20 000, money left $21 045

Stretch zone: Check that the amounts are totalled correctly and fall below $200 000. Students might choose, for example, a Basic, Standard and Mid-range totalling $177 200. This would leave $22 800.

2D Written methods of adding and subtracting

Explore 2 Student Book page 39 • Practice Book page 40

Specific learning focus
- Find the totals and differences of 5-digit numbers using a written method.

Global skills
- **Creative skills:** problem solving
- **Self-development skills:** reflecting on learning

Key vocabulary
- total, difference, estimate

Resources
- digit cards (one set per pair)

Language support

Continue to explain your thinking clearly and concisely as you model how to find totals and difference for students to copy. Encourage them to explain their own thinking to their partner.

Introductory activity

Write on the board the following numbers: 7943, 10 058 and 14 273.

Ask students, in pairs, to make three different addition calculations using pairs of these numbers, and three different subtraction calculations. For each calculation, they should estimate the answer by rounding first. As they work, ask them to describe their methods.

Main activity

Set the following problem.

The Dodgers baseball game had a crowd of 12 945. The Lions game had a crowd of 8961.

What was the crowd total of both games?

How many more fans were at the Dodgers game than the Lions game?

Ask students to discuss with a partner what they need to do to answer these problems, and how they can do this. Ask them to make an estimate before they calculate the answer. Take feedback. Share ways to add to find the total and ways to subtract to find how many more fans were at the Dodgers game. Together, demonstrate how you can use each strategy to solve the problem.

Write three 4- and/or 5-digit numbers on the board for students to add and find differences. Discuss and compare the strategies they use for adding three or more numbers.

Ask them to make up some problems using these numbers.

Explain the activity on page 39 of the Student Book and work through an example. Each pair will need a set of digit cards. Encourage them to use what they think is the most efficient strategy for each set of numbers.

Unit 2 Addition and subtraction

Ask students questions to support their learning, for example:

- Can you tell me one of the methods you used to add? Can you tell me another?
- Can you complete this sentence? To subtract two numbers, I can … To subtract 3 numbers, I can …
- Can you complete this sentence? I can check my addition is correct by …

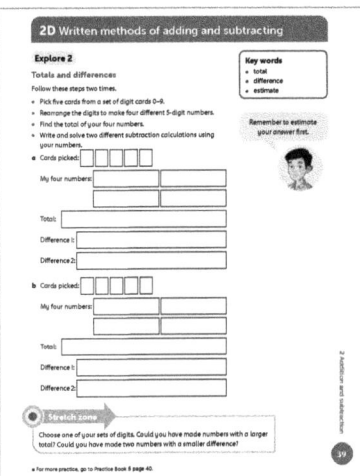

Differentiation

Supporting: Model for students how to find sums and differences of pairs of 3-digit numbers.

Consolidating: Ask students to find sums and differences of pairs of 5-digit numbers and explain their methods.

Extending: Challenge students to find sums and differences of pairs of 6-digit numbers, using more than one method.

Stretch zone: *Choose one of your sets of digits. Could you have made numbers with a larger total? Could you have made two numbers with a smaller difference?*

Check that students have given examples to justify their answers.

Reflection time

Review the questions on page 39 of the Student Book. Take feedback from the activity. Invite students to share their strategies for answering the calculations they made up. Encourage students to estimate by rounding first and then checking the calculations using the inverse operation.

Practice Book: Students complete Practice Book page 40. They can do this directly after the Main activity, as homework, or as the focus of a separate mathematics session to help students consolidate their learning and build fluency.

Give students digit cards to complete this activity or suggest that they make a set of their own.

Differentiated outcomes	
All students	should find totals and differences of pairs of 5-digit numbers independently and three or more 5-digit numbers with support.
Most students	will find totals of four 5-digit numbers and differences of pairs of 5-digit numbers.
Some students	may find totals and differences of pairs of 6-digit numbers.

Answers

Student Book page 39

Answers will vary because students make their own 5-digit numbers. Check that they have calculated the totals and differences for each set of numbers correctly. While they complete the activity, ask students to explain why they have chosen to use the strategies and methods that they have.

Practice Book page 40

Answers will vary because students make their own 5-digit numbers. Check that they have calculated the totals and differences for each pair of numbers correctly.

Stretch zone: Check that students' estimates are reasonable and that the calculation is correct.

Unit 2 Addition and subtraction

2E Adding and subtracting to solve problems

Discover Student Book page 40 • Practice Book page 41

Specific learning focus
Write addition and subtraction problems.

Global skills
- **Creative skills:** problem solving
- **Real-world skills:** financial literacy

Key vocabulary
- word problem, estimate, inverse operation

Resources
- mini whiteboards and markers

Language support
Model for students an example of a one-step and two-step word problem and display on the working wall, for example:
- *How much more does the red car cost than the blue car?*
- *When you buy two red cars, you get a $2500 discount. How much do two red cars cost with the discount?*

Introductory activity

Write on the board the numbers 36 821, 68 445 and 92 743. Ask students to add 6997 to each of the numbers and also subtract 6997 from each of the numbers, using their whiteboards. They should estimate their answers for each as well as check them using a method of their choice. Encourage them to consider which calculation strategy would be most efficient for each calculation.

Ask students to share four prices of information: what their estimates were for each calculation; their answers; how they calculated them; how they checked them to see whether they were correct.

Main activity

Give students the following word problem.

Johnson goes to the showroom and buys a new motorcycle for $24 995 and a set of accessories for $4745. The showroom owner gives Johnson £3750 off the price because there is a sale on. How much does Johnson pay for his goods?

Ask students to discuss in pairs how they would solve this two-step problem. Ask them to consider the following questions.
- *What is the first calculation you will do?*
- *Can you estimate the total before calculating?*
- *Which method will you choose to add the numbers?*
- *Can you estimate the amount Johnson has to pay when the sale price is calculated?*
- *Which method will you use to subtract?*

Discuss the answer to the problem as well as strategies for solving it.

Tell students that they will be thinking more about **word problems** involving addition and subtraction in the activity on page 40 of the Student Book. Ask students to work individually to write three problems about the car prices, then swap with a partner and calculate each other's answers.

Ask students questions to support their learning, for example:
- *Can you explain why you chose that method? Could you have used a different method?*
- *Which method will you use to subtract two numbers?*
- *How can you check whether your subtraction is correct?*

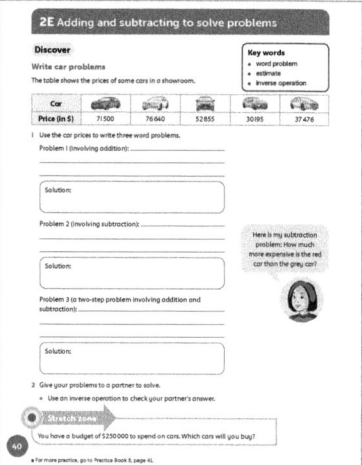

Differentiation

Supporting: Help students to word their problems by modelling addition and subtraction calculations.

Consolidating: Ask students to explain how they made their word problems and share the solutions.

Extending: Challenge students to make a three-step problem using 6-digit numbers.

Stretch zone: *You have a budget of $250 000 to spend on cars. Which cars will you buy?*

Students should be encouraged to justify their choices. Do they spend as much as possible of their $250 000? How much will they have left over?

Unit 2 Addition and subtraction

Reflection time

Ask different pairs of students to share one of their problems. Ask them to describe the strategy they used to solve the calculation. Ask other students whether they would have done it a different way and why.

Practice Book: Students complete Practice Book page 41. They can do this directly after the Main activity, as homework, or as the focus of a separate mathematics session to help students consolidate their learning and build fluency.

Revise how to round to the nearest 10 000, if appropriate. Encourage students to draw number lines for support.

Differentiated outcomes	
All students	should make word problems using prices from the table and solve them with support.
Most students	will make word problems and explain their methods for calculating.
Some students	will make three-step problems using 6-digit numbers and show alternative strategies for calculating.

Answers

Student Book page 40

Answers will vary because students choose their own prices from the table to form word problems. Check that they have written suitable word problems and solved them correctly.

Practice Book page 41

Town	Population	Nearest 10 000
Silao	74 242	70 000
Ticul	32 796	30 000
Garcia	93 641	90 000
Marfil	29 375	30 000
Apizaco	49 506	50 000

1 Estimate 120 000, total 123 748.
2 Ticul and Garzia together is larger – estimate 3000, difference 2689.
3 Estimate 4000, difference 3421.
4 New population estimate 80 000, total 79 742. Estimate of how many more people now live in Silao than Apizaco: 30 000, difference 30 236.

Stretch zone: Check that the estimate and answer is appropriate for the calculation chosen.

2E Adding and subtracting to solve problems

Explore Student Book page 41 • Practice Book page 42

Specific learning focus
- Use addition and subtraction to compare 4-digit numbers.

Global skills
- **Creative skills:** problem solving
- **Real-world skills:** interpreting information
- **Interpersonal skills:** communication

Key vocabulary
- word problem, estimate, inverse operation

Resources
- mini whiteboards and markers
- base-10 equipment (optional)

Language support

Provide sentence frames for students to refer to, to support them to write their answers to each word problem as a full sentence, for example: 'Everest is ____ m higher than Lhotse.'

 ### Introductory activity

Model an example of subtraction involving 5-digit and 4-digit numbers. At each stage of recording, explain what is happening. Can students tell you what should happen next? Practise putting the calculations into context to help students with the context used in the Student Book. For example:

8320 − 8187 = 133

A mountain is 8320 m high. A nearby mountain is 8187 m high. How much higher is the first mountain? The first mountain is 133 m higher.

 ### Main activity

Ask students to look at the information on Student Book page 41. Ask, *What do you notice? What is the information telling you?* Remind students that height is measured in metres and that the highest mountains in the world are all above 8000 m. They can then compare this figure with the heights given for the tallest mountains.

Unit 2 Addition and subtraction

Ask, *What is the difference between the heights of Everest and Lhotse?* Write the two heights on the board (8848 m and 8516 m). Ask students to write these on their whiteboards in columns and find the difference in height by subtracting using column subtraction. You could model this for some students with base-10 equipment to help them with the regrouping. Check that they agree that the difference is 332 m. Choose one student to talk through the method.

Students now complete the questions on page 41 of the Student Book. Make sure that students understand that the questions are asking them to find the totals or difference in heights of two mountains at a time. They could write their answers in full sentences on a separate sheet of paper or as part of their calculation in the table.

Ask further questions relating to the mountain heights, for example:

- *Is Lhotse higher than K2? How do you know?*
- *Is Kangchenjunga more than 100 m higher than Lhotse?*
- *Is the height of K2 between 8600 and 8610 m?*
- *If you climb to the top of Lhotse and back down, how far will you have climbed?*

Differentiation

Supporting: Model the calculations using base-10 equipment and encourage students to use strategies they know to help them complete the subtractions.

Consolidating: Ask students to continue to estimate and use a range of strategies when doing calculations. They should model the written subtraction method to a partner.

Extending: Ask students to use the most efficient method and check their answers using an **inverse operation**.

Stretch zone: *Which two mountains are closest in height? What is the difference between their heights?*

Students should give examples to show how they calculated their answers and checked them.

Reflection time

Select two of the problems from the Main activity to go through. Students should explain their methods and the rest of the class can check their work. Reinforce that estimating the answers before working them out is vital, especially when working with larger numbers. It allows students to see whether their answer looks 'about right'. They can also use estimates when checking with inverse operations.

Practice Book: Students complete Practice Book page 42. They can do this directly after the Main activity, as homework, or as the focus of a separate mathematics session to help students consolidate their learning and build fluency.

Explain that students will give the questions they write to another child or an adult but they will need to check their answers so they should be confident they can solve their own problems.

Differentiated outcomes	
All students	should solve the word problems using different strategies with support.
Most students	will solve the word problems using written and mental methods as appropriate.
Some students	may make and solve two- or three-step word problems using written or mental methods, estimating and checking.

Unit 2 Addition and subtraction

Answers

Student Book page 41

Problem	Calculation	Estimate	Method
How much higher is Everest than Lhotse?	8848 – 8516	8800 – 8500 = 300	Check that students have used a suitable method and obtained the answer 332 m.
I climbed both K2 and Kangchenjunga. What total height did I climb?	8611 + 8586	8600 + 8600 = 17 200	Check that students have used a suitable method and obtained the answer 17 197 m.
What is the difference in height between K2 and Makalu?	8611 – 8485	8600 – 8500 = 100	Check that students have used a suitable method and obtained the answer 126 m.
I climbed K2 and then Lhotse. How much higher than 17 000 m have I climbed?	(8611 + 8516) – 17 000	8600 + 8500 – 17 000 = 100	Check that students have used a suitable method and obtained the answer 127 m.

Practice Book page 42

Town	Population	Nearest 1000
Zefta	92 667	93 000
Safaga	32 944	33 000
Hurghada	95 622	96 000
Qotour	23 842	24 000
Basyun	55 523	56 000

Check that students have written suitable word problems and provided the correct answers.

Stretch zone: Check that students have correctly calculated using the rounded figures and the actual figures.

2 Addition and subtraction

Connect Student Book page 42

Big idea

I can add and subtract large numbers using a written method or a mental method. The strategy I use depends on the numbers in the calculation.

Global skills

- **Creative skills:** problem solving
- **Real-world skills:** presenting information
- **Interpersonal skills:** teamwork/communication

Key vocabulary

- addition, subtraction, estimate, rounding, mental method, strategy, written method, inverse operation, rounding, compensation, partitioning, near multiples, counting on or back

Resources

- flipchart paper and markers

Language support

Support students to write vocabulary-rich word problems by translating in some instances and providing bilingual dictionaries.

Introductory activity

Organise students into groups. Ask them to think of as many situations they can think of where we use 4- and 5-digit numbers in real life. Ask them to jot these down on a of piece of paper.

Allow groups a few minutes to come up with ideas and then come back together as a class. Record their ideas on the board, including any additional vocabulary that could relate to their ideas. You may also want to add examples of your own.

Students can then refer to this class list for ideas for their word problems during the main activity.

Unit 2 Addition and subtraction

Main activity

Arrange the class into mixed–attainment groups of three students. They should use flipchart paper to support collaborative work and only record answers in their book when they have completed the investigation. Refer them to page 42 of the Student Book. Discuss the context of football supporters in the example. Encourage them to think of their own contexts for the two problems they will work on, drawing on the list from the Introductory activity, if necessary. Remind them to set the problem in a realistic context, estimate first and check their answers using an inverse operation.

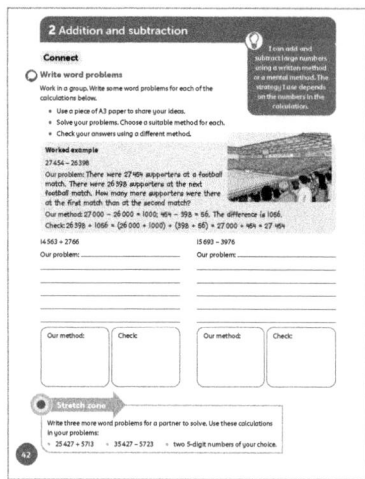

Differentiation

Supporting: Encourage students to use concrete resources or number lines to support them in their calculations.

Consolidating: Ask students to explain their choice of method to the whole group.

Extending: Encourage students to think of more than one possible method of calculation and present to groups why one method is most efficient.

Stretch zone: *Write three more word problems for a partner to solve. Use these calculations in your problems:*

- *25 427 + 5713*   • *35 427 – 5723*
- *two 5-digit numbers of your choice.*

 Reflection time

Each group should present their findings on flipchart paper. Compare the answers to the problems in context. Encourage students to share and compare the strategies they used to carry out the calculations.

Differentiated outcomes	
All students	should contribute to the work of the group.
Most students	will check the accuracy of the calculations.
Some students	will explain a range of methods to others.

Answers

Student Book page 42

Students choose their own contexts for the problems.

The numerical answers are 17 329 and 11 717.

2 Addition and subtraction

Review   Student Book page 43 • Practice Book page 43

Global skills
- **Real-world skills:** presenting information/interpreting information
- **Interpersonal skills:** communication
- **Self-development skills:** reflecting on learning

Student Book

With young students, assessment activities are most effective when carried out as an everyday classroom activity. Watch as students partition, exchange and regroup when adding or subtracting up to 6-digit numbers, using both mental and written methods. Provide additional paper for students to use for jotttings. It may also be appropriate to continue to provide students with concrete resources such as counters and a place-value tables, to support them with their calculations.

Listen to check that students understand how to estimate and to check answers using inverse operations.

Ask them additional questions to assess their understanding, for example:
- *Which mental calculations can you tell me about?*
- *What is the best strategy for adding 345 927 and 42 877?*
- *When do you use rounding and adjusting?*

Ask students to work in independently to complete the mind map on page 43 of the Student Book to show what they now know about mental calculation strategies. Encourage them to give examples for each.

For question 2, remind students of the steps to follow when solving a word problem, if necessary.

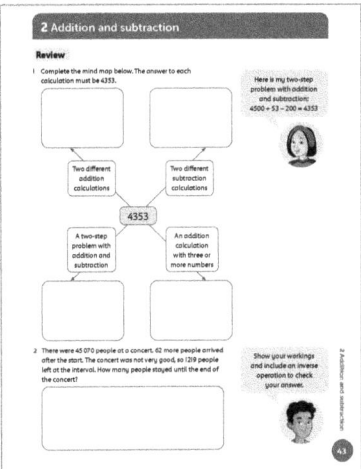

Answers

Student Book page 43

1 Answers will vary because students choose their own examples to complete the mind map.

2 43 913

Practice Book

With students in the upper primary years, it is appropriate to complete this as a whole-class discussion. You may choose to keep a record of the class discussion or a copy of the Review page for your own records. Use the Student Book to briefly remind students of the areas of mathematics that they have worked on in this unit.

Ensure that students have a copy of the Student Book to support them as they discuss and answer the questions in the Practice Book.

Allow students plenty of time for discussion before asking them to share their responses with the rest of the class. If students complete this assessment at home, encourage them to discuss this with adults.

Make a note of areas that students still feel unsure about. As addition and subtraction are essential in many of the other units, you can review and revisit this regularly. You can also build addition and subtraction of larger numbers into everyday practice, for example in discussions about distances between countries, depths of oceans and populations.

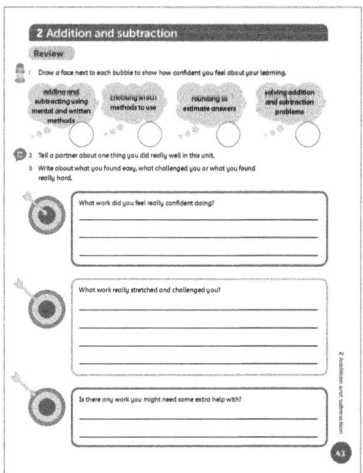

Additional material

There are additional end-of-unit assessment available on the *Oxford Owl for School* website.

3 Multiplication and division

Overview

Big idea

The Big idea for this unit is that multiplication can be understood as a repeated aggregation structure ('repeated addition') but also more broadly as a scaling structure, using the 'times as many' concept.

Multiplication and division are supported through understanding of arrays, through secure knowledge of multiplication and division facts ('times tables') and confident use of the place-value system.

A focus of this unit is the development of written methods for multiplication and division. Familiarity with the variety of multiplication and division strategies in the unit will equip students with the skills needed to interpret multiplication and division calculations in different ways and enable them to apply the methods, both mental and written, to a range of contexts.

Look out for

- **Students who sometimes make errors when they use compact column written methods (algorithms) for calculating.** This is because they do not understand the 'rules' and need plenty of practice using, for example, base-10 equipment to appreciate the effect on the numbers of each step in the method.
- **Students who make errors because they have a poor understanding of place value and have remembered the 'rules' incorrectly.** They should not treat the individual digits as numbers on their own, but work with the full range of digits in the number and appreciate the value of each digit.
- **Students who confuse factors and multiples.** Give students plenty of experience of listing the multiples of a number, these multiples being that number's times tables and all being bigger than the number, and experience of finding the factors of a number, which will all be smaller than the number.

Possible misconceptions

- **Students may think that multiplying by 10 is the same as 'adding a zero'.** Work with students on seeing multiplication and division by 10 (or 100, 1000 and so on) as moving digits to the left or right in the place-value column structure.

Key vocabulary

- addition, altogether
- partitioning grid method, column method
- subtraction, difference, compensation, halve
- how many more?, multiplication, groups of, product, written method, double
- divide, division, share, remainder, quotient, divisor, dividend, multiple, factor, common factor, common multiple, bracket, estimate, inverse operation, division fact, multiplication fact
- place-value table, chunking
- round, round up, round down, decimal point, decimal place, power of 10, prime number, composite numbers, prime factor, factor pair, square numbers, cube numbers, superscript, proportion, scaling problem
- which operation?

Coverage in lessons

Learning objective	E	3A	3B	3C	3D	3E	3F	3G	3H	3I	3J	C	R
Identify multiples and factors, including finding all factor pairs of a number, and common factors of two numbers.	✓	✓	✓	✓	✓								✓
Know and use the vocabulary of prime numbers, prime factors and composite (non- prime) numbers.			✓						✓				✓
Establish whether a number up to 100 is prime and recall prime numbers up to 19.									✓				✓
Multiply numbers up to 4 digits by a one- or two-digit number using a formal written method, including long multiplication for two-digit numbers.						✓					✓	✓	✓
Multiply and divide numbers mentally drawing upon known facts.	✓	✓	✓	✓	✓	✓	✓	✓		✓	✓	✓	✓
Divide numbers up to 4 digits by a one-digit number using the formal written method of short division and interpret remainders appropriately for the context.		✓							✓		✓	✓	✓
Multiply and divide whole numbers and those involving decimals by 10, 100 and 1000.				✓	✓		✓						✓
Recognise and use square numbers and cube numbers, and the notation for squared (^{2}) and cubed (^{3}).										✓			✓
Solve problems involving multiplication and division including using their knowledge of factors and multiples, squares and cubes.	✓	✓	✓	✓	✓	✓		✓		✓	✓	✓	✓
Solve problems involving addition, subtraction, multiplication and division and a combination of these, including understanding the meaning of the equals sign.	✓	✓		✓	✓	✓		✓			✓	✓	✓
Solve problems involving multiplication and division, including scaling by simple fractions and problems involving simple rates.			✓		✓	✓		✓			✓	✓	✓

3 Multiplication and division

Engage Student Book page 44

Big question
- How do I use what I know about multiplication and division to calculate with large numbers?

Global skills
- **Creative skills:** problem solving/exploring
- **Interpersonal skills:** communication
- **Self-development skills:** reflecting on learning

Key vocabulary
- partitioning, column method, compensation, grid method, product, quotient, divisor, inverse, brackets

Resources
- mini whiteboards and markers
- large sheets of paper and coloured pens

Language support
Set up this lesson by introducing any key or unfamiliar vocabulary that students will need to use in their discussion of the Engage page. This might include 'bunches' of bananas, 'market stall', 'crates' and 'awning'. Support with visual examples and model how to pronounce any new vocabulary accurately.

 **Introductory activity**

Write this problem on the board.

There are:

8 boxes with 12 apples each

6 boxes with 9 oranges each

5 boxes with 20 satsumas each.

How many pieces of fruit are there altogether?

Ask students, arranged in mixed-attainment groups, to work this out and to show their methods on a large sheet of paper. Methods may include multiplying using the formal written method (column method), grid method or mental methods. Each group should present their results to the class, describing their strategies.

 Main activity

Students continue to work in their mixed-attainment groups. Look together at page 44 of the Student Book. If you have access to an IWB, display the page. Look at the picture and ask them to describe what they can see. Encourage them to use mathematical language, for example: 'Each tray of eggs is arranged in a 6 × 4 array.'

Students should work in their groups to answer the questions in the speech bubbles. Once they have answered these, they can create their own word problems and work out the answers on their whiteboards.

As you move between groups, ask questions to assess their understanding, for example:
- *How many … are there altogether?*
- *How much will they cost if they are 25p each?*
- *How many … are in each box?*
- *How many would that be if we have double the amount?*
- *What calculation could you use to find out how many bananas there are? What strategy would you use?*

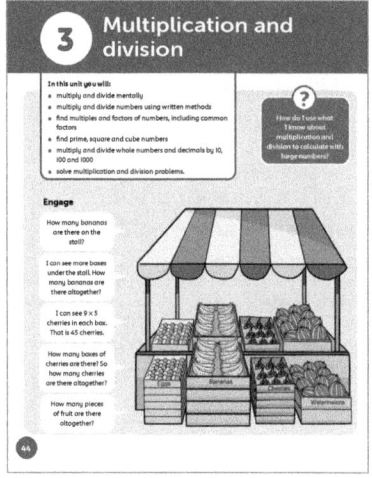

Differentiation
Supporting: Ask students to respond to the questions you put to each group.

Consolidating: Ask students to offer strategies to solve the problems.

Extending: Ask students to explain the strategies clearly to the rest of the group.

 Reflection time

Discuss the answers to each of the speech-bubble questions as well as the methods groups used to carry out the calculations. Ask them why they chose each method and see whether any other groups would have chosen a different strategy.

Groups can then share any word problems they made for the rest of the class to solve.

3A Multiplication and division facts

Discover Student Book pages 45–46 • Practice Book page 44

Specific learning focus
- Use multiplication and division facts for the two times to ten times tables to solve word problems.

Global skills
- **Creative skills:** problem solving
- **Interpersonal skills:** communication/teamwork

Key vocabulary
- division facts, multiplication facts, inverse operations, partitioning

Resources
- mini whiteboards and markers
- timers or stopwatches
- base-10 equipment

Language support
Remind students of what 'how many?' questions are asking them to find out. Encourage students to use the artwork in the Student Book to support their understanding of the word problems they are answering. Clarify, as necessary.

 **Introductory activity**

Write this problem on the board.

Feng scored 15 points in a chess tournament. Her friend Hua scored three times as many points. How many points did Hua score?

Ask students to discuss with a partner how to answer this problem. When they have solved the problem, ask pairs to share their strategies with the class. If no pair has used **partitioning**, demonstrate the following:

$15 \times 3 = 10 \times 3 + 5 \times 3 = 30 + 15 = 45$

You can also model this strategy using base-10 equipment.

Now tell students that the champion at the chess tournament scored 60 times as many points as Feng. Ask, *How can you work out how many points the champion scored?*

Pairs work to solve the problem and then share strategies with the class. If no pair offers this strategy, show them that you can find six lots of 15 and then make the answer 10 times bigger:

$(10 \times 6) + (5 \times 6) = 90$

$90 \times 10 = 900$

Again, use base-10 equipment to model this strategy.

 Main activity

Students should take it in turns to answer the multiplication facts questions on page 45 of the Student Book while their partner times them. Once they have both completed the questions, they can check their answers together. They should tell their partner how they remember the facts, with a particular focus on any that their partner found challenging.

They should support each other to solve the word problems on pages 45–46 of the Student Book, sharing the strategies they are using.

Ask pairs of students who complete all the questions to create additional word problems that can be shared in Reflection time.

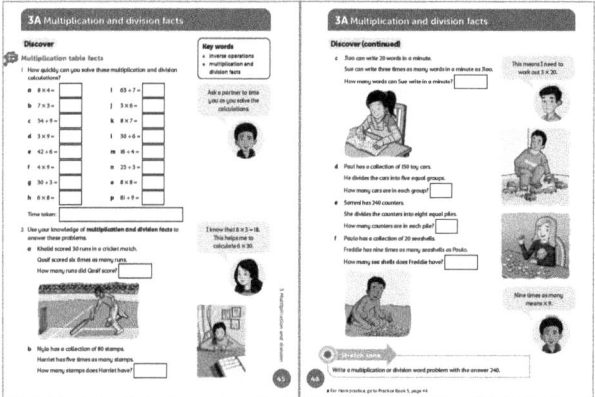

Differentiation

Supporting: Take students through the problem-solving steps and encourage them to use concrete resources such as base-10 equipment to help them solve the word problems.

Consolidating: Ask students to share with you their strategies for solving the word problems.

Extending: Ask students to create similar problems for the rest of the class to solve.

Stretch zone: *Write a multiplication or division word problem with the answer 240.*

Challenge students to choose a realistic context and check that their calculations work out correctly to 240.

Reflection time

Discuss students' answers to the word problems and ask them to share their strategies for calculating the answers. Ask pairs to work on the problems created by the pairs who completed all the questions during the Main activity. Students who created the questions should model the solutions when pairs have had sufficient time to solve them. Ask whether any pairs used a different strategy and ask them to model it.

Practice Book: Students complete Practice Book page 44. They can do this directly after the Main activity, as homework, or as the focus of a separate mathematics session to help students consolidate their learning and build fluency.

Recap the relationship between division and multiplication, as necessary.

Differentiated outcomes	
All students	should solve word problems with support.
Most students	will solve word problems working in pairs and sharing strategies.
Some students	may quickly use their knowledge of multiplication facts to solve and create word problems.

Answers

Student Book pages 45–46

1. a $8 \times 4 = 32$ g $30 \div 3 = 10$ m $16 \div 4 = 4$
   b $7 \times 3 = 21$ h $6 \times 8 = 48$ n $25 \div 5 = 5$
   c $54 \div 9 = 6$ i $63 \div 7 = 9$ o $8 \times 8 = 64$
   d $3 \times 9 = 27$ j $3 \times 6 = 18$ p $81 \div 9 = 9$
   e $42 \div 6 = 7$ k $8 \times 7 = 56$
   f $4 \times 9 = 36$ l $30 \div 6 = 5$

2. a 180   c 60   e 30
   b 400   d 30   f 180

Practice Book page 44

	×	7	3	6	9	10
1	8	56	24	48	72	80
2	4	28	12	24	36	40
3	2	14	6	12	18	20
4	5	35	15	30	45	50
5	10	70	30	60	90	100
6	11	77	33	66	99	110
7	6	42	18	36	54	60
8	12	84	36	72	108	120

9. Students choose their own multiplication facts to write statements.

Stretch zone: Using $6 \times 9 = 54$, $12 \times 9 = 108$ (double 54) and so $12 \times 90 = 1080$.

3A Multiplication and division facts

Explore Student Book page 47 • Practice Book page 45

Specific learning focus
- Use multiplication and division facts for the two times to ten times tables to play the divisors game.

Global skills
- **Creative skills:** exploring
- **Interpersonal skills:** communication/teamwork

Key vocabulary
- division facts, multiplication facts, divisor

Resources
- mini whiteboards and markers
- 0–9 digit cards

Language support

Create a poster showing the divisibility rules for 2, 5, 10, 3 and 8 in words, including examples.

Students can refer to these throughout the rest of the unit. Add any other divisibility rules to the poster as they are learned.

 **Introductory activity**

Write this number sequence on the board: 5, 10, 15, 20. As a class, continue to chant the sequence to 100. Can they recall what we call all these numbers? (multiples of 5). Ask students what they notice about all multiples of 5. Agree that multiples of 5 end with a 5 or 0.

Ask students whether 43 will divide exactly by 5. *What about 28? What about 35? How can you tell whether a number will divide exactly by 5?* Point out that numbers that divide exactly by 5 are also multiples of 5.

Repeat for multiples of 2 with this sequence: 10, 12, 14, 16, 18. Agree that multiples of 2 are even, and that numbers that divide by 2 must end in 0, 2, 4, 6 or 8.

Unit 3 Multiplication and division

Repeat for multiples of 10 (50, 60, 70, 80) and 100 (500, 600, 700). Agree that multiples of 10 end with zero and multiples of 100 end with two zeros.

Ask students to work with a partner and write on their whiteboards six numbers that are multiples of:

- 2 and 5
- 2 and 10
- 2, 5, 10 and 100.

They should write the **multiplication facts** for each number they choose, for example:

200 is a multiple of 2, 5, 10 and 100.

$200 = 2 \times 100 = 5 \times 40 = 10 \times 20 = 100 \times 2$.

 **Main activity**

Write on the board the following divisibility rules for 3 and 8.

A number is divisible by 3 if the sum of its digits is a multiple of 3.

A number is divisible by 8 if either the number made by the last 3 digits is divisible by 8, or if that number can be halved three times and still be a whole number.

Ask students to work in pairs and check whether these rules are true by trying a selection of numbers up to 100.

Choose a student to come to the front of the class and model how to play the game on page 47 of the Student Book with you.

Students should then play the game in pairs. As they play, circulate asking questions, for example:

Are any of your numbers divisible by both 2 and 5? How do you know? Can you think of a number less than 50 that would be divisible by 2, 5 and 8? What about a 3-digit number?

Using the divisibility rule for 3, can you choose three digit cards to make three different 3-digit numbers divisible by 3?

Differentiation

Supporting: Give students the list of rules with examples to help them.

Consolidating: Ask students to explain how they allocated points for one or more of their numbers. Can they say why the number is not divisible by, for example, 3?

Extending: Challenge students to try to find numbers less than 100 that are divisible by three of the following: 2, 3, 5 and 8. Ask them to explain how they know.

Stretch zone: *Write four 3-digit numbers that each score a total of 10 points.*

Students could choose, for example, 120, 240, 360, 840. Students can play the game, making 4-digit numbers. Ask, *Is it easier to get a higher score with more digits?* Students might realise that the numbers that might benefit from more digits are those that divide by 3 because the extra digits might help the sum of the digits make a multiple of 3. For other divisors, only the final three digits or fewer are needed to check the divisibility.

 Reflection time

Ask individual students to tell you some of the numbers they made that are divisible by 3, then 8, 10 and 100. Ensure that they explain how they know. Ask each pair to complete these sentences.

- A number is divisible by 3 if … .
- A number is divisible by 8 if … .
- A number is divisible by 10 if … .
- A number is divisible by 100 if … .

Practice Book: Students complete Practice Book page 45. They can do this directly after the Main activity, as homework, or as the focus of a separate mathematics session to help students consolidate their learning and build fluency.

Look at the worked example together, discussing how each is a related multiplication fact. Point out, for example, if they doubled the multiplier, how that changes the product. Ask students to give you more examples for $8 \times 7 = 56$, for example $8 \times 70 = 560$ and $16 \times 14 = 224$.

Differentiated outcomes	
All students	should recognise numbers that are divisible by 2 and 5.
Most students	will use the divisibility rules for 3 and 8 to recognise whether a number is a multiple of 3 or 8.
Some students	may use divisibility rules to make 3-digit numbers that are multiples of 2 and 3 numbers and choose their numbers strategically to gain more points.

Unit 3 Multiplication and division 59

Answers

Student Book page 47

Answers will vary according to the numbers made. While students are playing, ask them to explain how they can make a number that they know is divisible by 2, 3, 5 or 8 without having to do the division. Ask them to tell you the divisibility rules they know.

Practice Book page 45

1. $4 \times 7 = 28$, $7 \times 4 = 28$, $28 \div 4 = 7$, $28 \div 7 = 4$
2. $8 \times 6 = 48$, $6 \times 8 = 48$, $48 \div 8 = 6$, $48 \div 6 = 8$
3. $3 \times 7 = 21$, $7 \times 3 = 21$, $21 \div 3 = 7$, $21 \div 7 = 3$
4. $5 \times 9 = 45$, $9 \times 5 = 45$, $45 \div 5 = 9$, $45 \div 9 = 5$
5. $4 \times 12 = 48$, $12 \times 4 = 48$, $48 \div 4 = 12$, $48 \div 12 = 4$

Stretch zone: Examples could be $9 \times 18 = 162$ and $90 \times 3 = 270$.

3B Factors and multiples

Discover Student Book page 48 • Practice Book page 46

Specific learning focus
- Recognise multiples of 3, 4 and 5 and factors of numbers up to 120.

Global skills
- **Creative skills:** investigating
- **Interpersonal skills:** communication/teamwork
- **Self-development skills:** reflecting on learning

Key vocabulary
- multiple, factor, common multiple, common factor

Resources
- mini whiteboards and markers

Language support
Add definitions of 'factors' and 'multiple', including examples, to your working wall.

Introductory activity

Sit the class in a circle. Ask each student to think of their birthdate. Play a game of 'Cross the circle'. Say:
- *Cross the circle if your birthdate is an odd number.*
- *Cross the circle if your birthdate is a 2-digit number.*
- *Cross the circle if your birthdate is a multiple of 3.*
- *Cross the circle if your birthdate is factor of 24.*

Check that students can recall what a **factor** is. Ask them for a definition and then ask, for example, *Is 4 a factor of 24? Why? Is 5 a factor of 24? Why not?* Remove one chair before you make the last statement. This will leave one student standing in the middle. They should be the next person to make a statement so they can sit down when students move across the circle.

Main activity

Ask students to work in pairs. Look together at the second column and first row of the table on page 48 of the Student Book. *What type of number can you put in this section?* (an even number that is also a **multiple** of 3). Ask pairs to write possible numbers on their whiteboards. Repeat, looking at each column in turn and at least one section in each row.

Go through the instructions for the activity. Make sure that students understand that they may need to move the numbers to different sections as they go along so that they are sure to be able to use each number.

As you move between groups, ask questions to assess students' understanding and use of the vocabulary, for example:
- *How can you explain what a multiple is?*
- *What is a factor? Is there another way to explain?*
- *Is it possible for a number to be both a multiple of 3 and 5? How do you know?*

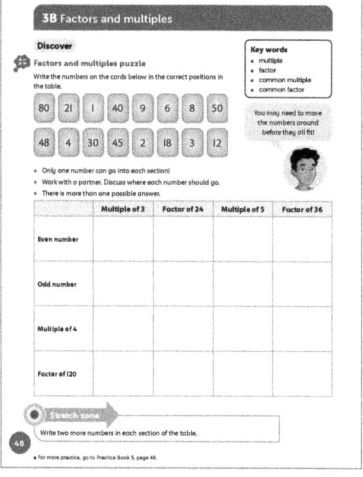

Differentiation

Supporting: Remind students of the divisibility rule for 3. Work together to find and list all factor pairs for 24 and 36 so that they can refer to these as they complete the puzzle.

Consolidating: Ask students to justify their choices for which section they put a number in. Can they tell you why they ruled out other sections?

Extending: Challenge students to add an additional row and column to the table, choosing their own multiple and factor criteria.

Stretch zone: *Write two more numbers in each section of the table.*

Students could also make up their own grid puzzle similar to the one on page 48 of the Student Book. They could choose their own multiples and factors as well as numbers to choose from.

Reflection time

Invite pairs of students to share the numbers they put into each section of the grid and also any extra numbers they added. They should be encouraged to explain how they knew to put the numbers where they did.

Practice Book: Students complete Practice Book page 46. They can do this directly after the Main activity, as homework, or as the focus of a separate mathematics session to help students consolidate their learning and build fluency.

Differentiated outcomes	
All students	should find examples to fit into each section with support.
Most students	will match the cards to the correct section.
Some students	may find additional examples to fit into the sections.

Answers

Student Book page 48

There are several answers. One possible answer is:

	Multiple of 3	Factor of 24	Multiple of 5	Factor of 36
Even number	18	2	50	4
Odd number	21	3	45	9
Multiple of 4	48	8	80	12
Factor of 120	30	1	40	6

Practice Book page 46

Answers will vary because students choose their own numbers to match the statements. Possible answers include:

1 3, 9, 15, 21, 27
2 2, 4, 8, 16, 32
3 1, 2, 4, 5, 8
4 36, 45, 54, 63, 72
5 10, 20, 30, 40, 50
6 21, 35, 49, 63, 77

Stretch zone: 3, 12 and 18 are all multiples of 3 and all factors of 36, for example.

3 is odd but 12 and 18 are even. 12 is a multiple of 4 and 18 is a multiple of 9, for example.

3B Factors and multiples

Explore Student Book pages 49–50 • Practice Book page 47

Specific learning focus
- Find factor pairs and common factors.

Global skills
- **Creative skills:** investigating
- **Real-world skills:** presenting information

Key vocabulary
- multiple, factor, common multiple, common factor, factor pair

Resources
- mini whiteboards and markers
- counters or cubes

Language support

Ask questions to further support students' understanding and use of the vocabulary, for example:
- *How can you explain what a common factor is?*
- *What is a factor pair? Is there another way to explain?*

 Introductory activity

Start by looking together at the Think back example on page 49 of the Student Book. Ask students to recall what a factor of a number is, and to say what the factors of 18 are. Can they see how the **factors pairs** are displayed in the spider diagram? Can they explain why they are called factor pairs?

Ask students to draw a similar spider diagram with 24 in the centre. Tell them to work in pairs to find all the factor pairs of 24 and label the legs of the spider with these. They should find: 1, 2, 3, 4, 6, 8, 12, 24.

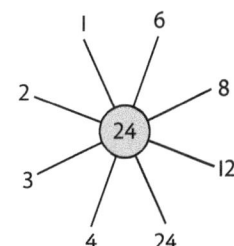

Unit 3 Multiplication and division 61

 **Main activity**

Ask students to write multiplication facts that make 12 to find the factor pairs of 12, and so find all the factors of 12. They should find 1 and 12, 2 and 6 and 3 and 4, making the factors of 12: 1, 2, 3, 4, 6 and 12.

Repeat for finding the factors of 18: the factor pairs are 1 and 18, 2 and 9, 3 and 6, making the factors of 18: 1, 2, 3, 6, 9 and 18.

Ask, *How many factors do 12 and 18 each have? Does one number have more factors than the other? Which factors are in the list for both 12 and for 18?* (1, 2, 3, 6). Explain that these are called **common factors** as they are factors of both numbers.

Students then work in their pairs to complete the activities on pages 49–50 of the Student Book to find factor pairs, factors and common factors.

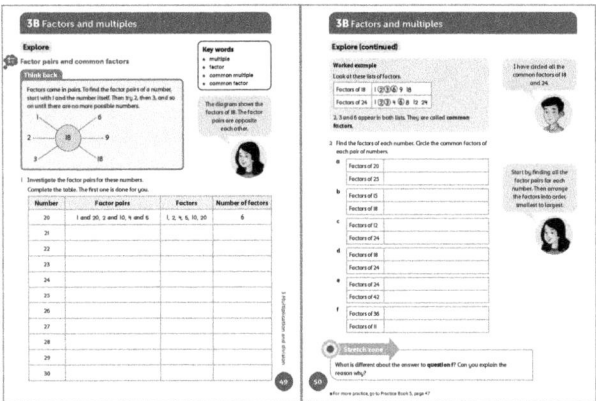

Differentiation

Supporting: Help students to find factors by arranging counters or cubes in arrays. For example, 12 counters will make an array of 3 by 4, so 3 and 4 are factors of 12.

Consolidating: Ask students to explain how they found the factors of a number.

Extending: Ask students to describe how to find the common factors of a pair of numbers.

Stretch zone: *What is different about the answer to question f? Can you explain the reason why?*

Students may notice that the only common factor of 36 and 11 is 1, because 11 only has factors 1 and 11, and 11 is not a factor of 36.

Reflection time

Look back at the completed table on page 49 and ask students to focus on the last column, which gives the total number of factors for each number. Ask them to describe anything they notice. They might notice, for instance, that just because one number is larger than another, this doesn't mean that it has more factors.

Ask some pairs to describe how they found the common factors of numbers on page 50. You could challenge the class to work out the common factors of 14 and 70 and choose a student to explain how they did it.

Practice Book: Students complete Practice Book page 47. They can do this directly after the Main activity, as homework, or as the focus of a separate mathematics session to help students consolidate their learning and build fluency.

Differentiated outcomes	
All students	should find most factor pairs for a number using concrete resources for support.
Most students	will find all factor pairs for 30 or any smaller number as well as common factors for a pair of numbers 30 or smaller.
Some students	may find common factors for larger numbers and explain why some pairs of numbers only have one common factor.

Answers

Student Book pages 49–50

1

Number	Factor pairs	Factors	Number of factors
20	1 and 20, 2 and 10, 4 and 5	1, 2, 4, 5, 10, 20	6
21	1 and 21, 3 and 7	1, 3, 7, 21	4
22	1 and 22, 2 and 11	1, 2, 11, 22	4
23	1 and 23	1, 23	2
24	1 and 24, 2 and 12, 3 and 8, 4 and 6	1, 2, 3, 4, 6, 8, 12, 24	8
25	1 and 25, 5 and 5	1, 5, 25	3
26	1 and 26, 2 and 13	1, 2, 13, 26	4
27	1 and 27, 3 and 9	1, 3, 9, 27	4
28	1 and 28, 2 and 14, 4 and 7	1, 2, 4, 7, 14, 28	6
29	1 and 29	1, 29	2
30	1 and 30, 2 and 15, 3 and 10, 5 and 6	1, 2, 3, 5, 6, 10, 15, 30	8

2 a Factors of 20: 1, 2, 4, 5, 10, 20
Factors of 25: 1, 5, 25
Common factors: 1, 5

b Factors of 15: 1, 3, 5, 15
Factors of 18: 1, 2, 3, 6, 9, 18
Common factors: 1, 3

c Factors of 12: 1, 2, 3, 4, 6, 12
Factors of 24: 1, 2, 3, 4, 6, 8, 12, 24
Common factors: 1, 2, 3, 4, 6, 12

d Factors of 18: 1, 2, 3, 6, 9, 18
Factors of 24: 1, 2, 3, 4, 6, 8, 12, 24
Common factors: 1, 2, 3, 6

e Factors of 24: 1, 2, 3, 4, 6, 8, 12, 24
Factors of 42: 1, 2, 3, 6, 7, 14, 21, 42
Common factors: 1, 2, 3, 6

f Factors of 36: 1, 2, 3, 4, 6, 9, 12, 18, 36
Factors of 11: 1, 11
Common factors: 1

Practice Book page 47

1 Factors of 12: 1, 12, 2, 6, 3, 4
Factors of 15: 1, 15, 3, 5
Common factors: 1, 3

2 Factors of 24: 1, 24, 2, 12, 3, 8, 4, 6
Factors of 30: 1, 30, 2, 15, 3, 10, 5, 6
Common factors: 1, 2, 3, 6

3 Factors of 16: 1, 16, 2, 8, 4
Factors of 40: 1, 40, 2, 20, 4, 10, 5, 8
Common factors: 1, 2, 4, 8

4 Factors of 18: 1, 18, 2, 9, 3, 6
Factors of 26: 1, 26, 2, 13
Common factors: 1, 2

5 Factors of 22: 1, 22, 2, 11
Factors of 33: 1, 33, 3, 11
Common factors: 1, 11

6 Factors of 25: 1, 25, 5
Factors of 50: 1, 50, 2, 25, 5, 10
Common factors: 1, 5, 25

Stretch zone: 15, 30 30, 45 45, 60 are possible examples.

3C Using known facts to multiply

Discover Student Book pages 51–52 • Practice Book page 48

Specific learning focus
- Multiply multiples of 10 to 90, and multiples of 100 to 900, by a single-digit number.
- Multiply by 25 by multiplying by 100 and dividing by 4.

Global skills
- **Creative skills:** problem solving/exploring
- **Interpersonal skills:** communication/teamwork

Key vocabulary
- multiples of 10, known facts, mental strategies, division fact

Resources
- mini whiteboards and markers
- counting stick (or metre rule)
- base-10 equipment
- coloured counters (two colours per pair)
- 0–9 digit cards (two sets per pair – even numbers only)

Language support

Draw students' attention to the first speech bubble on page 52 of the Student Book to help them structure explanations of their calculation strategy. Make a similar sentence frame to support students who choose to halve and halve again and then multiply by 100 as their strategy instead.

 **Introductory activity**

Say, *I know the multiplication fact 9 × 12 = 108. What **division facts** do I know then?* (108 ÷ 12 = 9 and 108 ÷ 9 = 12) *How do we describe this relationship between multiplication and division?* (Division and multiplication are inverses.)

Show students a counting stick. Explain that 0 is at one end and 90 is at the other. Agree to count in nines from 0.

Unit 3 Multiplication and division

As you point to various divisions, ask students to write down the multiplication facts and corresponding division facts on their whiteboards. Focus initially on the nine times table and then repeat for other multiplication tables.

Ask pairs to discuss how to multiply 9 × 50. Agree that you can multiply 9 by 5 and then make the answer 10 times bigger: 9 × 5 = 45, and then 45 × 10 = 450. Model with base-10 equipment, if appropriate, to reinforce. Repeat with 90 × 5 and see what students notice.

 **Main activity**

Write 48 × 5 on the board and ask pairs to calculate this and share the **mental strategies** they use. If they do not share the following method, introduce it as an alternative strategy.

48 × 10 = 480 so 48 × 5 = 240 (as 5 is half of 10).

Ask students to use this method for the following:

52 × 5   88 × 5

36 × 5   31 × 5

35 × 5

Now write 48 × 25 and ask pairs to calculate this and share the strategies they use. If they do not share the following, introduce it as an alternative strategy.

Establish that, because 25 is a quarter of 100, you can multiply by 100 and then divide by 4 or halve your answer and halve it again. Alternatively, you can find a quarter of the number first and then multiply by 100.

Students works in pairs to play the games on pages 51–52 of the Student Book. As students play the games, circulate and ask questions to encourage them to use **known facts** to solve their calculations.

- How can you multiply 40 by 7? What times table fact could you use to help you?
- I can multiply by 25 by …
- How can you find quarters of numbers? Is there another way?

While students are working, observe who is recalling the multiplication facts easily and who is struggling and will require more practice learning their multiplication tables.

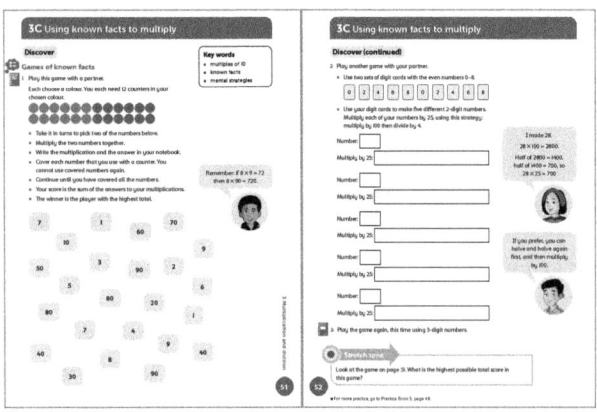

Differentiation

Supporting: Provide students with times table grids.

Consolidating: Encourage students to explain to their partner or an adult their calculation strategy for specific numbers.

Extending: Ask students to explore alternative strategies for multiplying by multiples of 10, for example multiply by 50 by multiplying by 100 then halving.

Stretch zone: *Look at the game on page 51. What is the highest possible total score in this game?*

The highest possible total is 22 306. This can be achieved by multiplying the largest numbers together in pairs at each step: 90 × 90, 80 × 80, 70 × 60, 50 × 40, 40 × 30, 20 × 10, 9 × 9, 8 × 7, 7 × 6, 5 × 4, 3 × 2 and 1 × 1, then adding these together.

Reflection time

Invite pairs to share the strategies they used to multiply pair of numbers from page 51 of the Student Book. Play the game together, either you against the class or divide the class into two teams.

Invite students to share some of the numbers they made on page 52 of the Student Book. Ask, *How did you multiply by 25?* Check students' confidence in using the strategy of multiplying by 100 and then dividing by four.

Practice Book: Students complete Practice Book page 48. They will need digit cards. They can do this directly after the Main activity, as homework, or as the focus of a separate mathematics session to help students consolidate their learning and build fluency.

Differentiated outcomes	
All students	should multiply two single-digit numbers, multiply by multiples of 10 and multiply by 25 with support.
Most students	will multiply multiples of 10 to 90 by a single-digit number or a multiple of 10 and multiply 2-digit numbers by 25 using the strategy of multiplying by 100, then dividing by 4.
Some students	may extend the activity to multiples of 100 and find and use different calculation strategies.

Answers

Student Book pages 51–52

Answers will vary because students choose which numbers to multiply. Check that students' multiplications are correct.

2 Answers will vary because students make their own 2-digit numbers and multiply them by 25. While students are working, ask them to explain why they used a particular method for a particular number. For example, when multiplying 36 by 25, they might remember that 36 is a multiple of 4 so they could do this:

$4 \times 25 = 100$

$36 \div 4 = 9$

$9 \times 100 = 900$

$36 \times 25 = 900$

3 Answers will vary because students make their own 3-digit numbers and multiply them by 25.

3C Using known facts to multiply

Explore Student Book pages 53–54 • Practice Book page 49

Specific learning focus
- Multiply by 19 or 21 by multiplying by 20 and adjusting.
- Use factors to multiply, for example multiply by 3, then double to multiply by 6.

Global skills
- **Creative skills:** exploring/investigating
- **Interpersonal skills:** communication/teamwork

Key vocabulary
- multiples of 10, known facts

Resources
- mini whiteboards and markers
- 1–9 digit cards

Language support
Model phrases and questions to support students with the necessary language, for example:
- *I can multiply by 21 by …*
- *I can multiply by 19 by …*
- *A factor is …*
- *What is an easy way to multiply by 4?*
- *An easy way to multiply by 8 is to …*
- *How can you multiply by 6 and 9?*

Practice Book page 48
Answers will vary because students make their own numbers to multiply by 25. Check that students' calculations are correct, and that they have made ten different calculations.

Stretch zone: 3 and 5 to make the number 53.

Introductory activity
This lesson looks at two different strategies for multiplication in detail. You may prefer to spread this learning across two mathematics sessions, focusing on the Introductory activity, the corresponding Student Book page and the Practice Book page activity in the first session.

Begin the lesson by practising multiplication and division facts. Call out an answer from a multiplication table. Ask students to write the multiplication facts and corresponding divisions. For example, for 56:

$7 \times 8 = 56, 56 \div 7 = 8, 8 \times 7 = 56, 56 \div 8 = 7$

Write 7×19 on the board. Ask pairs to discuss how to calculate this. Take feedback. Accept any strategies that give the correct answer. Focus on the idea of multiplying by 20 and adjusting, linking to the work on multiplying by multiples of 10 in 3C Discover.

Demonstrate how you can do this for 7×19 by using these number sentences:

7×20

$7 \times 2 \times 10 = 140$

$140 - 7 = 133$

Explain that we subtract 7 at the end because we multiplied by 20 instead of 19 and so we had one too many sevens.

Repeat for other calculations such as 5×19 and 9×19. Ask students to work these out. Ask them to show you their answers on their whiteboards.

Now ask them to use a similar strategy for multiplying 7 by 21:

7×20

$7 \times 2 \times 10 = 140$

$140 + 7 = 147$

Explain that we add 7 at the end because we multiplied by 20 instead of 21 and so we had one too few sevens.

Unit 3 Multiplication and division

Ask, *What is the same and what is different about the two strategies?*

Agree that for both you multiply by 20 and adjust. The difference is that you subtract a number when multiplying by 19 and you add a number when multiplying by 21. Call out some other numbers for students to multiply by 21.

You can use 5 and 9 again so that students may see patterns in the answers. Again, ask them to show their answers on their whiteboards.

At this point in the lesson, you may choose to ask students, in pairs, to play the game on page 53 of the Student Book. Explain and demonstrate the game. As students make numbers from their multiplications, they can place the number anywhere on the ladder as long as it keeps the numbers in order from bottom to top. If they cannot place an answer, they discard it. Ask students to work in pairs of similar mathematical experience so that they are evenly matched. Some pairs may need your support to play the game.

Main activity

Write on the board and say, *Tony says that if you know your 3 times table and know how to double you can complete the 6 times table. What do you think he means? Is he correct? Explain your thinking.*

Students work in pairs to discuss, then share ideas as a class. Agree that you can multiply by 6 by multiplying by 3 and doubling your answer. Work through some calculations to confirm.

Next, ask students to think about the eight and four times table. What statement could they make about these? Students should come up with a statement similar to: 'I can multiply by 4 by multiplying by 2 and doubling my answer.' Ask pairs to check that this strategy works.

Students should then work on the activity on page 54 of the Student Book in pairs.

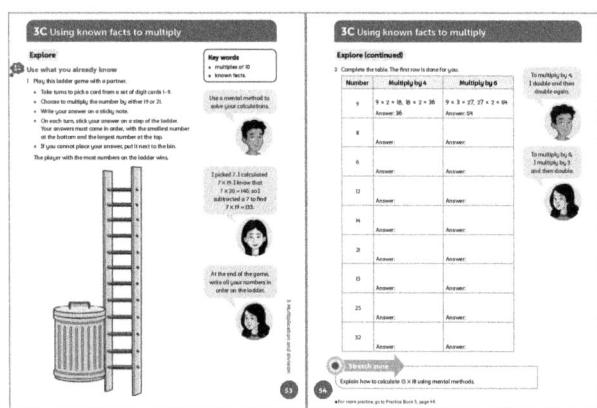

Differentiation

Supporting: Play the game with students so that you can support them and model the strategies.

Consolidating: Ask students to explain their thinking carefully.

Extending: Ask students to try to extend the game to 2-digit multiples of 10 multiplied by 19 and 21 and explore alternative strategies for multiplying.

Stretch zone: *Explain how to calculate 15 × 18 using mental methods.*

Prompt students if necessary to think about calculating 15 × 20 and then adjusting.

 Reflection time

Discuss strategies for winning the game on page 53 of the Student Book. For example, students could place lower numbers they made at the bottom of the ladder.

Ask students to work on the calculations below by drawing on the strategies they have been using or by using other strategies. They should share the strategies they are using with their partner and then discuss each as a class.

15 × 6

25 × 8

23 × 4

49 × 5

61 × 6

Practice Book: Students complete Practice Book page 49. They can do this directly after the Main activity, as homework, or as the focus of a separate mathematics session to help students consolidate their learning and build fluency.

Differentiated outcomes	
All students	should use the multiplying and doubling strategy with support.
Most students	will use the multiplying and doubling strategy as well as rounding and adjusting to complete calculations.
Some students	may extend the game to 2-digit numbers and understand and develop new strategies.

Answers

Student Book pages 53–54

1. Answers will vary because students make their own numbers to multiply by 19 or by 21. While students are working, observe who is using the 'multiply by 20 and adjust' strategy correctly. Ask students to explain why this strategy is useful. How could they use a similar strategy to multiply by 29 or 31?

2

Number	Multiply by 4	Multiply by 6
8	32	48
6	24	36
12	48	72
14	56	84
21	84	126
15	60	90
25	100	150
32	128	192

Practice Book page 49

	Starting number	× 20	× 19	× 21
1	9	180	171	189
2	6	120	114	126
3	4	80	76	84
4	5	100	95	105
5	2	40	38	42
6	8	160	152	168
7	7	140	133	147
8	30	600	570	630
9	20	400	380	420
10	50	1000	950	1050

Stretch zone: 25 × 19 = 475

3D Doubling and halving

Discover
Student Book page 55 • Practice Book page 50

Specific learning focus
- Double any number up to 100 and halve even numbers to 200.

Global skills
- **Creative skills:** investigating
- **Interpersonal skills:** communication/teamwork

Key vocabulary
- double, halve, partitioning, inverse operation

Resources
- mini whiteboards and markers

Language support
Continue to listen to how students read numbers, focusing particularly on the 3-digit numbers. Correct any errors in pronunciation, modelling how to say numbers correctly as necessary.

 Introductory activity

Students work in pairs. One of the pair picks a number between 1 and 20 and the other has to say its double. They then swap roles. They should keep note of the numbers they use and eventually double each number from 1 to 20.

Then, ask them to repeat, using the numbers 5, 10, 15, 20, 25, … up to 100.

Main activity

Ask pairs to double 78 and then halve 78. Ask pairs to share their strategies. Draw these diagrams on the board:

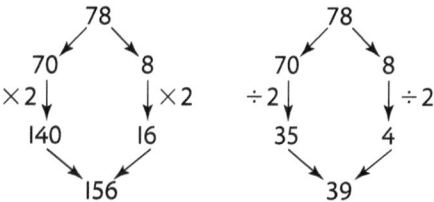

Encourage students to picture partitioning, doubling and recombining like this in their heads as they double. Ask them to double other numbers to 200 in this way.

Students work in pairs on the activity on page 55 of the Student Book so that they can take it in turns to match a number with its double or half. They should explain to their partner how they know that the answer is correct each time.

Once they have completed the activity, pairs should write a double or half 'riddle' to swap with another pair to solve. Give them an example, to help them get started, for example: 'Half of me is less than 5 but more than 2. I am a multiple of 3.'

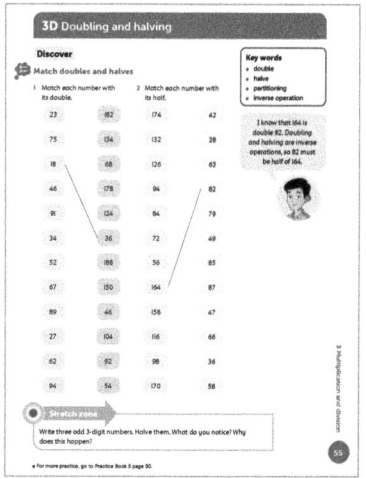

Unit 3 Multiplication and division

Differentiation

Supporting: Model using base-10 equipment to double and halve numbers.

Consolidating: Ask students to share their strategies with other students or an adult. Can they tell you what the relationship is between doubling and halving?

Extending: Challenge students to extend this strategy to 4-digit numbers.

Stretch zone: *Write three odd 3-digit numbers. Halve them. What do you notice? Why does this happen?*

Students write their chosen numbers, then halve them, on their whiteboards. They should notice that halving odd numbers always gives a number ending in $\frac{1}{2}$ or .5. Students might be able to say that this is because when halving an odd number, the 'odd' 1 gives $\frac{1}{2}$ when halved.

Reflection time

Ask pairs to think of other numbers they can double and halve. Ask, *What is the most difficult double or halve calculation you can carry out mentally?* Ask students who completed the Stretch zone question to share what they noticed with the class. Take examples of 2-digit and 3-digit odd numbers for all students to halve in order to check this.

Practice Book: Students complete Practice Book page 50. They can do this directly after the Main activity, as homework, or as the focus of a separate mathematics session to help students consolidate their learning and build fluency.

Look at the worked examples together. Ask, *Can you explain why one half is a whole number and the other is a decimal number? Looking at the numbers in the first column, which do you think will give decimal halves? Why?*

Differentiated outcomes	
All students	should double numbers to 100 and halve numbers to 200 with support.
Most students	will double numbers to 100 and halve numbers to 200 independently.
Some students	may double and halve numbers with up to four digits.

Answers

Student Book page 55

Check that students have doubled and halved correctly as they play the game, and that their explanations of their strategies for doubling and halving make sense and seem reasonable.

1		2	
23	46	174	87
75	150	132	66
18	36	126	63
46	92	94	47
91	182	84	42
34	68	72	36
52	104	56	28
67	134	164	82
89	178	158	79
27	54	116	58
62	124	98	49
94	188	170	85

Practice Book page 50

	Number	Double the number	Half the number
1	198	396	99
2	199	398	99.5
3	82	164	41
4	182	364	91
5	50	100	25
6	150	300	75
7	66	132	33
8	166	332	83
9	13	26	6.5
10	113	226	56.5
11	27	54	13.5
12	327	654	163.5

Stretch zone: Double 999 = 1998. Check for accurate reasoning, for example 'double 1000 = 2000, so doubling 1 less than 1000 gives 2 less than 2000'.

3D Doubling and halving

Explore Student Book page 56 • Practice Book page 51

Specific learning focus
- Double and halve multiples of 10 and 100.

Global skills
- **Creative skills:** exploring
- **Real-world skills:** presenting information

Key vocabulary
- double, halve, multiple of 10

Resources
- mini whiteboards and markers
- place-value table and counters or Gattegno chart (optional)

Language support
Read the speech bubbles on page 56 of the Student Book together. Ask students to explain their strategies, for example:
- *How can you halve 540?*
- *How can you double 7600?*

They can use the speech-bubble text to help them structure their explanation of their strategies.

Introductory activity
Write 230 on the board. Ask students to double this number in pairs. Take feedback on how they did this. If no one offers this strategy, model how to double 23 and then make their answer 10 times bigger. Say, *There are 23 tens in 230. Double 23 is 46 so double 23 tens is 46 tens. 46 tens is the same as 460.*

Call out multiples of 10, such as 170, 590. Ask students to double them using the strategy described above and record their answers on their whiteboards. Focus on halving in the same way. For example, ask them to halve 390 by halving 39 (19.5) and making the number ten times bigger (195).

Main activity
Develop the Introductory activity for doubling and halving multiples of 100, using numbers such as 2300 and 8700. Encourage students to make the number 100 times smaller, double or halve and then to make the number 100 times bigger or smaller again.

For example: 2300: 23, double: 46, make 100 times bigger: 4600; 23, halve: 11.5, make 100 times bigger: 1150. Emphasise that these two strategies rely on the ability to multiply and divide by 10 and 100 and to understand the effect. Support this understanding using a place-value table and counters or a Gattegno chart.

As you introduce each of these strategies, ask pairs to carry out the calculation first and then share their strategies. In this way, you may find they already know the strategy, or you may find that their strategy is more efficient. Encourage them to use whichever strategy is more efficient.

Students should work on their whiteboards to find the answers for the questions on page 56 of the Student Book and then record them in their books. If they are uncertain, they can check in pairs.

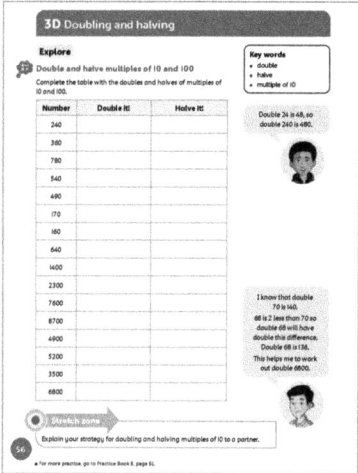

Differentiation
Supporting: Support students by modelling the use of place-value counters for doubling and halving up to 1000 to complete calculations with 3-digit multiples of 10.

Consolidating: Ask students to share their answers with their partner and describe their strategy.

Extending: Ask students to extend this strategy to working with 5-digit numbers.

Stretch zone: *Explain your strategy for doubling and halving multiples of 10 to a partner.*

Students can then also describe how they could use the strategy with 5-digit numbers and give examples.

Reflection time
Ask pairs to think of other numbers they can double and halve. What is the most difficult double or halve calculation they can carry out mentally?

Pairs find the biggest number they can double and halve.

Practice Book: Students complete Practice Book page 51. They can do this directly after the Main activity, as homework, or as the focus of a separate mathematics session to help students consolidate their learning and build fluency.

Before they begin, you may choose to work through a similar question to support them in coming up with different types of related facts.

Differentiated outcomes	
All students	should double and halve 3-digit numbers that are multiples of 10.
Most students	will double and halve 3- and 4-digit numbers that are multiples of 10.
Some students	may double and halve some 5-digit numbers.

Answers

Student Book page 56

Number	Double it!	Halve it!
240	480	120
360	720	180
780	1560	390
540	1080	270
490	980	245
170	340	85
160	320	80
640	1280	320
1400	2800	700
2300	4600	1150
7600	15 200	3800
8700	17 400	4350
4900	9800	2450
5200	10 400	2600
3500	7000	1750
6800	13 600	3400

Practice Book page 51

Check that the facts students derive for each diagram is correct. For example, for 'half of 180 = 90', students might have: 'double 180 = 360', 'half of 1.8 = 0.9', 18 ÷ 2 = 9, 180 ÷ 20 = 9.

Stretch zone: Double 97 = 194 can give double 970 = 1940, or double 98 = 196 and so on.

3E Written methods for multiplying

Discover Student Book page 57 • Practice Book page 52

Specific learning focus
- Multiply 3-digit numbers by single-digit numbers.

Global skills
- **Creative skills:** problem solving/exploring
- **Real-world skills:** presenting information
- **Interpersonal skills:** communication

Key vocabulary
- written methods, estimate

Resources
- 0–9 digit cards

Language support

Review words of position such as 'row', 'column', 'place', '1st', '2nd', '3rd', 'under', and 'above' to support students in being able to describe how to use a written method for multiplication.

 **Introductory activity**

Write this problem on the board.

Serena has five sheets of stamps. On each sheet, there are 85 stamps. How many stamps does she have altogether?

Ask students to discuss with a partner how they could work out the answer. Take feedback. Invite students to demonstrate their methods on the board. Aim for these strategies: multiply by 10 and halve, partition and multiply both by 5, the grid method. Model any of these methods that are not mentioned by students.

Main activity

Give students the following problem.

Samir collects shells. He has 6 bags. Each bag contains 346 shells. How many shells does he have altogether?

Ask students to estimate their answer first, then work it out using the column method. Invite a student to model the method or model it yourself.

```
     3 4 6
   ×     6
   -------
   2 0 7 6
     2 3
```

Explain the task on page 57 of the Student Book, using the worked example and the text in the speech bubbles. Each pair will need a set of digit cards. Encourage them to use short multiplication for each set of numbers and then use a mental method to check their answer. Remind them that they should estimate before they work out their answer. You may wish to pair students by prior experience. This way, you can ask some pairs to focus on multiplying 2-digit numbers by single-digit numbers, then move on to working with 3-digit numbers. Other pairs could start with 3-digit numbers and move on to working with 4-digit numbers.

While students complete the activity, ask them to explain why they have chosen to use the strategies and methods that they have by asking questions such as:

- How could you make a reasonable estimate for this calculation?
- How do you record this calculation using a **written method**?
- Is there another strategy that you can use to find the answers to a multiplication calculation?
- Which method did you prefer for this calculation? Why?
- Which operation and strategy did you use to check your answer?

After students have made two calculations, they can make more, recording their working on a separate sheet of paper. They could also use the digit cards to make up a multiplication word problem including a 3-digit and single-digit number and swap it with another pair to solve.

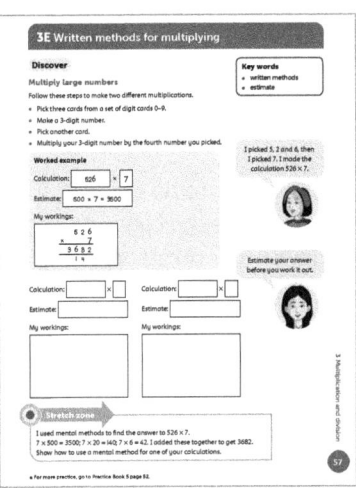

Differentiation

Supporting: Model how to multiply 3-digit numbers by single-digit numbers using the column method.

Consolidating: Ask students to explain how to multiply 3-digit numbers by single-digit numbers using the column method.

Extending: Challenge students to multiply 4-digit numbers by single-digit numbers using the column method.

Stretch zone: *I used mental methods to find the answer to 526 × 7.*

7 × 500 = 3500; 7 × 20 = 140; 7 × 6 = 42. I added these together to get 3682.

Show how to use a mental method for one of your calculations.

Check that students have chosen an appropriate mental method and have explained it clearly.

 Reflection time

Take feedback from the activity, working through some students' answers as a class, including students sharing their **estimates** and how they reached them. Were their estimates reasonable? Invite students to describe the column method in their own ways. This will support other students.

Practice Book: Students complete Practice Book page 52. They can do this directly after the Main activity, as homework, or as the focus of a separate mathematics session to help students consolidate their learning and build fluency.

Encourage students to continue to estimate their answers before they work them out.

Differentiated outcomes	
All students	should multiply 3-digit numbers by single-digit numbers with support.
Most students	will multiply 3-digit numbers by single-digit numbers using short multiplication.
Some students	may multiply 4-digit numbers by single-digit numbers using short multiplication.

Answers

Student Book page 57

Answers will vary because students make their own 3-digit numbers. Check that they have calculated the products for each pair of numbers correctly.

Practice Book page 52
1. 318
2. 322
3. 432
4. 336
5. 2916
6. 17 472
7. 8325
8. 2556

Stretch zone:
1. 300
2. 350
3. 500
4. 320
5. 3000
6. 18 000
7. 9000
8. 3000

3E Written methods for multiplying

Explore Student Book pages 58–59 • Practice Book page 53

Specific learning focus
- Multiply 3-digit numbers by single-digit numbers.

Global skills
- **Creative skills:** problem solving
- **Real-world skills:** interpreting information
- **Self-development skills:** reflecting on learning

Key vocabulary
- grid method, column method

Resources
- 1–9 digit cards

Language support
Explain any context vocabulary in the word problems that is unfamiliar to students as they work through the questions.

Introductory activity

Ask a student to choose a digit card, for example 4.

The class should all count up together in fours to 40: 4, 8, 12, 16, 20, 24, 28, 32, 36, 40.

Then they count up in 40s to 400: 40, 80, 120, 160, 200, 240, 280, 320, 360, 400.

Finally, they count in 400s to 4000: 400, 800, 1200, 1600, 2000, 2400, 2800, 3200, 3600, 4000.

Repeat with different single-digit numbers.

Main activity

Ask students to look at the worked example in question 1 on page 58 of the Student Book. Read the word problem, then discuss with students how the answer can be estimated, before explaining each stage of the **grid method** to get to the final answer as you work through it on the board.

Now model how to calculate the answer using the **column method** on the board, beside the grid method workings. Ask students at each stage to mentally calculate the sub-products before reaching the final answer.

$$\begin{array}{r} 3\ 6 \\ \times\ 9\ 8 \\ \hline 2\ 8\ 8 \\ 3\ 2\ \overset{4}{4}\ 0 \\ \hline 3\ \overset{5}{5}\ 2\ 8 \end{array}$$

Ask students to look at the calculations using the two methods. They can discuss in pairs what is the same and what is different about the two methods, then come together for a discussion.

Students then work on the remaining questions on pages 58–59 of the Student Book.

Ask questions to encourage students to explain their workings, for example:
- *What is the grid method?*
- *How can you know whether an answer is reasonable?*
- *If you know that 6 × 7 is 42, then what else do you know?*
- *What is the column method? How is it different from the grid method?*

Encourage most students to do the Stretch zone activity. Explain that they may use the contexts from the word problems they have just worked on, if they are struggling to come up with their own.

72 Unit 3 Multiplication and division

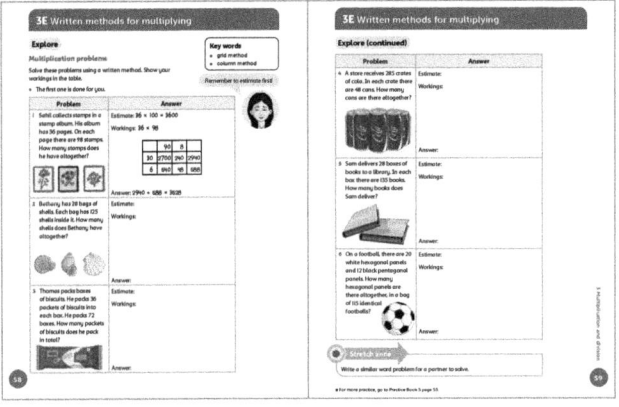

Differentiated outcomes	
All students	should multiply 2-digit numbers by 2-digit numbers using the grid method with support.
Most students	will multiply 2-digit numbers by 2-digit numbers using the column method.
Some students	may multiply 3-digit numbers by 2-digit numbers using the column method.

Answers

Student Book pages 58–59

1 (Provided): Estimate 36 × 100 = 3600, answer = 3528.

2 Estimate 30 × 120 = 3600, answer = 3500.

3 Estimate 40 × 70 = 2800, answer = 2592.

4 Estimate 300 × 50 = 15 000, answer = 13 680.

5 Estimate 30 × 130 = 3900, answer = 3780.

6 Estimate 20 × 120 = 2400, answer = 2300.

Practice Book page 53

1 432

2 332

3 464

4 3276

5 581

6 4584

7 6885

Stretch zone: 9 × 765 = 10 × 765 − 765

Differentiation

Supporting: Focus on 2-digit number × 2-digit number and using the grid method only. Give students a times table grid as additional support, if appropriate.

Consolidating: Ask students to explain how they made their estimate and to describe the calculation method they used for one of the questions.

Extending: Challenge students to make up a word problem that requires them to multiply a 4-digit number by a 2-digit number.

Stretch zone: *Write a similar word problem for a partner to solve.*

Students should use 2- and 3-digit numbers in their word problem and focus on multiplication. They should calculate the answer to their own problem before sharing it with their partner. After their partner has solved the problem, they can compare strategies, discussing which is more efficient.

Reflection time

Take feedback from the activity. Invite students to explain the column method in their own ways. This will support other students. Students who have written their own problems can share these for the rest of the class to solve.

Practice Book: Students complete Practice Book page 53. They can do this directly after the Main activity, as homework, or as the focus of a separate mathematics session to help students consolidate their learning and build fluency.

Look at the worked example together, taking students through each step of the calculation. Say, for example, *First I multiply 2 ones by 5. That's 10 ones or 10. I record the 0 in the same column, but the 1 goes in the tens column because that's how many tens there are. Next, I multiply 4 tens by 5 …*

3F Multiplying and dividing by 10, 100 and 1000

Discover Student Book page 60 • Practice Book page 54

Specific learning focus
- Multiply and divide whole numbers and those involving decimals by 10, 100 and 1000.

Global skills
- **Creative skills:** exploring
- **Real-world skills:** presenting information
- **Interpersonal skills:** communication

Key vocabulary
- place-value grid, decimal point, decimal place, powers of 10

Resources
- mini whiteboards and markers
- 0–9 digit cards for each student

Language support
Use a place-value grid with counters as part of a classroom display to show how numbers change when we multiply and divide by powers of 10. Include text in speech bubbles to explain these changes, for example: 'If I move the digits one place to the left, the number is 10 times bigger.'

 **Introductory activity**

Write on the board:

24   138   9.32   74.39

Ask students to use a **place-value grid** and counters to set up each number in turn and then, with a partner, discuss what happens to the numbers as they are multiplied by 10, 100 or 1000. Take feedback. Ask, *What happens to the digits in each number as they are multiplied?* Discuss the mental calculation strategy of moving the digits two places to the left.

 Main activity

Look at the worked example for question 1 on page 60 of the Student Book together. Ask students to explain how the start number has changed as it is multiplied by 10, 100 and 1000. Each student needs a set of digit cards. You may wish to support some pairs of students. This activity involves multiplying a 2-digit number by 10, 100 and 1000, so is relatively straightforward if students are reminded of the value of each column and how to move digits to the left when multiplying by powers of 10. Continue to support students as they work in pairs. Pairs can continue to work with a place-value grid.

Ask questions to encourage them to develop their thinking. For example,

- *I multiply a 2-digit number by 1000. How many digits will my answer have?*
- *My number is 24 and I multiply by 100. What is the value of the 2 digit now?*
- *I multiply 452 by 100. What column will the 5 digit be in now?*

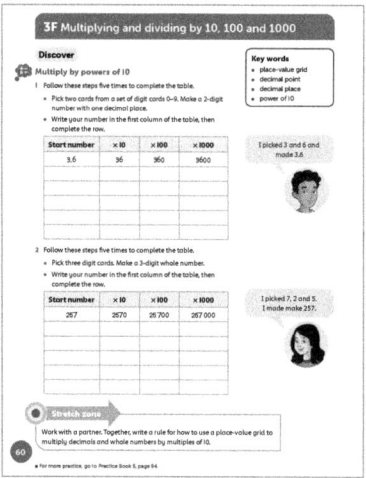

Differentiation

Supporting: Work with students to model multiplying by 10 by moving the digits to the left on a place-value grid.

Consolidating: Ask students to explain how to multiply by 1000.

Extending: Ask students to extend the calculations to multiplying by 10 000.

Stretch zone: *Work with a partner. Together, write a rule for how to use a place-value grid to multiply decimals and whole numbers by multiples of 10.*

Ask pairs to write their explanations on their whiteboards, then share them with another pair or student. Can the other pair or student suggest any changes to make the explanation clearer?

 Reflection time

Ask students to share the methods they used to multiply by 10, 100 or 1000. Ask, *When do we need to do this in real life?* Ask students to think of examples such as 'A crate contains 48 tins of fruit. How many tins will be in 100 crates?'

Practice Book: Students complete Practice Book page 54. They can do this directly after the Main activity, as homework, or as the focus of a separate mathematics session to help students consolidate their learning and build fluency.

Revise how as numbers get 10 times larger each digit moves one place to the left.

Unit 3 Multiplication and division

Differentiated outcomes	
All students	should carry out the calculations using a place-value grid and with support.
Most students	will carry out the calculations independently.
Some students	may do the calculations mentally with confidence and extend the calculations to multiplying by 10 000.

Answers

Student Book page 60

Answers will vary because students make their own numbers. Check that they have calculated the products for each pair of numbers correctly.

Practice Book page 54

		TTh	Th	H	T	O	.	t
	5.8					5	.	8
	× 10				5	8		
1	× 100			5	8	0		
2	× 1000		5	8	0	0		
	11.6				1	1	.	6
3	× 10			1	1	6		
4	× 100		1	1	6	0		
5	× 1000	1	1	6	0	0		
	27.3				2	7	.	3
6	× 10			2	7	3		
7	× 100		2	7	3	0		
8	× 1000	2	7	3	0	0		
	10.1				1	0	.	1
9	× 10			1	0	1		
10	× 100		1	0	1	0		
11	× 1000	1	0	1	0	0		

Stretch zone: 4.327 × 100 000 = 432 700. Students can set out place-value counters on a grid to represent 4.327, then move the counters five places to the left as they are multiplying by 100 000, which is the same as 10 × 10 × 10 × 10 × 10.

3F Multiplying and dividing by 10, 100 and 1000

Explore Student Book page 61 • Practice Book page 55

Specific learning focus
- Multiply and divide whole numbers and those involving decimals by 10, 100 and 1000.

Global skills
- **Creative skills:** exploring
- **Real-world skills:** presenting information
- **Interpersonal skills:** communication

Key vocabulary
- place-value grid, decimal point, decimal place, power of 10

Resources
- mini whiteboards and markers
- 0–9 digit cards

Language support
Ask questions to support the methods, for example:
- *How can you divide 473 by 100? Is there another way?*
- *The place-value grid helps you divide by …*

Introductory activity

Ask a student to choose three cards from a set of 0–9 digit cards and make a 3-digit number with them, using decimals if they wish.

The rest of the students should write down the number and then multiply it by 10, and by 100, and by 1000. Check by asking students to hold up their whiteboards.

Repeat with other 3-digit numbers drawn from the digit cards.

Main activity

Write on the board the number 47 500.

Ask students to divide this by 10 and then by 100 and then by 1000. Share answers.

Using a place-value grid, emphasise that when you divide by 10, digits move one place to the right; when you divide by 100, they move two places; when you divide by 1000, they move three places.

Discuss what is the same and what is different about multiplying by 10, 100 or 1000 and dividing by 10, 100 or 1000.

Look at the worked example for question 1 on page 61 of the Student Book together. Ask students to explain how the start number has changed as it is divided by 10, 100 and 1000. Each student needs a set of digit cards. You may wish to support some groups of students. This activity involves dividing numbers by 10, 100 and 1000, so is relatively straightforward if students are reminded of the value of each column and how to move digits to the right when dividing by powers of 10. Continue to remind them of this as they work in their pairs. Pairs can continue to work with a place-value grid.

Ask questions to encourage them to develop their thinking. For example:

- I divide a 4-digit number by 1000. How many digits will my answer have?
- My number is 249 and I divide by 100. What is the value of the 4 digit now?
- I divide 4520 by 100. What column will the 5 digit be in now?

Once students have completed their tables, they can use one of their calculations to make a word problem, which they can give to another pair to solve.

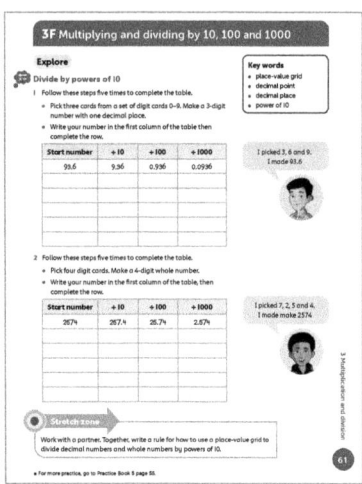

Differentiation

Supporting: Model dividing by 10 and moving the digits to the right using a place-value grid.

Consolidating: Ask students to explain how to divide by 10, 100 and 1000. *How are they the same? How are they different?*

Extending: Ask students to extend the calculations to dividing by 10 000.

Stretch zone: Work with a partner. Together, write a rule for how to use a place-value grid to divide decimal numbers and whole numbers by powers of 10.

Students can also make up word problems that involve dividing decimal numbers by 10, 100 or 1000 to share with the class.

Reflection time

Ask students to share the answers in the tables on page 61 of the Student Book. Ask, *Do you feel confident with dividing by moving digits to the right? How could you check your answers?*

Practice Book: Students complete Practice Book page 55. They can do this directly after the Main activity, as homework, or as the focus of a separate mathematics session to help students consolidate their learning and build fluency.

Revise how numbers get 10 times smaller as each digit moves one place to the right.

Differentiated outcomes	
All students	should divide whole and decimal numbers by 10, 100 and 1000 using a place-value grid with support.
Most students	will divide whole and decimal numbers by 10, 100 and 1000 independently using a place-value grid.
Some students	may extend the pattern to dividing whole and decimal numbers by 10 000.

Answers

Student Book page 61

Answers will vary because students use digit cards to find which number they will divide. Check that they have divided the numbers correctly. While they complete the activity, ask students to explain how they used the strategy.

Practice Book page 55

		TTh	Th	H	T	O	.	t	h	th
	2856		2	8	5	6	.			
	÷ 10			2	8	5	.	6		
1	÷ 100				2	8	.	5	6	
2	÷ 1000					2	.	8	5	6
	28 560	2	8	5	6	0				
3	÷ 10		2	8	5	6				
4	÷ 100			2	8	5	.	6		
5	÷ 1000				2	8	.	5	6	
	35 405	3	5	4	0	5				
6	÷ 10		3	5	4	0	.	5		
7	÷ 100			3	5	4	.	0	5	
8	÷ 1000				3	5	.	4	0	5
	3007		3	0	0	7				
9	÷ 10			3	0	0	.	7		
10	÷ 100				3	0	.	0	7	
11	÷ 1000					3	.	0	0	7

Stretch zone: 941 000 ÷ 10 000 = 94.1

Students may be able to describe how dividing by 10 000 is the same as moving digits four places to the right.

3G Written methods for dividing

Discover Student Book page 62 • Practice Book page 56

Specific learning focus
- Divide 3-digit numbers by single-digit numbers using a written method, including those with a remainder.

Global skills
- **Creative skills:** exploring
- **Real-world skills:** presenting information
- **Interpersonal skills:** communication

Key vocabulary
- chunking, quotient, divisor, dividend, remainder

Resources
- mini whiteboards and markers
- 0–9 digit cards for each pair

Language support
Display on the working wall a division calculation and label the parts as: 'quotient', 'divisor', 'dividend' and 'remainder'. Include a definition of each term in speech bubbles.

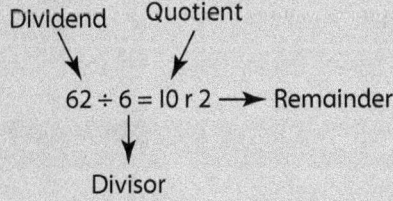

Introductory activity

Write this problem on the board.

Ryan picks 148 apples to sell in bags of 7 apples. How many bags does he have?

Ask students to discuss with a partner how they could work this out, estimating first. They can note ideas on their whiteboards. Share strategies as a class, encouraging students to demonstrate their ideas on the board. Focus on **chunking**. Ask them to work with their partner to take away the greatest number of groups of 7. Take feedback. Invite a student to demonstrate if appropriate, otherwise you can model the calculation:

148 ÷ 7

140 (2 lots of 7 is 14, so 20 lots of 7 is 140)

8

7 (1 lot of 7)

Agree that Ryan can sell 21 bags. Ask: *What can he do with the leftover apple?* Agree that it won't fill a bag, so he can eat it! Establish that students rounded their answer down. Check that the answer is correct by multiplying 21 by 7 and adding the **remainder**.

Main activity

Write on the board the calculation 853 ÷ 6. Talk through the example and model how the division is worked out using long division.

```
        1 4 2   r1
    6 | 8 5 3
        6 0 0
        ―――――
        2 5 3
        2 4 0
        ―――――
          1 3
          1 2
          ―――
            1
```

Tell students that the answer to a division calculation is called the **quotient** and that the number you divide by is called the **divisor**. The number you are dividing is called the **dividend**.

Demonstrate the game on page 62 of the Student Book, choosing a student to be your partner. Each pair needs a set of digit cards. As previously, you may wish to pair students by prior experience for the Student Book activity. You can ask some students to focus on dividing 2-digit numbers by single-digit numbers in the first instance; some on 3-digit numbers only and some on both 3- and 4-digit numbers.

Support the key language, by asking questions such as:
- *Can you explain how to divide using chunking?*
- *What is a remainder?*
- *Can you show me the quotient?*

If students complete the game quickly, ask them to extend it to play to best out of three turns each.

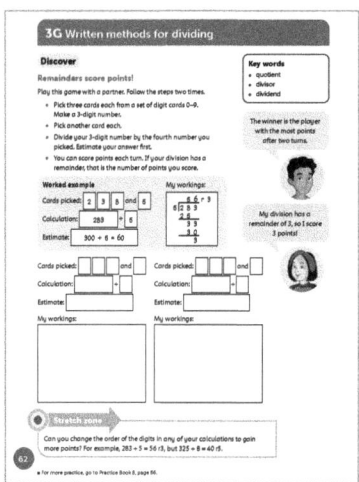

Differentiation

Supporting: Use place-value counters or base-10 equipment to model the division.

Unit 3 Multiplication and division

Consolidating: Ask students to explain how to divide 3-digit numbers by single-digit numbers.

Extending: Challenge students to divide 4-digit or 5-digit numbers by single-digit numbers.

Stretch zone: *Can you change the order of the digits in any of your calculations to gain more points? For example, 283 ÷ 5 = 56 r3, but 325 ÷ 8 = 40 r5.*

Check that students have successfully changed one of their calculations to gain more points.

You could also ask students to make up their own game to help them practise division.

Reflection time

Take feedback from the game. Invite pairs to share. Ask, *How did you divide your numbers?* Ask them to record their methods on the board. Play the game together, either you against the class or divide the class into two teams.

Practice Book: Students complete Practice Book page 56. They can do this directly after the Main activity, as homework, or as the focus of a separate mathematics session to help students consolidate their learning and build fluency.

Students will be looking at short division in the next lesson. You may prefer to ask them to complete this activity after they have had more experience with this written method. Alternatively, you could also ask students to use long multiplication to find the answers if they prefer. Students could use place-value counters or base-10 equipment here to help with the calculations. Remind students to estimate the answer before calculating.

Differentiated outcomes	
All students	should be able to divide 2-digit numbers by single-digit numbers.
Most students	will divide 3-digit numbers by single-digit numbers.
Some students	may divide 4-digit numbers by single-digit numbers.

Answers

Student Book page 62

Answers will vary because students use digit cards to find the numbers they will use in their divisions. Check that they have divided the numbers correctly. While they complete the activity, ask students to explain why they have chosen to use the strategies and methods that they have.

Practice Book page 56

Accept any reasonable estimates.

1. Estimate: 360 ÷ 6 = 60. Answer: 58
2. Estimate: 550 ÷ 5 = 110. Answer: 110 r 1
3. Estimate: 480 ÷ 4 = 120. Answer: 120 r 2
4. Estimate: 800 ÷ 8 = 100. Answer: 105 r 6
5. Estimate: 150 ÷ 3 = 50. Answer: 50 r 2
6. Estimate: 450 ÷ 9 = 50. Answer: 51 r 2
7. Estimate: 900 ÷ 3 = 300. Answer: 296

Stretch zone: Students should notice that 150 ÷ 3 = 50 so there are 2 left over in 152.

3G Written methods for dividing

Explore Student Book pages 63–64 • Practice Book page 57

Specific learning focus
- Decide whether to round an answer up or down after division using a written method, depending on the context.

Global skills
- **Creative skills:** problem solving
- **Real-world skills:** interpreting information

Key vocabulary
- round up, round down

Resources
- mini whiteboards and markers

Language support

Display on the working wall an example of short division. Label the parts as: 'quotient', 'divisor', 'dividend' and 'remainder'. Write clear steps on how to work through the calculation next to it. For example:

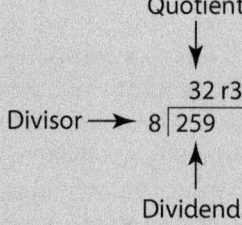

Quotient ↓

Divisor → 8)259 32 r3

↑ Dividend

1. First, I think about how many lots of 8 are in 2. None.
2. Next, I think about how many lots of 8 are in 25. 3.
3. That leaves 1 remainder. I record that in the next column, and so on.

Unit 3 Multiplication and division

Introductory activity

Write this problem on the board.

Demi receives a delivery of 375 apples at her shop. She decides to put all of the apples in bags of 8. How many bags does she fill?

Ask students to work with a partner to find the answer, writing their workings on their whiteboards. Then take suggestions from the pairs. Agree that she fills 46 bags with 7 apples left over. Ask, *Do you round the answer up or down? Why?* Agree down because the 7 left over are not enough to fill another bag to sell. Establish that deciding whether to **round up** or **round down** depends on the context of the problem.

Write this problem on the board.

Harry has 796 trading cards. He wants to put them into a folder. Each page has slots for 9 cards. How many pages will he fill?

Ask pairs to estimate the answer and share estimates. Show students how to find the answer using the short division method, taking them through the calculation step by step.

$$9 \overline{)79^{7}6} 88 \text{ r}4$$

Show students how they could record the remainder as a fraction, a decimal or by rounding. Discuss how to deal with the remainder in the context of the problem. Agree that 88 pages will be filled with 4 remaining cards so you could partially fill an 89th page with 4 cards. Establish that deciding whether to round up or down depends on the context of the problem.

Explain the activities on pages 63–64 of the Student Book, noting that for this activity students will need to round all remainders. Emphasise that students need to explain why they made the decision to round up or down. For this activity, use mixed-attainment pairs. One of the pair can model the division to their partner who may be uncertain and then the partner can decide whether to round up or down. For question 5 on page 64, students should think about making problems that have a remainder, and they should use their knowledge of divisibility tests for this.

Ask students questions to monitor their understanding, for example:

- *When do you round an answer up?*
- *When do you round an answer down?*
- *Can you make up a problem where you need to round the answer up?*
- *Can you make up a problem where you need to round the answer down?*

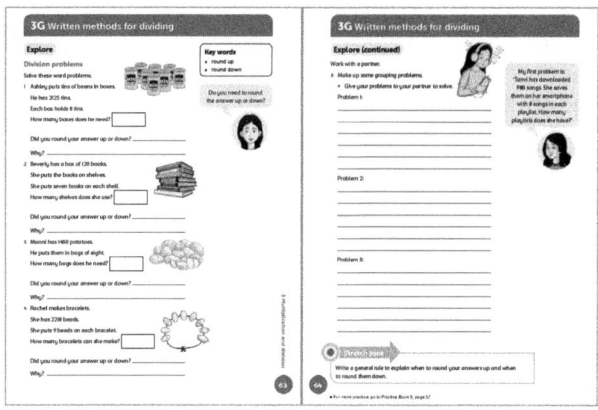

Differentiation

Supporting: Help students by explaining why you should round up or down.

Consolidating: Ask students to model their division method.

Extending: Ask students to explain how they created their own problems.

Stretch zone: *Write a general rule to explain when to round your answers up and when to round them down.*

Students might say that rounding up happens when the remainder is used (for example for tins to go into boxes), rounding down is used when the remainder is discarded (for example for beads on bracelets).

Reflection time

Select four different pairs of students to share their solutions for each of the problems. They should model the strategy for dividing and explain why they decided to round up or down.

Practice Book: Students complete Practice Book page 57. They can do this directly after the Main activity, as homework, or as the focus of a separate mathematics session to help students consolidate their learning and build fluency.

This activity provides further practice in solving division problems that involve rounding up or down depending on the context.

Differentiated outcomes	
All students	should understand whether to round up or down when using written methods of division with support.
Most students	will carry out division using a written method and decide whether to round up or down.
Some students	may create their own problems that involve division by a written method and rounding up or down.

Unit 3 Multiplication and division

Answers

Student Book pages 63–64

1. 266. Round the answer up for the remaining 5 tins to go into a box.
2. 18. Round the answer up so the remaining book can go on a shelf.
3. 182. Round the answer down because the remaining 2 won't fill a bag.
4. 246. Round the answer down because the remaining 2 won't make a bracelet.

Practice Book page 57

1. 378 boxes. Round up.
2. 195 cakes. Round down.
3. 95 necklaces. No rounding.
4. 7 packets. Round down.

Stretch zone: Check that students have rounded to a number that can be divided easily.

3H Prime numbers

Discover
Student Book page 65 • Practice Book page 58

Specific learning focus
- Identify all prime numbers up to 100.

Global skills
- **Creative skills:** investigating
- **Interpersonal skills:** communication

Key vocabulary
- prime number, composite numbers, factors

Resources
- 100-squares
- counters or cubes

Language support
Add definitions of 'prime number' and 'composite numbers', including examples, to your working wall.

Introductory activity

Give students a pile of counters. Ask them to count out 12 counters and arrange them in an array. Ask them to write down how many rows and columns are in the array. Can they make a different array with 12 counters? Compare answers around the class. Students may have found 3 rows of 4, 6 rows of 2 and so on. Explain that the arrays all show factors of 12.

Now ask students to take an extra counter so that they have 13. Ask them to try to make different arrays with 13 counters. What do they notice? (They can only make one 1 × 13 array.) Explain that a number that only has one array, or only two factors, is called a **prime number**.

So, 13 is a prime number, but 12 is called a **composite number** (it is composed of different arrays).

Main activity

Ask students to work with the counters and test numbers from 2 up to 11. They should try to make arrays with each number of counters. If they can only find one array for a number, they have found a prime number.

Discuss which numbers are prime (2, 3, 5, 7, 11) and which are composite (4, 6, 8, 9, 10).

Now ask students to look at page 65 of the Student Book. Read together the first section of text and then talk them through the steps of how to find the prime numbers up to 100 using Eratosthenes' sieve. They should all observe how step 1 was completed, circling 2 and crossing out remaining multiples of 2. Then they complete the remaining steps individually.

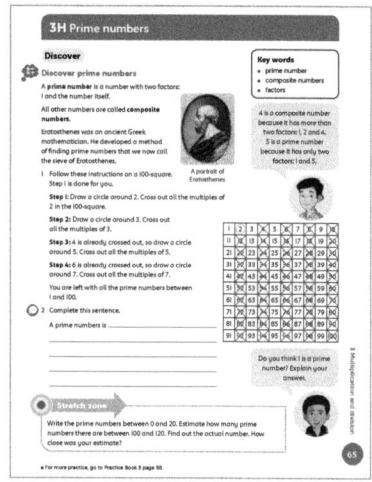

Differentiation

Supporting: Help students with finding arrays and completing the steps of the prime number sieve.

Consolidating: Ask students to explain how they found prime numbers beyond 11.

Extending: Challenge students to find prime numbers from 101 to 120.

Stretch zone: *Write the prime numbers between 0 and 20. Estimate how many prime numbers there are between 100 and 120. Find out the actual number. How close was your estimate?*

There are five prime numbers between 100 and 120: 101, 103, 107, 109 and 113.

Unit 3 Multiplication and division

Reflection time

Choose students to share their definitions of a prime number with the class. Look together at the completed prime sieves. Ask, *Can you tell me why the multiples of 4 and 6 were already completed on the sieve? Why was it only necessary to continue up to multiples of 7?* Ask individual students to choose a number on the 100-square and say how they know it is either prime or composite.

Discuss the question in the second speech bubble: *Do you think 1 is a prime number?* Ask students to explain their thinking. If there is some disagreement, go back to the definition in the first speech bubble on Student Book page 65 and agree that it has only one factor so it is not prime.

Practice Book: Students complete Practice Book page 58. They can do this directly after the Main activity, as homework, or as the focus of a separate mathematics session to help students consolidate their learning and build fluency.

Review how to sort using Carroll and Venn diagrams as needed.

Differentiated outcomes	
All students	should find prime numbers to 100 with support.
Most students	will find prime numbers to 100 independently.
Some students	may find prime numbers beyond 100.

Answers

Student Book page 65

1. Prime numbers to 100 are: 2, 3, 5, 7, 11, 13, 17, 19, 23, 29, 31, 37, 41, 43, 47, 53, 59, 61, 67, 71, 73, 79, 83, 89, 97

Practice Book page 58

1. Odd, prime: 3, 5, 7, 11, 13, 17, 19, 23

   Not odd, prime: 2

   Odd, not prime: 1, 9, 15, 21, 25

   Not odd, not prime: 4, 6, 8, 10, 12, 14, 16, 18, 20, 22, 24

2. Even numbers: 2, 4, 6, 8, 10, 12, 14, 16, 18, 20, 22, 24

   Prime numbers: 2, 3, 5, 7, 11, 13, 17, 19, 23

   Even, prime: 2

Stretch zone: There is only one even prime number, 2, the rest are odd.

3H Prime numbers

Explore Student Book page 66 • Practice Book page 59

Specific learning focus
- Discover prime numbers.

Global skills
- **Creative skills:** investigating
- **Real-world skills:** presenting information

Key vocabulary
- prime number, composite number, factor, prime factor

Resources
- counters or interlocking cubes

Language support

Encourage students to refer to the speech bubble on page 66 of the Student Book to help them recall the key words relating to the learning focus and to use as a model for describing their thinking.

 **Introductory activity**

Write on the board the numbers from 2 to 10. Ask students to write down, for each number, the factors of that number and any multiplication facts that give that number. For example, for 6 they can write the factors 1, 2, 3, 6, and the multiplication facts 1×6, 2×3, 3×2, 6×1.

Compare answers and ask students what they notice about the numbers and their factors. *Which numbers are prime numbers? How many factors do they have?*

 **Main activity**

Recall with students that numbers are either prime (two factors, one array) or composite (more than two factors, more than one array). Introduce the notion of **prime factors**, using 12 as an example.

Ask students to write a multiplication fact that has 12 as an answer, for example 3×4, and confirm that 3 and 4 are factors of 12. Notice that 3 is a prime factor, but that 4 is not prime. So, now write a multiplication fact that has 4 as an answer: 2×2. As 2 is a prime number, we can say that 2 and 2 are prime factors of 4, so they are prime factors of 12 as well.

The prime factors of 12 are therefore 2, 2 and 3, because $2 \times 2 \times 3$ gives 12.

Unit 3 Multiplication and division

Refer students to the table on page 66 of the Student Book. Ask them to complete the table for the numbers 1 to 20, identifying each number's factors, whether it is prime or composite and the prime factors that multiply together to make the number.

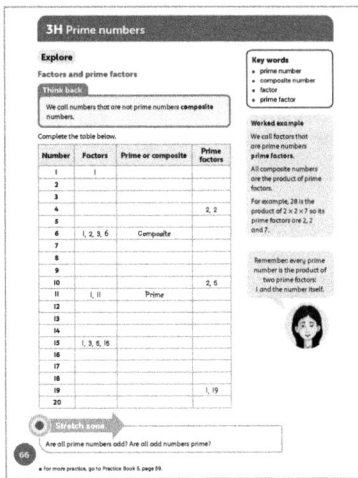

Differentiation

Supporting: Help students to form arrays using counters or cubes to find the factors.

Consolidating: Ask students to explain how they found the prime factors for each number.

Extending: Challenge students to find the prime factors for numbers 21–30.

Stretch zone: *Are all prime numbers odd? Are all odd numbers prime?*

Students should notice that only 2 is both even and prime, and that many odd numbers are not prime. One example is 15 because it has more than 2 factors (1, 3, 5, 15).

Reflection time

Work through the table together and ask students to explain how they found the factors for each number. Check that they understand that prime factors can be repeated. For example, the prime factors for 16 are 2, 2, 2, 2 because 2 × 2 × 2 × 2 = 16.

Practice Book: Students complete Practice Book page 59. They can do this directly after the Main activity, as homework, or as the focus of a separate mathematics session to help students consolidate their learning and build fluency.

Some students may find this activity challenging. Encourage them to work systematically to identify all factors.

Differentiated outcomes	
All students	should find factors of numbers to 20 independently and prime factors of numbers to 20 with support.
Most students	will find prime factors of numbers to 20 independently.
Some students	may find prime factors of numbers beyond 20.

Answers

Student Book page 66

Number	Factors	Prime or Composite	Prime factors
1	1		
2	1, 2	Prime	2
3	1, 3	Prime	3
4	1, 2, 4	Composite	2, 2
5	1, 5	Prime	5
6	1, 2, 3, 6	Composite	2, 3
7	1, 7	Prime	7
8	1, 2, 4, 8	Composite	2, 2, 2
9	1, 3, 9	Composite	3, 3
10	1, 2, 5, 10	Composite	2, 5
11	1, 11	Prime	11
12	1, 2, 3, 4, 6, 12	Composite	2, 2, 3
13	1, 13	Prime	13
14	1, 2, 7, 14	Composite	2, 7
15	1, 3, 5, 15	Composite	3, 5
16	1, 2, 4, 8, 16	Composite	2, 2, 2, 2
17	1, 17	Prime	17
18	1, 2, 3, 6, 9, 18	Composite	2, 3, 3
19	1, 19	Prime	19
20	1, 2, 4, 5, 10, 20	Composite	2, 2, 5

Practice Book page 59

Number	Factors	Prime or Composite	Prime factors
60	1, 2, 3, 4, 5, 6, 10, 12, 15, 20, 30, 60	Composite	2, 2, 3, 5
61	1, 61	Prime	61
62	1, 2, 31, 62	Composite	2, 31
63	1, 3, 7, 9, 21, 63	Composite	3, 3, 7
64	1, 2, 4, 8, 16, 32, 64	Composite	2, 2, 2, 2, 2, 2
65	1, 5, 13, 65	Composite	5, 13
66	1, 2, 3, 6, 11, 22, 33, 66	Composite	2, 3, 11
67	1, 67	Prime	67
68	1, 2, 4, 17, 34, 68	Composite	2, 2, 17
69	1, 3, 23, 69	Composite	3, 23
70	1, 2, 5, 7, 10, 14, 35, 70	Composite	2, 5, 7
71	1, 71	Prime	71
72	1, 2, 3, 4, 6, 8, 9, 12, 18, 24, 36, 72	Composite	2, 2, 2, 3, 3,

Stretch zone: 96 has more factors.

Factors of 92: 1, 2, 4, 23, 46, 92

Factors of 96: 1, 2, 3, 4, 6, 8, 12, 16, 24, 32, 48, 96

31 Square and cube numbers

Discover Student Book page 67 • Practice Book page 60

Specific learning focus
- Identify square and cube numbers

Global skills
- **Creative skills:** investigating
- **Real-world skills:** presenting information

Key vocabulary
- square numbers, cube numbers, superscript

Resources
- square tiles
- interlocking cubes
- counters (optional)

Language support
Create visual examples of squared and cubed numbers for your working wall and add speech bubbles, for example:
- '3 squared means $3 \times 3 = 9$.'
- 'We can show square numbers using square arrays.'
- '2 cubed means $2 \times 2 \times 2 = 8$.'
- 'We can build cube numbers using small cubes to make larger cubes.'

Introductory activity

Write the following on the board and ask students to answer them.

$1 \times 1 =$

$2 \times 2 =$

$3 \times 3 =$

$4 \times 4 =$

$5 \times 5 =$

$6 \times 6 =$

$7 \times 7 =$

$8 \times 8 =$

$9 \times 9 =$

$10 \times 10 =$

Go through the answers. Ask, *Do you see any patterns? Can you describe them?* Students might notice how the numbers in the calculations increase by one each time, and how the products increase by successive odd numbers.

Explain that there is another way you can describe each product: they are all **square numbers**. Tell students that they will be investigating how square numbers get their name.

Main activity

Give students, working in pairs, a pile of square tiles and ask them to make 10 different-size squares using the tiles. They can start with a single tile, then make the next possible size square using four small tiles, then continue making larger and larger squares with more and more tiles.

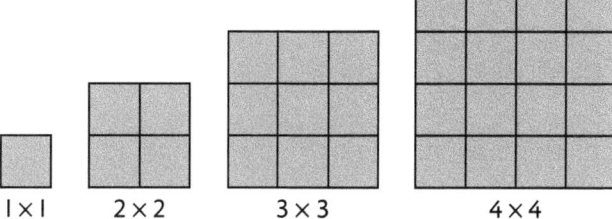

1×1 2×2 3×3 4×4

As they make each new square, they should write down how many tiles there are along each side, and how many in total. Explain that the number of tiles needed to make each square is called a square number.

Then give students some interlocking cubes and ask them to make a cube that is twice the size of a single cube. Ask, *What do I mean by twice the size?* Agree you mean twice as long, twice as wide and twice as high as a single cube. *How many small cubes do you need?* Give pairs some time to investigate to discover that it is 8 cubes. Then repeat and make a cube that is 3 times as wide, 3 times as long, and 3 times as high as a single cube. Students should record the new height, width and length. *How many small cubes do you need?* Explain that the number of small cubes needed to make each larger cube is called a **cube number**.

Students should then complete the activities on page 67 of the Student Book, which consolidates the earlier activity. Introduce the superscript here by writing on the board:

$2 \times 2 = 2$ squared, written as 2^2

$2 \times 2 \times 2 = 2$ cubed, written as 2^3

Explain that the small number is called a **superscript**, and that it is telling us the power of a number, or how many times the number is multiplied together.

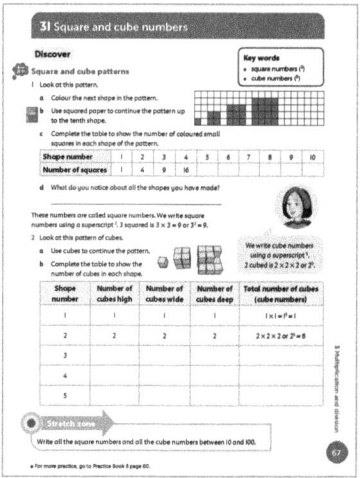

Unit 3 Multiplication and division

Differentiation

Supporting: Help students to organise their work and approach building squares and cubes systematically.

Consolidating: Ask students to describe any patterns they can see in the square and cube numbers in more than one way, for example: $3 \times 3 \times 3$, a cube 3 cubes high, 3 cubes long, and 3 wide, and 3^3.

Extending: Challenge students to find square numbers above 100 and cube numbers up to 1000.

Stretch zone: *Write all the square numbers and all the cube numbers between 10 and 100.*

Students should also write the numbers and the calculations that make them.

 Reflection time

Look together at the pattern of square numbers and the differences between them. For example, as they count up 1, 4, 9, 16, 25 … the differences between the numbers are 3, 5, 7, 9 … .

Ask, *How do you find the square of a number without making it with tiles or drawing it?*

Write on the board 121, 144 and 169. Ask whether these are square numbers. How do they know? Students should be able to say which number has been multiplied by itself to make these examples.

Practice Book: Students complete Practice Book page 60. They can do this directly after the Main activity, as homework, or as the focus of a separate mathematics session to help students consolidate their learning and build fluency.

Students may find using counters for this investigation useful.

For the Stretch zone question, students may find it challenging to describe and explain their thinking in words. Once they have worked out what happens when you add pairs of consecutive triangle numbers, suggest they use diagrams to show why. They can rearrange each triangle number diagram into right-angled triangles and then combine consecutive triangles to make a square.

Differentiated outcomes	
All students	should make a list of square and cube numbers with support.
Most students	will make a list of square and cube numbers independently.
Some students	may find square and cube numbers beyond 100.

Answers

Student Book page 67

1

c

Shape number	1	2	3	4	5	6	7	8	9	10
Number of squares	1	4	9	16	25	36	49	64	81	100

d Students may notice, for example, that the difference between the number of squares needed for each consecutive shape is a consecutive odd number.

2

Shape number	Number of cubes high	Number of cubes wide	Number of cubes deep	Total number of cubes (cube numbers)
1 (Provided)	1	1	1	$1 \times 1 \times 1 = 1^3$ $= 1$
2 (Provided)	2	2	2	$2 \times 2 \times 2 = 2^3$ $= 8$
3	3	3	3	$3 \times 3 \times 3 = 3^3$ $= 27$
4	4	4	4	$4 \times 4 \times 4 = 4^3$ $= 64$
5	5	5	5	$5 \times 5 \times 5 = 5^3$ $= 125$

Practice Book page 60

	Square number	Sum of two prime numbers?
1	4	$4 = 2 + 2$
2	9	$9 = 2 + 7$
	16	$16 = 5 + 11$
3	25	$25 = 2 + 23$
4	36	$36 = 5 + 31$
5	49	$49 = 2 + 47$
6	64	$64 = 5 + 59$
7	81	$81 = 2 + 79$
8	100	$100 = 47 + 53$

Stretch zone: $1 + 3 = 4$

$3 + 6 = 9$

$6 + 10 = 16$

Adding consecutive triangle numbers makes square numbers.

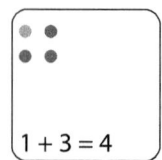 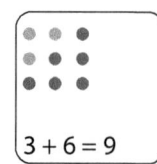 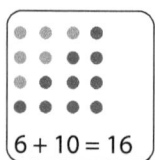

Unit 3 Multiplication and division

31 Square and cube numbers

Explore Student Book pages 68–69 • Practice Book page 61

Specific learning focus
- Solve problems using square and cube numbers.

Global skills
- **Creative skills:** investigating
- **Interpersonal skills:** communication/teamwork

Key vocabulary
- square numbers, cube numbers

Resources
- cm-squared paper
- mini whiteboards and markers

Language support
Support students with phrases and questions, for example:
- A square number is …
- What is 7 squared?
- 7 squared is …
- Can you tell me any other square numbers?

 **Introductory activity**

Write this pattern on the board:

$1 =$

$1 + 3 =$

$1 + 3 + 5 =$

$1 + 3 + 5 + 7 =$

$1 + 3 + 5 + 7 + 9 =$

Ask students to copy it to their whiteboards and write the answers to the calculations.

They should look at these calculations made from odd numbers and say what they notice about the answers. They should notice that the answers are all square numbers. Ask, *Can you work out what calculation you need to write to continue the pattern? What square number will the answer be?*

 **Main activity**

Ask pairs of students to refer to their list of all the square numbers to 100 to help them with the activities on pages 68–69 of the Student Book. They should work in pairs. They check one another's answers and then discuss the patterns they are seeing.

For question 1, check that students remember what the superscript means, for example $5^2 = 5 \times 5$, and then ask them to describe the sorts of numbers that result from these subtractions to their partner. In question 4, remind students that, for example, 4^3 means $4 \times 4 \times 4$.

Before students begin working on the activity on page 69 of the Student Book, ask them to read the speech bubbles. Ask, *Do you agree that the first two characters are wrong? Can you explain why?* You might need to prompt students to notice that as well as one large square and 64 small ones, there are also some squares that are 2×2 (how many?), and 3×3, 4×4 and so on.

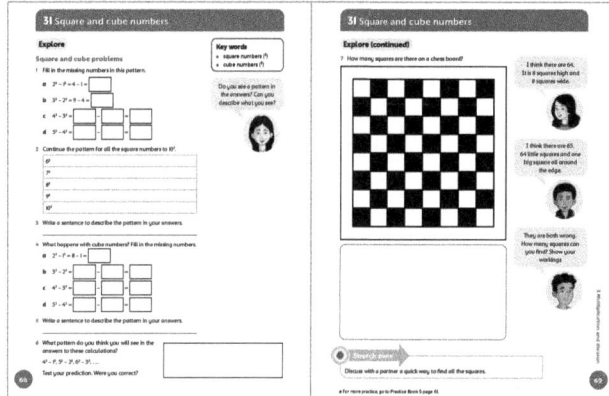

Differentiation

Supporting: Help students to list all the square numbers to 10×10.

Consolidating: Ask students to describe the patterns they are noticing.

Extending: Challenge students to predict and explain any patterns they are noticing.

Stretch zone: *Discuss with a partner a quick way to find all the squares.*

Students might work systematically, for example counting how many 7×7 squares they can see, then 6×6, then noticing the pattern of how many squares there are of each size.

 Reflection time

Invite students to share the patterns they made and their ideas about what happens each time. For example: the first pattern (questions 1 and 2) increases in twos. The differences between square numbers that differ by 3 increase by consecutive multiples of 6 (question 6).

Unit 3 Multiplication and division

Write $5^2 - 3^2$ and $6^2 - 4^2$ on the board. Ask students to find the answers and predict the answers on the sequence that continues. Test it to see whether they are correct. They should find that the answers, when finding the difference between squares of numbers that differ by 2, increase by 4 each time.

In the chess board problem, there is 1 square of 8×8, 4 squares of 7×7, 9 squares of 6×6 and so on, leading to a total of 204 squares.

Practice Book: Students complete Practice Book page 61. They can do this directly after the Main activity, as homework, or as the focus of a separate mathematics session to help students consolidate their learning and build fluency.

Discuss how to cube a number by multiplying it by itself 3 times, for example, $7^3 = 7 \times 7 \times 7 = 343$. Prompt students to look at the numbers being used in the additions in the third column. Ask, *Can you see a sequence? What is the rule?*

Differentiated outcomes	
All students	should find square numbers to 10×10 and begin to calculate cube numbers with support.
Most students	will notice the patterns that are formed by finding the differences between square numbers.
Some students	may explore other patterns with square numbers.

Answers

Student Book pages 68–69

1. a $2^2 - 1^2 = 4 - 1 = 3$
   b $3^2 - 2^2 = 9 - 4 = 5$
   c $4^2 - 3^2 = 16 - 9 = 7$
   d $5^2 - 4^2 = 25 - 16 = 9$
2. a $6^2 - 5^2 = 36 - 25 = 11$
   b $7^2 - 6^2 = 49 - 36 = 13$
   c $8^2 - 7^2 = 64 - 49 = 15$
   d $9^2 - 8^2 = 81 - 64 = 17$
   e $10^2 - 9^2 = 100 - 81 = 19$
3. The answer is the same as the total of the original numbers.
4. a (Provided) $2^3 - 1^3 = 8 - 1 = 7$
   b $3^3 - 2^3 = 27 - 8 = 19$
   c $4^3 - 3^3 = 64 - 27 = 37$
   d $5^3 - 4^3 = 125 - 64 = 61$
5. The answers go up 12, 18, 24 and so on.
6. $4^2 - 1^2 = 15$, $5^2 - 2^2 = 21$, $6^2 - 3^2 = 27$. The answers go up in sixes.
7. There is 1 large square of 8×8, 4 squares of 7×7, 9 squares of 6×6, and so on. The total is $1 + 4 + 9 + 16 + 25 + 36 + 49 + 64 = 204$.

Practice Book page 61

	Number cubed	Product	Addition
1	4^3	$= 64$	$= 13 + 15 + 17 + 19$
2	5^3	$= 125$	$= 21 + 23 + 25 + 27 + 29$
3	6^3	$= 216$	$= 31 + 33 + 35 + 37 + 39 + 41$
4	7^3	$= 343$	$= 43 + 45 + 47 + 49 + 51 + 53 + 55$
5	8^3	$= 512$	$= 57 + 59 + 61 + 63 + 65 + 67 + 69 + 71$
6	9^3	$= 729$	$= 73 + 75 + 77 + 79 + 81 + 83 + 85 + 87 + 89$
7	10^3	$= 1000$	$= 91 + 93 + 95 + 97 + 99 + 101 + 103 + 105 + 107 + 109$

Stretch zone: The answers are made by adding a growing series of consecutive odd numbers.

3J Multiplying and dividing to solve problems

Discover Student Book page 70 • Practice Book page 62

Specific learning focus
- Multiplying and dividing to solve problems with recipes involving scaling.

Global skills
- **Creative skills:** problem solving
- **Real-world skills:** presenting information/interpreting information
- **Interpersonal skills:** communication/teamwork

Key vocabulary
- proportion, scaling problem

Resources
- mini whiteboards and markers

Language support
Have visual examples of the ingredients listed for the dahl recipe on page 70 of the Student Book. Explain what the abbreviation 'tsp' means (teaspoon) and explain that $\frac{1}{2}$ teaspoon is equivalent to a bit less than 3 grams.

 **Introductory activity**

Write this problem on the board.

To make a fruit syrup drink, mix 100 ml of syrup with 500 ml of water. How much of each will be needed to make 3 litres of this drink, and how much of each to make 60 litres?

Ask students to work in pairs on this problem, writing their answers on their whiteboards. They should choose multiplication or division strategies and be able to explain how they worked out the answers.

Discuss their strategies and agree on the answer as a class. For example, one drink is 600 ml, or 0.6 litres. 60 litres is 0.6 × 100 so to make 60 litres they will need 100 times as much syrup and water as one drink. To make 3 litres, students could divide the amounts for 60 litres by 20.

 **Main activity**

Point out the recipe on page 70 of the Student Book. Discuss the dish, clarifying what the ingredients are as necessary. Review the quantities and units of measure used.

Ask students how many people the original recipe is for. (4) *What will you need to do to make it for 2 people? What about 8 people?* Agree that they need to make it smaller for 2 people and bigger for 8. Explain that these are **scaling problems**. They need to make the recipe half as big and twice as big. For each ingredient, they need to scale it down by a half for 2 people. For example, instead of 3 tsp curry powder, this will need to be $1\frac{1}{2}$ tsp for 2 people. For 8 people, instead of 250 g tomatoes, they need double, which is 500 g tomatoes, and so on.

Explain that if the number of people is not an exact multiple of 4, or a simple fraction of 4, then students need to use **proportioning** to find the amount of ingredients. To make the recipe for 10 people, the multiple needed for each ingredient is $\frac{10}{4}$, or 2.5. So, for example, instead of 1 onion, you need $2\frac{1}{2}$ onions.

Students should work in pairs to complete the table showing the ingredients for each different number of people, using scaling or proportioning as appropriate.

Encourage students to use all the facts from the unit. Ask questions to prompt them, for example:
- *How can I find the recipe for 2 people by dividing the amounts?*
- *How do I divide 1.2 litres by 4? Does converting to millilitres make it easier?*
- *How can I find the amounts for 10 people if I have a recipe for 2 people and 8 people?*

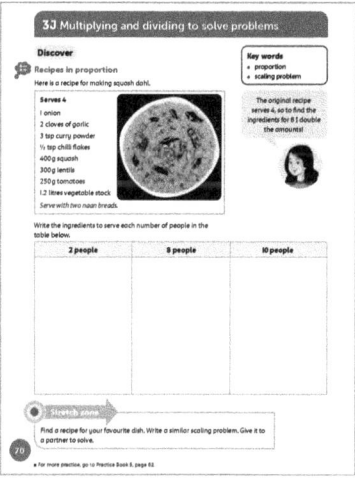

Differentiation

Supporting: Help students with multiplying and dividing the amounts that appear as fractions.

Consolidating: Ask students to explain their strategies for calculating the amounts.

Extending: Challenge students to calculate a recipe for 5 people. What do they already know that could help?

Stretch zone: *Find a recipe for your favourite dish. Write a similar scaling problem. Give it to a partner to solve.*

Students can base their problem on a recipe found from home or on the internet.

Unit 3 Multiplication and division

Reflection time

Talk through each column in the table to check the amounts for each number of people. Ask students to explain how they calculated the ingredients for each number of people. Ask, *Did you use the amounts you knew to get the amounts for 10 people? How can you use what you now know to find the amounts for 6 people?*

Practice Book: Students complete Practice Book page 62. They can do this directly after the Main activity, as homework, or as the focus of a separate mathematics session to help students consolidate their learning and build fluency.

Discuss the ingredients list, using photographs to explain ingredients if necessary, so students can concentrate on scaling amounts rather than focusing on unfamiliar vocabulary.

Differentiated outcomes	
All students	should calculate the recipe quantities by scaling up or down with support.
Most students	will calculate the recipe quantities by scaling up or down independently.
Some students	may calculate the recipe quantities for other numbers of people.

Answers

Student Book page 70

2 people	8 people	10 people
½ onion	2 onions	2½ onions
1 clove of garlic	4 cloves of garlic	5 cloves of garlic
1½ tsp curry powder	6 tsp curry powder	7½ tsp curry powder
¼ tsp chilli flakes	1 tsp chilli flakes	1¼ tsp chilli flakes
200 g squash	800 g squash	1000 g squash
150 g lentils	600 g lentils	750 g lentils
125 g tomatoes	500 g tomatoes	625 g tomatoes
0.6 litres vegetable stock	2.4 litres vegetable stock	3 litres vegetable stock
1 naan bread	4 naan breads	5 naan breads

Practice Book page 62

1

1 pizza	5 pizzas	8 pizzas	
150 g	750 g	1200 g	flour
½ tsp	2½ tsp	4 tsp	yeast
½ tbsp	2½ tbsp	4 tbsp	olive oil
50 ml	250 ml	400 ml	passata
½ handful	2½ handfuls	4 handfuls	basil
½ clove	2½ cloves	4 cloves	garlic
62.5 g	312.5 g	500 g	mozzarella
2½	12½	20	cherry tomatoes

2 3750 grams flour

3 4 jars

Stretch zone: Check that students have made a suitable word problem based on the recipe.

3J Multiplying and dividing to solve problems

Explore Student Book page 71 • Practice Book page 63

Specific learning focus
- Multiply and divide to solve word problems involving scaling.

Global skills
- **Creative skills:** problem solving
- **Real-world skills:** interpreting information
- **Interpersonal skills:** communication/teamwork

Key vocabulary
- scaling, proportion, which operation?

Resources
- mini whiteboards and markers

Language support
Check that students understand any unfamiliar vocabulary used in each word problem so that poor understanding does not distract or interfere with them solving the problems.

Introductory activity

Share this problem with students.

A newspaper shop sells three types of game magazines. Puzzles for Fun, Challenge Your Mind and Can You Solve It?

    Puzzles for Fun has 20 pages and costs 20¢

    Challenge Your Mind has 40 pages and costs 50¢

    Can You Solve It? has 25 pages and costs 35¢

On Saturday the newsagent sold some of each magazine and took $41.25. She sold 15 Can You Solve It?, and twice as many Puzzles for Fun as Challenge Your Mind. How many Puzzles for Fun and Challenge Your Mind were sold?

Encourage students to think carefully about how to work out the answer and choose operations and strategies that will solve the problem. Students may decide, for example, to subtract from the total amount the sales of *Can You Solve It?*, then try different numbers for the remaining magazines to get to the amount remaining. They should show their workings on their whiteboards.

Main activity

Ask students to read the word problems on page 71 of the Student Book. Remind them to think about what the key information is for each problem and to decide which operation and which strategies to use to calculate the answers.

Students can work together in pairs or small groups, and they should set out their workings to help them explain their methods later. Encourage them to estimate their answers first and check them using an inverse operation or a different calculation strategy.

Ask questions as students are working, for example:
- *How can I find the total cooking time?*
- *How many packets of stickers do I need to buy?*
- *What do you know about the litres and millilitres that will help you answer this question about lemonade?*
- *How can I find how much drink I need for 10 people?*
- *We use multiplication when we cook. When would you use division?*

Once students have completed the problems, they can make up their own real-life problems involving multiplication and division.

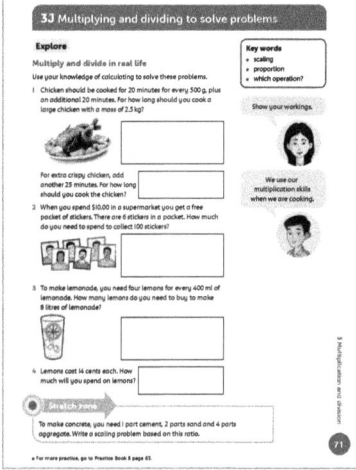

Differentiation
Supporting: Help students draw out the information they need to solve each problem and choose the correct operations.

Consolidating: Students should set out their workings clearly and be able to describe their method to you.

Extending: Challenge students to write their own similar two-step word problems including more than one type of operation.

Stretch zone: *To make concrete, you need 1 part cement, 2 parts sand and 4 parts aggregate. Write a scaling problem based on this ratio.*

Students should work in pairs to solve each other's problems.

Reflection time
Discuss together the approaches to solving the word problems on page 71. How did students identify which information, operations and strategies to use? Select students to share their methods. Ask the rest of the students to say whether they used different methods from those shared and to explain their choice.

Unit 3 Multiplication and division

Practice Book: Students complete Practice Book page 63. They can do this directly after the Main activity, as homework, or as the focus of a separate mathematics session to help students consolidate their learning and build fluency.

Explain that fruit squash is a concentrated fruit flavoured sugar that you mix with water to make a drink, and is often drunk by children.

Differentiated outcomes	
All students	should solve the word problems with support.
Most students	will solve the word problems independently.
Some students	may make their own similar two-step word problems using more than one type of operation.

Answers

Student Book page 71

1 120 minutes (2 hours); 145 minutes (2 hours 25 minutes)
2 $170
3 80 lemons
4 $11.20

Practice Book page 63

1 3 bottles
2 19.25 litres water
3 $4.50

Stretch zone: Based on the information above, 70 glasses cost $4.50, which makes about 6¢ per glass. To raise money for charity, you could charge 10¢ per glass.

3 Multiplication and division

Connect   Student Book page 72

Big idea

I can multiply and divide large numbers using a written method or a mental method, using known facts. The strategy I use depends on the numbers in the calculation.

Global skills

- **Creative skills:** investigating
- **Interpersonal skills:** teamwork
- **Self-development skills:** reflecting on learning

Key vocabulary

- multiplication, division, population, estimate

Resources

- flipchart paper and markers

Language support

Support students with the language required to work out the answers to the word problems, for example:

- There were more people in the first town than the second.
- The total population of the three towns is ... .
- I worked out the difference in the populations using a subtraction method.

 Introductory activity

Write a selection of 3-, 4- and 5-digit numbers on the board and ask students to read them aloud and say which number is 1 more and 1 less, for example.

 Main activity

Refer students to page 72 of the Student Book and discuss the context of populations of three towns in the example. Read through the questions in the speech bubbles to start students thinking about the type of problems they could create. In the second speech bubble it says that, on average, five people live in each home. Explain to students that some homes may have fewer than five people, others may have more, but if the people were shared equally between all the homes, there would be five in each.

Ask students to work in mixed–attainment groups. They should use flipchart paper to support collaborative work and only record the problems in their book when they have completed them as a group. They should be able to solve the problems they create and choose an efficient strategy to do so.

Differentiation

Supporting: Encourage students to take an active role in their group. Assign them a set role or task in the group, if appropriate.

Consolidating: Ask students to explain how they created their word problems and what operations and strategies they used to solve them.

Extending: Ask students to use more than one strategy to solve other groups' word problems. Can they add an additional step to it to make it more challenging and then solve it?

Stretch zone: *Write a two-step word problem about one of the towns that involves both multiplication and division.*

Remind students to set the problem in a realistic context, estimate first and check their answers using an inverse operation.

Reflection time

Each group should present their problems using the flipchart paper. Encourage students to share the strategies they used to carry out the calculations. Can other groups suggest another more efficient strategy?

Students could work with a partner to make up similar problems in local contexts.

Differentiated outcomes	
All students	should contribute to the work of the group.
Most students	will explain their calculation strategies and check the accuracy of the calculations.
Some students	may explain a range of methods for solving problems to others.

Answers

Student Book page 72

Answers will vary because students make up their own problems using the context of the towns' populations.

---

3 Multiplication and division

Review   Student Book page 73 • Practice Book page 64

Global skills

- **Real-world skills:** presenting information/interpreting information
- **Interpersonal skills:** communication
- **Self-development skills:** reflecting on learning

Student Book

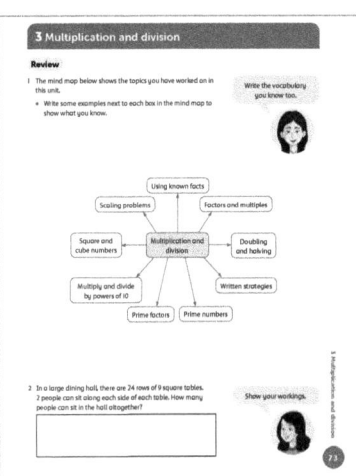

With young students, assessment activities are most effective when carried out as an everyday classroom activity. Students should have digit cards and place-value cards to help them generate numbers. Some students may also benefit from having base-10 equipment or counters available to support them in their calculations.

Ask students to work individually to complete the mind map on page 73 of the Student Book to show you what they now know about multiplication and division from this unit. Encourage students to use larger numbers where possible in their examples. Check that students can recognise and calculate square and cube numbers and know what prime and composite numbers are.

Ask students further questions to assess their understanding, for example:

- *Which mental calculations for multiplication can you tell me about?*
- *Which is the best strategy for finding the factors of 48?*
- *How do you know if a number is prime or not? Can you give me an example greater than 20?*

Answers

Student Book page 73

1 Answers will vary because students choose their own examples to complete the mind map.

2 1728

Practice Book

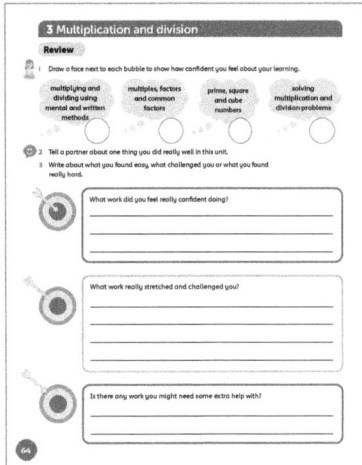

With students in the upper primary years, it is appropriate to complete this as a whole-class discussion. You may choose to keep a record of the class discussion or a copy of the Review page for your own records. Use the Student Book to briefly remind students of the areas of mathematics that they have worked on in this unit.

Ensure that students have a copy of the Student Book to support them as they discuss and answer the questions in the Practice Book.

Allow students plenty of time for discussion before asking them to share their responses with the rest of the class. If students complete this assessment at home, encourage them to discuss this with adults.

Make a note of areas that students still feel unsure about. You can also build multiplication and division into everyday practice, for example looking for large numbers in the media.

Answers will vary as students choose their own calculations to match the statements and strategies given. Check that the answers given are correct, and that the calculations do require the requested strategy.

Additional material

There are additional end-of-unit assessment available on the *Oxford Owl for School* website.

4 Fractions, decimals and percentages

Overview

Big idea

The Big idea when teaching fractions is the idea of equality. For example, taking a fraction of a shape involves dividing the shape into equal areas. Similarly, when calculating fractions of quantities, we divide the quantity into equal parts.

It is also important that students develop an understanding of fraction notation. A fraction is written as follows:

$\frac{2}{9}$ (This is the numerator and tells you the number of parts in this fraction.)
(This is the denominator and tells you how many pieces the whole is divided into.)

In this unit, students learn about equivalent fractions. These are fractions that may be written differently but which are equivalent. For example:
$$\frac{1}{2} = \frac{2}{4} = \frac{3}{6} = \frac{20}{40}$$

Students learn about mixed numbers. These are fractions that are greater than 1 and so are a mix of whole numbers and parts of a whole. For example, we would more usually write $1\frac{1}{2}$ rather than $\frac{3}{2}$ and say 'one and a half' rather than 'three halves'.

Finally, students learn to convert fractions into decimal fractions, using place value and a decimal point to separate whole numbers from the decimal fractions – tenths, hundredths, thousandths and so on. A special kind of decimal, a percentage (which represents hundredths), is also considered. All of these types of fractions are used in problems about proportions and ratios between amounts.

Look out for

- **Students who confuse the operations on fractions.** Use a good range of visual representations so that each calculation is made meaningful in a diagram and students can relate the operation to the context.
- **Students who find the notion of a ratio difficult.** Provide students with examples that represent 'part of a whole' and the ratio aspect of 'a part to a different part'.

Possible misconceptions

- **Students may say that the number after 4.9 (4 wholes and 9 tenths) is 4.10 (4 wholes and 10 tenths).** Practise counting with number lines to overcome this.
- **Students may think that 0.3 is greater than 0.4 because $\frac{1}{3}$ is greater than $\frac{1}{4}$.** Provide good representations of shaded shapes and number lines to overcome this.

Key vocabulary

- equal parts, multiple, whole
- fraction, proper/improper fraction, unit fraction, tangram
- mixed number, numerator, denominator, (lowest) common denominator, equivalent fraction, equivalences
- reduced to, ascending order, multiplier, bar model
- half, quarter, eighth, third, sixth, fifth, tenth, hundredth
- proportion, ratio
- one in every … , one for every …
- decimal, decimal fraction, decimal point, decimal place, rounding
- percentage, per cent, %
- nearest whole number, nearest tenth

Coverage in lessons

Learning objective	E	4A	4B	4C	4D	4E	4F	4G	4H	4I	4J	4K	C	R
Compare and order fractions whose denominators are all multiples of the same number.	✓	✓					✓							
Identify, name and write equivalent fractions of a given fraction, represented visually, including tenths and hundredths.	✓	✓	✓											✓
Recognise mixed numbers and improper fractions and convert from one form to the other and write mathematical statements > 1 as a mixed number (for example $\frac{2}{5} + \frac{4}{5} + \frac{6}{5} = 1\frac{1}{5}$).				✓										✓
Add and subtract fractions with the same denominator and denominators that are multiples of the same number.	✓				✓									
Multiply proper fractions and mixed numbers by whole numbers, supported by materials and diagrams.						✓								✓
Read and write decimal numbers as fractions (for example $0.71 = \frac{71}{100}$).	✓		✓											✓
Recognise and use thousandths and relate them to tenths, hundredths and decimal equivalents.			✓					✓	✓					✓
Round decimals with two decimal places to the nearest whole number and to one decimal place.								✓	✓					✓
Read, write, order and compare numbers with up to three decimal places.								✓	✓					
Solve problems involving number up to three decimal places.								✓	✓					
Recognise the per cent symbol (%) and understand that per cent relates to 'number of parts per hundred', and write percentages as a fraction with denominator 100, and as a decimal.	✓										✓			✓
Solve problems that require knowing percentage and decimal equivalents of $\frac{1}{2}, \frac{1}{4}, \frac{1}{5}, \frac{2}{4}, \frac{4}{5}$ and those fractions with a denominator of a multiple of 10 or 25.	✓	✓			✓	✓				✓	✓	✓	✓	✓
Understand that proportion compares the part to the whole and use this knowledge to solve problems and find proportions of amounts.											✓			
Understand that ratio shows how much there is of one thing compared to another thing or things and use this knowledge to solve ratio problems.												✓		

Unit 4 Fractions, decimals and percentages

4 Fractions, decimals and percentages

Engage Student Book page 74

Big question
- What do we know about fractions, decimals and percentages?

Global skills
- **Creative skills:** exploring
- **Interpersonal skills:** communication/teamwork
- **Self-development skills:** reflecting on learning

Key vocabulary
- fraction, percentage, decimal fraction, tangram

Resources
- mini whiteboards and markers
- Resource sheet 4.1: set of tangram pieces – one per student with the shapes cut out (and laminated, optionally)

Language support
Start a fractions and decimals wall display for the classroom and add examples of numbers and images as the unit progresses.

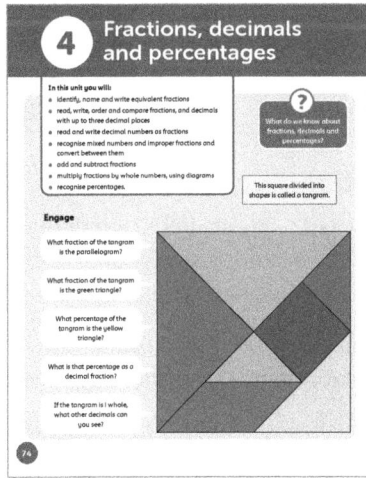

Introductory activity

Ask students to work in pairs to list on their whiteboards all the fractions they know. They should write each fraction and draw an image to illustrate it. Share the fractions and images as a class.

Decide which is the smallest fraction that anyone has listed and which is the largest. Ask, *How do you know that is the smallest? How do you know that is the largest?*

Main activity

Look together at page 74 of the Student Book. If you have access to an IWB, display the page. Give students time to look at the tangram design and see what they notice about it. Can they identify any fractions immediately by looking? Ask, *What fraction of the whole square is each of the light blue and pink triangles? How do you know?* Students should work in pairs. Give each pair a set of tangram pieces cut out from Resource sheet 4.1. Pairs work out the fraction each piece represents by laying pieces on top of each other to compare. For example, in the colour version of the design the yellow triangle will cover half of one of the large triangles, so yellow must be half of a half, which is $\frac{1}{4}$. They should look at each speech-bubble question in turn and discuss and investigate in their pairs.

Differentiation
Supporting: Help students to compare the tangram pieces to work out the fractions.

Consolidating: Ask students to explain how they found out the fraction for each piece.

Extending: Challenge students to label the pieces as percentages or decimals.

Reflection time

Discuss the results of the tangram investigation as a class. Ask pairs of students to say something about what they found out about the size of each piece as a fraction of the whole. Some students will be able to make statements comparing pieces to each other. For example, they may say that the red square is the same fraction as the yellow triangle.

Work through the speech bubbles together and see how students respond, asking them to justify their answers.

4A Equivalent fractions

Discover 1 Student Book page 75 • Practice Book page 65

Specific learning focus
- Recognise equivalences between halves, quarters and eighths, thirds and sixths, fifths and tenths.

Global skills
- **Creative skills:** investigating
- **Real-world skills:** presenting information

Key vocabulary
- equivalent fractions, equivalences, numerator, denominator

Resources
- mini whiteboards and markers
- six 3 cm-wide strips of A4 paper, cut across the width, for each student

Language support
Check that students are pronouncing the names of fractions correctly, for example saying 'one sixth' and not 'one six'. Help them with the '-th' endings on many fractions and point out that some fractions have their own word, namely 'half' instead of 'one twoth' and 'quarter' instead of 'one fourth' (although some countries do also use 'one fourth').

Introductory activity

Ask pairs of students to draw three rectangles the same size on their whiteboards. Ask them to divide the first into two equal parts, the second into four equal parts and the third into eight equal parts. Asks pairs to discuss what the fractions should be called and how they should be written.

Ask, *Which is the smallest fraction?* Establish that eighths are the smallest because the shape is divided into more parts and therefore these parts are smaller. Remind students of the terms **numerator** and **denominator** and agree that we would write $\frac{1}{2}, \frac{1}{4}, \frac{1}{8}$.

Main activity

Ask, *How many halves are the same as one whole?* Invite a student to write a number sentence to show this: $\frac{2}{2} = 1$ whole.

Agree that this means $\frac{2}{2}$ and 1 are equivalent. Repeat for quarters and eighths. Reinforce by referring back to their rectangles from the Introductory activity.

Work on the first part of the activity on page 75 of the Student Book together. Give students their strips of paper. Ensure that they all keep one of their strips whole. Check that they label each section of the folded strips with the appropriate fraction. Ask questions related to the **equivalences**, for example, *How many eighths are the same as $\frac{1}{4}$?* Ask students to work with their partner to explore thirds and sixths. They can then continue to work in pairs to complete questions 2–4 on page 75 of the Student Book.

Ask students questions to check their equivalences. Encourage students to use full sentences to answer, for example:

- *How many quarters are the same as a whole? Can you explain why?*
- *My friend says that $\frac{75}{75}$ is equal to one whole. Are they correct?*
- *Four eighths are the same as …*
- *Two eighths is equivalent to …*

Students need to keep their strips for use in 4A Explore 2, or they can use strips cut out from Resource sheet 4.2 in the Explore 2 lesson.

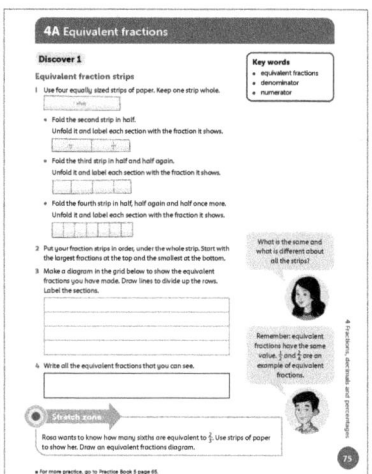

Differentiation

Supporting: Help students think of different **equivalent fractions** for $\frac{1}{2}$ and to explain how they know what they are. Ask them to show you, using their fraction strips.

Consolidating: Ask students to explain why two fractions are equivalent and to give examples.

Extending: Challenge students to create lists of other equivalent fractions beyond those they have shown with their fraction strips.

Stretch zone: *Rosa wants to know how many sixths are equivalent to $\frac{2}{3}$. Use strips of paper to show her. Draw an equivalent fractions diagram.*

You could also ask students to make a fraction wall poster for the classroom showing whole, halves, thirds, quarters, sixths and eighths in descending order. Around their poster, they can write different fraction equivalences.

Unit 4 Fractions, decimals and percentages

Reflection time

Ask pairs to show their fraction strips and to describe some of the equivalences. Ensure that they include $\frac{1}{2} + \frac{1}{4} = \frac{3}{4}$ and $\frac{3}{6} = \frac{1}{2}$.

Work together to make a poster that illustrates some of these equivalent fractions using images of the strips of paper. Add the poster to your wall display.

Practice Book: Students complete Practice Book page 65. They can do this directly after the Main activity, as homework, or as the focus of a separate mathematics session to help students consolidate their learning and build fluency.

Students will extend their learning to look at ninths, tenths and twelfths. Once students have completed the activity, give them an opportunity to discuss, compare and check their answers to the Stretch zone question in pairs or small groups.

Differentiated outcomes	
All students	should understand equivalences for $\frac{1}{2}$.
Most students	will understand a wider range of equivalences.
Some students	may generalise to find more equivalent fractions.

4A Equivalent fractions

Discover 2 Student Book page 76 • Practice Book page 66

Specific learning focus
- Recognise equivalence between fractions.

Global skills
- **Creative skills:** investigating
- **Real-world skills:** presenting information

Key vocabulary
- equivalent fractions, numerator, denominator

Resources
- small square sheets of paper
- coloured pencils
- interlocking cubes
- squared paper

Answers

Student Book page 75

Check that students have labelled their fraction strips correctly and used them to write the required equivalents.

Practice Book page 65

Some answers may vary but check that equivalents have been used. Example answers include:

1 $\frac{2}{6}$ 5 $\frac{2}{3}$ 9 $\frac{1}{2}$
2 $\frac{4}{10}$ 6 $\frac{6}{8}$ 10 $\frac{1}{4}$
3 $\frac{1}{2}$ 7 $\frac{4}{5}$ 11 $\frac{2}{3}$
4 $\frac{2}{5}$ 8 $\frac{6}{10}$ 12 $\frac{1}{3}$

Stretch zone: Check that students' definitions of equivalent fractions include references to the fractions being of equal size.

Language support

It is important for students to say the fractions out loud, especially halves, quarters and eighths, to get used to pronouncing them correctly. Ask students direct questions that will require them to do so, for example:

- *What does the numerator/denominator show?*
- *Tell me two fractions that are equivalent to one half …*

 ### Introductory activity

Give each student a small square of paper. Ask them to fold it in half. Then ask them to show you the fold. Then, ask them to fold the paper in half in a different way. (They may fold horizontally, vertically or along a diagonal.) Ask, *How do you know that you folded the paper exactly in half?* (They should notice equal areas of each half.)

Ask students to fold their square into quarters. Then ask them to think of as many different ways as they can of shading in one quarter of the square. Finally, ask them to work in groups of four to produce as many different fractions as they can by folding squares of paper.

Unit 4 Fractions, decimals and percentages

 **Main activity**

Talk to students about equivalent fractions, drawing on their work in the previous lesson. Discuss how fractions can be represented in different ways by referring to the chocolate bar example on page 76 of the Student Book. Ask, *For example, how could you divide this bar in half? Is there more than one way?*

Ask students to work on the activity on page 76 of the Student Book in pairs. As they work, ask questions such as:

- *How many squares of chocolate are there altogether? How could you describe this bar as an array? What does that tell us about how many equal parts it can be divided into?* (It can be divided into 3 and 8 equal parts.)
- *What fraction is one row of the bar?*
- *What fraction is two columns of the bar?*

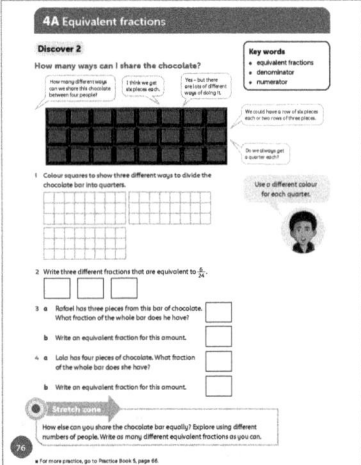

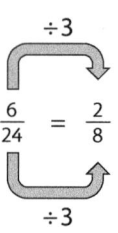

Then give students another fraction, perhaps $\frac{20}{32}$, and ask them to use this method to find equivalences and check by folding paper or sketching.

Practice Book: Students complete Practice Book page 66. They can do this directly after the Main activity, as homework, or as the focus of a separate mathematics session to help students consolidate their learning and build fluency.

Give students five minutes to silently write down everything they notice about the fraction wall. After five minutes, take one fact from each student. Then students can complete the activity.

Differentiated outcomes	
All students	should find a range of equivalent fractions.
Most students	will find a range of equivalent fractions and define an equivalent fraction.
Some students	will understand the rule for reducing fractions to their simplest form.

Differentiation

Supporting: Provide squared paper for students to shade or fold to get equivalent fractions. Students could also 'build' the bar using interlocking cubes.

Consolidating: Ask students to draw diagrams on separate sheets of squared paper to represent any additional equivalent fractions they find.

Extending: Challenge students to explain the process for reducing fractions to their simplest forms.

Stretch zone: *How else can you share the chocolate bar equally? Explore using different numbers of people. Write as many different equivalent fractions as you can.*

A bar of 24 squares could be shared equally into halves, thirds, quarters, sixths, eighths and twelfths. Students might give the equivalent fractions, $\frac{1}{8} = \frac{2}{16} = \frac{3}{24}$, or $\frac{3}{8} = \frac{6}{16} = \frac{9}{24}$ and so on. Students can carry out similar activities using other items that can be split into fractions in this way.

Reflection time

Writing some equivalent fractions to $\frac{6}{24}$ from question 2 and ask students what they notice. Take their ideas and then draw a diagram to show the multiplication and division to find the equivalent fractions.

Answers

Student Book page 76

1. Answers will vary because students choose their own ways to divide the chocolate bar into quarters. Any answer with six pieces is acceptable.
2. $\frac{6}{24} = \frac{12}{48} = \frac{3}{12} = \frac{1}{4}$, for example.
3. a $\frac{1}{8}$ b $\frac{2}{16}$
4. a $\frac{1}{6}$ b $\frac{2}{12}$

Practice Book page 66

1. $\frac{2}{3} = \frac{4}{6} = \frac{6}{9} = \frac{8}{12} = \frac{10}{15}$
2. $\frac{3}{5} = \frac{6}{10} = \frac{9}{15}$
3. $\frac{5}{6} = \frac{10}{12}$
4. $\frac{1}{8} = \frac{2}{16}$
5. $\frac{4}{10} = \frac{2}{5} = \frac{6}{15}$
6. $\frac{2}{12} = \frac{1}{6}$
7. $\frac{10}{15} = \frac{8}{12} = \frac{6}{9} = \frac{4}{6} = \frac{2}{3}$
8. $\frac{12}{15} = \frac{4}{5}$

Stretch zone: For example: $\frac{1}{2} = \frac{50}{100}, \frac{1}{4} = \frac{25}{100}, \frac{3}{4} = \frac{75}{100}, \frac{2}{10} = \frac{20}{100}, \frac{3}{5} = \frac{60}{100}$

Unit 4 Fractions, decimals and percentages

4A Equivalent fractions

Explore 1 Student Book page 77 • Practice Book page 67

Specific learning focus
- Shade 100-squares to show equivalent fractions as decimals in tenths and hundredths.

Global skills
- **Creative skills:** problem solving
- **Real-world skills:** presenting information

Key vocabulary
- equivalent fractions, tenths, hundredths

Resources
- 100-squares
- counting stick

Language support
Continue to encourage clear and accurate pronunciation of fractions. For example, for $\frac{1}{5}$, say *one-fifth* rather than *one over five*.

Introductory activity

Hold up a counting stick, marked in 10 sections. Explain that one end is 0 and the other end is 1. Ask students to count together in tenths as you point along the stick: *1 tenth, 2 tenths, 3 tenths* and so on, then count back from 10 tenths to 0.

Repeat, counting in hundredths from 0. Explain that the one end is still 0 and the other end is still 1. Ask students to count together in hundredths as you point to each division along the stick: *10 hundredths, 20 hundredths, 30 hundredths* and so on, then count back from 100 hundredths to 0.

Main activity

Give students blank 100-squares. Ask, *How many rows are there on the 100-square?* (10) *If there are 10 equal rows, what fraction of the square is each column?* $\left(\frac{1}{10}\right)$ *How many squares is that out of 100?* (10) *How can we describe that as a fraction out of 100?* $\left(\frac{10}{100}\right)$ *What does that tell us about $\frac{10}{100}$ and $\frac{1}{10}$?* (that they are equivalent).

Look together at the examples at the top of page 77 of the Student Book. Ask, *What fraction is the red area?* $\left(\frac{1}{10}\right.$ or $\left.\frac{10}{100}\right)$. *What fraction is the blue area?* $\left(\frac{5}{10}, \frac{50}{100}\right.$ or $\left.\frac{1}{2}\right)$

Explain that for grids labelled a–d you would like them to shade a fraction of each 100-square as each question asks but that they should record the equivalent hundredths, tenths or another fractional equivalent, as appropriate.

As students work in pairs, ask questions to help them shade each 100-square correctly.

If you want to find $\frac{1}{4}$ of something, what calculation do you need to do? (divide by 4) *So, if you want to find $\frac{1}{4}$ of 100-squares, what number do you need to divide by 4?* (100) *What will that tell us?* (how many squares to shade).

We know that one row or column shaded shows $\frac{1}{10}$. How many tenths equal $\frac{3}{5}$? Can that help us work out how many rows or columns we need to shade to show $\frac{3}{5}$?

If we know that we shade six columns to show $\frac{3}{5}$, how can we use this information to work out how many columns to shade to show $\frac{1}{5}$?

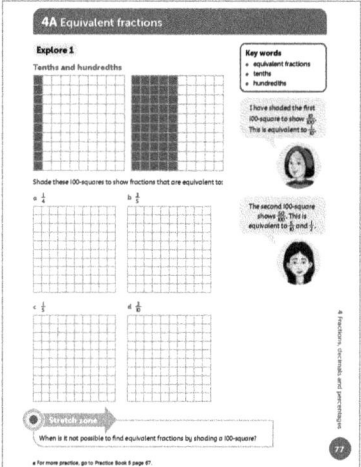

Differentiation
Supporting: Help students to identify how many columns to colour for each fraction by building on what they already know about tenths and hundredths.

Consolidating: Ask students to explain how they shaded each fraction.

Extending: Ask students to find other fraction equivalents on a 100-square.

Stretch zone: *When is it not possible to find equivalent fractions by shading a 100-square?*

Students may recognise that some fractions cannot be shaded easily, such as $\frac{1}{6}$, $\frac{1}{8}$ and so on (although $\frac{3}{6}$ and $\frac{4}{8}$ can be shaded).

 ### Reflection time

Discuss as a class the equivalences found during the Main activity. Which fractions were easier to shade on the 100-square? Why?

Show students a 100-square and ask whether someone would like to shade $\frac{1}{3}$. *How many complete columns should be shaded first?* (3) Then ask how many squares in the next column should be shaded. (3) *How much more of another square should be shaded?* $\left(\frac{1}{3}\right)$

Unit 4 Fractions, decimals and percentages

Practice Book: Students complete Practice Book page 67. They can do this directly after the Main activity, as homework, or as the focus of a separate mathematics session to help students consolidate their learning and build fluency.

Students may choose to fold strips of paper to find equivalent fractions, build or refer to a fraction wall, or multiply or divide the numerator and denominator by the same number to find equivalent fractions. They can include decimal and percentage equivalents if they feel comfortable doing so at this point.

Differentiated outcomes	
All students	should find equivalent fractions on a 100-square with support.
Most students	will find equivalent fractions on a 100-square.
Some students	may find and shade a wider range of equivalent fractions on a 100-square.

Answers

Student Book page 77

Check that students have correctly shaded the 100-square for each fraction.

a $\frac{1}{4}$ = any 25 squares   b $\frac{3}{5}$ = any 60 squares

c $\frac{1}{5}$ = any 20 squares   d $\frac{3}{10}$ = any 30 squares

Practice Book page 67

Answers will vary because students choose their own equivalent fractions to those given. Examples are:

1 $\frac{2}{5} = \frac{4}{10} = \frac{40}{100} = \frac{6}{15} = 40\% = 0.4$

2 $\frac{1}{4} = \frac{25}{100} = \frac{2}{8} = \frac{4}{16} = 25\% = 0.25$

3 $\frac{3}{4} = \frac{6}{8} = 75\% = 0.75 = \frac{12}{16}$

4 $\frac{8}{10} = 0.8 = 80\% = \frac{16}{20} = \frac{4}{5}$

Stretch zone: $\frac{55}{100} = \frac{11}{20} = \frac{110}{200} = 55\% = 0.55$

4A Equivalent fractions

Explore 2 Student Book page 78 • Practice Book page 68

Specific learning focus

- Calculate fractions of amounts and find an equivalent fraction of the same amount.

Global skills

- **Creative skills:** investigating

Key vocabulary

- equivalent fraction, numerator, denominator, unit fractions

Resources

- mini whiteboards and markers
- fraction strips from 4A Discover 1 or Resource sheet 4.2: photocopiable strips
- A4 paper, rulers, counters

Language support

Support students with questions and phrases, for example:

- An equivalent fraction is ….

 Introductory activity

Allow pairs of students two minutes to list on their whiteboards all the equivalent fractions they can remember. Take feedback and list these on the board.

Focus on fifths and tenths. Ask pairs of students to use a ruler to draw four rectangles, 10 cm × 2 cm, on a piece of A4 paper. Tell them to keep the first one whole and divide the others into halves, fifths and tenths. Ask them to find equivalences and write them down, for example:

$\frac{1}{5} = \frac{2}{10}$ $\frac{8}{10} = \frac{4}{5}$ $\frac{1}{2} = \frac{5}{10}$

Main activity

Ask students to take their halves, quarters and eighths strips from 4A Discover 1 and to lay them out, or give them new strips cut out from Resource sheet 4.2. Give each student 16 counters. Ask students to share the counters into each of the sections of the strips in turn and to make a note of the quantity for one half, one quarter and one eighth. Remind students that these are called **unit fractions**.

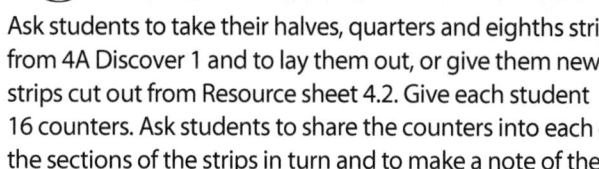

Draw 16 squares as a 4 × 4 array. Ask, *How many squares must be shaded to show $\frac{1}{4}$?* (4) Then ask, *How many squares would be $\frac{3}{4}$ of 16?* (3 × 4 = 12)

Repeat with 24 squares in a 6 × 4 array and find $\frac{1}{4}$ of 24 to help find $\frac{3}{4}$ of 24.

Look together at the worked example on page 78 of the Student Book. Draw a strip on the board split into 3 parts and share 24 'counters' equally between the parts to show that $\frac{2}{3}$ of 24 equals 16. Say, *Imagine that I have 20 counters. Could I share them equally to make a fraction of 20 that would equal 16?*

Start by drawing a strip divided in two and split the counters equally to show that they cannot do this with halves. Repeat for quarters. Then move on to fifths. Agree that $\frac{4}{5}$ of 20 equals 16.

Unit 4 Fractions, decimals and percentages

Repeat with other amounts of counters and fraction strips to show, for example, that half of 32 = 16 and $\frac{4}{6}$ of 32 = 16, $\frac{4}{6}$ of 24 = 16.

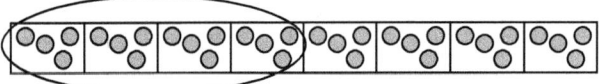

Organise students into groups of six, for example. Give each student a set number of counters (16, 20, 24, 32, 36 and 40) to work with and a set of fraction strips. For question 1, for example, the student who has 24 counters needs to find $\frac{5}{6}$ of 24 (20). Each student then uses their own number of counters and fraction strips to try to find one or more fractions that equal 20. Explain that the fraction strips are to help them, but they may be able to come up with other fractions as well. For example, they may suggest $\frac{5}{9}$ of 36 = 20. Tell students that, in some cases, they will not be able to find an equivalent fraction with their number of counters.

Once they have found an equivalent fraction, they should share their answers with the group to check.

Students may find this activity challenging. Emphasise its exploratory nature and provide support as necessary.

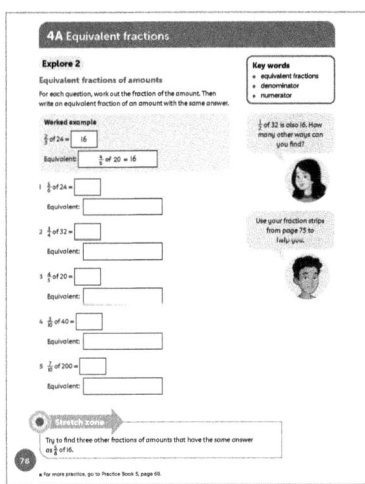

Differentiation

Supporting: Have students work with the smallest number of counters.

Consolidating: Ask students to explain to others in their group how they found an equivalent fraction.

Extending: Students work with the largest number of counters. Encourage them to try to find equivalent fractions beyond those possible to make with the fraction strips.

Stretch zone: *Try to find three other fractions of amounts that have the same answer as $\frac{5}{8}$ of 16.*

Students should work out that $\frac{5}{8}$ of 16 is 10. They may list fractions such as $\frac{1}{10}$ of 100, $\frac{2}{10}$ of 50, $\frac{1}{2}$ of 20. Encourage them to find equivalences mentally.

Reflection time

Discuss as a class the equivalences students found for each question. Discuss what they found easy and challenging about the activity.

Practice Book: Students complete Practice Book page 68. They can do this directly after the Main activity, as homework, or as the focus of a separate mathematics session to help students consolidate their learning and build fluency.

Students could use real counters or similar to find fractions of amounts.

Differentiated outcomes	
All students	should find equivalent fractions of amounts using counters and fraction strips with adult support.
Most students	will find equivalent fractions of amounts using counters and fraction strips with peer support.
Some students	may relate finding equivalent fractions of amounts to division.

Answers

Student Book page 78

Examples of possible answers:

1 $\frac{5}{6}$ of 24 = 20, $\frac{2}{3}$ of 30 = 20

2 $\frac{3}{4}$ of 32 = 24, $\frac{1}{2}$ of 48 = 24

3 $\frac{4}{5}$ of 20 = 16, $\frac{2}{3}$ of 24 = 16

4 $\frac{3}{10}$ of 40 = 12, $\frac{3}{8}$ of 32 = 12

5 $\frac{7}{10}$ of 200 = 140, $\frac{1}{5}$ of 700 = 140

Practice Book page 68

1 16
2 8
3 24
4 24
5 2
6 6
7 16
8 16
9 8
10 16
11 16
12 9

Stretch zone: $\frac{1}{2}$ of 16 = 8, $\frac{3}{4}$ of 16 = 12, $\frac{7}{8}$ of 16 = 14

Unit 4 Fractions, decimals and percentages   101

4B Fraction and decimal equivalents

Discover Student Book page 79 • Practice Book page 69

Specific learning focus
- Recognise and use the equivalence between decimal and fraction forms.

Global skills
- **Creative skills:** exploring

Key vocabulary
- fraction, decimal equivalent

Resources
- counting stick divided into ten sections
- calculators

Language support
The key phrase in this activity is, *What is the decimal equivalent?* Ask students to repeat it aloud as a class, checking their pronunciation of 'equivalent', in particular. Encourage them to ask their partner this question throughout the activity.

Introductory activity

Hold up a counting stick and explain to students that it starts from 0 at one end and finishes at 1. Ask, *How many divisions are there?* (10) *What fraction should we say at each one when we count up?* Give pairs a minute to consider the answer and then agree that they will count up in tenths. Count up and back in tenths as a class: *one tenth, two tenths, … ten tenths.*

What if we count up in decimals? Can you recall what decimal is equivalent to one tenth? Students may recall from previous work that it is 0.1. Count up and back in steps of 0.1 as a class: *zero point one, zero point two … 1.*

Ask some random questions to check that students understand the equivalence. For example, if you say six tenths, they should say zero point 6.

Main activity

Write the following fractions on the board:
$\frac{2}{10}, \frac{4}{10}, \frac{6}{10}, \frac{8}{10}, \frac{10}{10}$

Ask students what they notice about the pattern in these fractions. They might say that the numerators go up in twos, the denominators are all tens. Remind them of counting up in tenths and **decimals equivalents** and

then ask them to say what each fraction is as a decimal (0.2, 0.4, 0.6, 0.8, 1.0).

Ask, *Can you think of equivalent fractions to each of the tenths?* For example, $\frac{2}{10} = \frac{1}{5}$, so if $\frac{2}{10} = 0.2$, then $\frac{1}{5} = 0.2$. Can they match the rest of the tenths to equivalent fractions in fifths, and then match the fifths to decimal equivalents? For example, they might say that $\frac{6}{10} = 0.6 = \frac{3}{5}$. Students may find it useful to use number lines to help them find and visualise equivalences. For example:

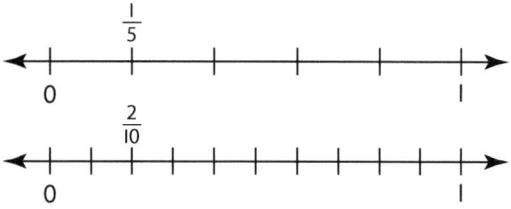

Ask students to complete the activities on page 79 of the Student Book. As they work, ask questions prompting them to explain how they found the equivalents.

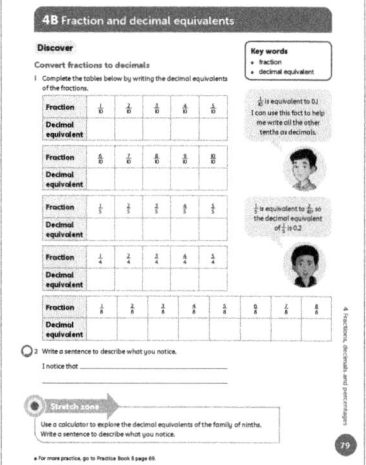

Differentiation

Supporting: Help students to tell you what patterns they are noticing by modelling the language.

Consolidating: Ask students to tell you which equivalent fractions they can see.

Extending: Challenge students to predict the answers before they carry out the calculations.

Stretch zone: *Use a calculator to explore the decimal equivalents of the family of ninths. Write a sentence to describe what you notice.*

Encourage students by looking at $\frac{1}{9}$. They should notice that the number in the equivalent decimal number repeats itself and is the same as the numerator.

Reflection time

Ask each student to share with the rest of the class something new they have learned. They could, for example, give two fractions or decimals they discovered are equivalent. See whether the class can find another that is equal to the two given in each case.

Practice Book: Students complete Practice Book page 69. They can do this directly after the Main activity, as homework, or as the focus of a separate mathematics session to help students consolidate their learning and build fluency.

Students will need a calculator to work out decimal equivalents. Take them through how to use a calculator to find decimal equivalents of fractions as a class, if necessary.

Differentiated outcomes	
All students	should complete the equivalence tables and notice patterns.
Most students	will complete the equivalence tables and notice patterns and equivalences.
Some students	may predict the answers before they find the equivalent fractions.

Answers

Student Book page 79

1. Tenths: 0.1, 0.2, 0.3, 0.4, 0.5, 0.6, 0.7, 0.8, 0.8, 1.0

   Fifths: 0.2, 0.4, 0.6, 0.8, 1.0

   Quarters: 0.25, 0.5, 0.75, 1.0, 1.25

   Eighths: 0.125, 0.25, 0.375, 0.5, 0.625, 0.75, 0.875, 1.0

2. Students may prefer to describe any patterns they notice in the different decimal equivalents orally.

Practice Book page 69

1

Fraction	$\frac{1}{7}$	$\frac{2}{7}$	$\frac{3}{7}$	$\frac{4}{7}$	$\frac{5}{7}$	$\frac{6}{7}$	$\frac{7}{7}$
Decimal	0.142	0.285	0.428	0.571	0.714	0.857	1.0

2

Fraction	$\frac{1}{9}$	$\frac{2}{9}$	$\frac{3}{9}$	$\frac{4}{9}$	$\frac{5}{9}$	$\frac{6}{9}$	$\frac{7}{9}$	$\frac{8}{9}$	$\frac{9}{9}$
Decimal	0.111	0.222	0.333	0.444	0.555	0.666	0.777	0.888	1.000

Stretch zone:

Fraction	$\frac{1}{11}$	$\frac{2}{11}$	$\frac{3}{11}$	$\frac{4}{11}$	$\frac{5}{11}$	$\frac{6}{11}$	$\frac{7}{11}$	$\frac{8}{11}$	$\frac{9}{11}$	$\frac{10}{11}$	$\frac{11}{11}$
Decimal	0.090	0.181	0.272	0.363	0.454	0.545	0.636	0.727	0.818	0.909	1.000

4B Fraction and decimal equivalents

Explore Student Book page 80 • Practice Book page 70

Specific learning focus
- Recognise and use the equivalence between decimal and fraction forms.

Global skills
- **Creative skills:** exploring
- **Real-world skills:** research/presenting information
- **Interpersonal skills:** communication

Key vocabulary
- fraction, decimal equivalent

Resources
- large cards with the digits 0 and 1 on them
- Resource sheet 4.3: place-value grid
- base-10 equipment

Language support

Focus on the difference between tenths, hundredths and thousandths. Read out loud, enunciating clearly, for example, $\frac{8}{10}$, $\frac{8}{1000}$ and $\frac{8}{100}$. Ask students to write down the fractions that you have said in the order you said them. Can they hear the difference? Swap roles.

 **Introductory activity**

Place a large '0' card at one side at the front of the classroom and a '1' on the opposite side. Ask half the class to choose a fraction (tenths, fifths, quarters). The other half must choose a decimal number between 0 and 1, and they should write their chosen number on their whiteboard.

Now ask a student to come to the front of the class and to stand in the correct place between the '0' and '1' that represents their number. Ask another student to come up and to stand in the correct place in relation to the first student. Repeat this until there are five students on the 'number line' at the front of the class in order. Repeat the activity.

 **Main activity**

Write on the board:

$\frac{4}{10}$ $\frac{4}{100}$ $\frac{4}{1000}$

Unit 4 Fractions, decimals and percentages

Ask students to say how each number is pronounced, making sure they notice the difference between tenths, hundredths and thousandths.

Ask them to write the numbers as decimal fractions on a place-value grid (Resource sheet 4.3). They should be able to write them as 0.4, 0.04 and 0.004. Then ask the questions from the second speech bubble. What is the same and what is different between the three numbers? (They all have a 4 and they all have a decimal point, but they differ by the position of the decimal point and the value that the 4 represents.)

Ten Thousands	Thousands	Hundreds	Tens	Ones	tenths	hundredths	thousandths
TTh	Th	H	T	O	t	h	th
				0	4		
				0	0	4	
				0	0	0	4

Now ask students to complete the tables on page 80 of the Student Book, working in pairs to compare answers and to share strategies. Pay particular attention to what students notice about the fractions and decimal equivalents in terms of tenths and hundredths. Make a list of all the responses. Write the list on the board for the whole class.

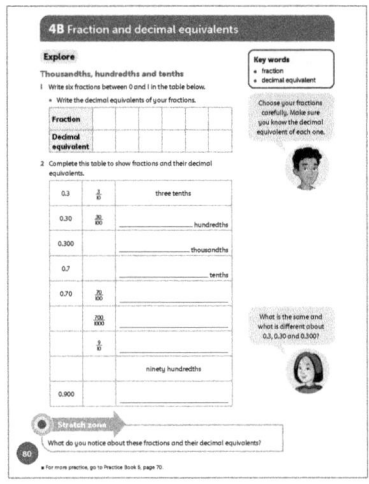

Differentiation

Supporting: Use base-10 equipment to represent fractional amount; that is, 1 ones-cube to represent $\frac{1}{1000}$, 1 tens-rod to represent $\frac{1}{100}$ and so on.

Consolidating: Ask students to explain their strategies for converting between fractions and decimals.

Extending: Ask students to tell you the decimal equivalents for $\frac{32}{100}$, $\frac{77}{100}$ and $\frac{59}{100}$, for example.

Stretch zone: *What do you notice about these fractions and their decimal equivalents?*

Students should make links between the numerators and where the digits appear in the decimal equivalents. When written as decimals, the numerators end in the place column indicated by the denominators.

Reflection time

Ask students to share their answers for the equivalents in the two tables. Ask them to explain how they found the equivalent decimal for a fraction and vice versa. Now ask pairs to write their own fractions in tenths, hundredths or thousandths for the rest of the class to convert to decimal equivalents.

Practice Book: Students complete Practice Book page 70. They can do this directly after the Main activity, as homework, or as the focus of a separate mathematics session to help students consolidate their learning and build fluency.

Encourage students to read their fractions and decimal numbers out loud to practise saying them correctly.

Differentiated outcomes	
All students	should understand the equivalences between tenths and hundredths and decimal forms with support.
Most students	will understand the equivalences between tenths, hundredths and thousandths in decimal forms.
Some students	may understand the equivalences between tenths, hundredths and thousandths and decimal forms and explain what is the same and different about them.

Answers

Student Book page 80

1 Answers will vary because students choose their own fractions. Check that students have written the correct decimal equivalent for each of their fractions.

2

0.30	$\frac{30}{100}$	thirty hundredths
0.300	$\frac{300}{1000}$	three hundred thousandths
0.7	$\frac{7}{10}$	seven tenths
0.70	$\frac{70}{100}$	seventy hundredths
0.700	$\frac{700}{1000}$	seven hundred thousandths
0.9	$\frac{9}{10}$	nine tenths
0.90	$\frac{90}{100}$	ninety hundredths
0.900	$\frac{900}{1000}$	nine hundred thousandths

Practice Book page 70

1 $\frac{68}{100} = 0.68$

2 $\frac{71}{100} = 0.71$

3 $\frac{17}{100} = 0.17$

4 $\frac{25}{100} = 0.25$

5 $\frac{64}{100} = 0.64$

6 $\frac{80}{100} = 0.8$

7 $\frac{8}{100} = 0.08$

8 $\frac{100}{100} = 1.0$

Stretch zone: $\frac{1}{3}, \frac{2}{3}, \frac{1}{7}, \frac{1}{9}, \frac{1}{11}$ cannot be shown because 3, 7, 9 and 11 do not divide into 100 exactly.

4C Improper fractions and mixed numbers

Discover Student Book pages 81–82 • Practice Book page 71

Specific learning focus
- Recognise mixed numbers and change an improper fraction to a mixed number.

Global skills
- **Creative skills:** investigating
- **Interpersonal skills:** communication/teamwork

Key vocabulary
- mixed number, numerator, denominator, proper fraction, improper fraction

Resources
- mini whiteboards and markers
- interlocking cubes

Language support
Add to your working wall display the sentence stems:
- An improper fraction is …
- A mixed number is …
- To change a mixed number to an improper fraction, you….

Model or support students to complete these sentences and add their responses to the working wall. Students can refer to them across the unit.

 **Introductory activity**

Write $1\frac{3}{8}$ on the board. Say, *Sketch me an image to represent this number.* (Do not read the fraction aloud.) Ask students to work in pairs to draw an image on their whiteboards that shows $1\frac{3}{8}$. Say, *Talk to your partner about how you might say this number.*

Share the images students have drawn on their whiteboards. Ask students how they think they should say the fraction and agree on 'one and three eighths'. Tell students it is called a **mixed number**, because is it a number made up of a whole number and a **proper fraction**. Explain that a proper fraction is a fraction where the numerator is smaller than the denominator.

Tell students to look at their images for $1\frac{3}{8}$ again, or display other appropriate examples. Ask, *How many eighths are there altogether?* (11) Write $\frac{11}{8}$ on the board. Agree that: $\frac{11}{8} = 1\frac{3}{8}$. Explain that $\frac{11}{8}$ is called an **improper fraction**. Ask, *Think about proper fractions. What makes this one improper?* Agree that it is because the numerator is larger than the denominator. Repeat for other mixed numbers with a variety of denominators such as $2\frac{1}{4}$ and $3\frac{3}{5}$.

 **Main activity**

Point out the worked example on page 81 of the Student Book. Students can use interlocking cubes (bricks) to model the example for themselves, starting with 14 cubes and making towers of 5 cubes. They will make 2 towers of 5 cubes and have 4 cubes left over. Ask, *What does each cube represent if 5 cubes make a tower?* Agree that each cube represents one fifth of a tower, so 14 cubes means 14 fifths, or $\frac{14}{5}$. The completed towers show that $\frac{14}{5} = 2\frac{4}{5}$.

Ask students to complete the activities in the Student Book in pairs. Make sure that they have enough cubes to make the towers for each example. As they work, ask them to read the mixed number to you. This allows you to check their pronunciation and understanding. You can give them more practice by asking them to make mixed-number towers using a different starting number of cubes.

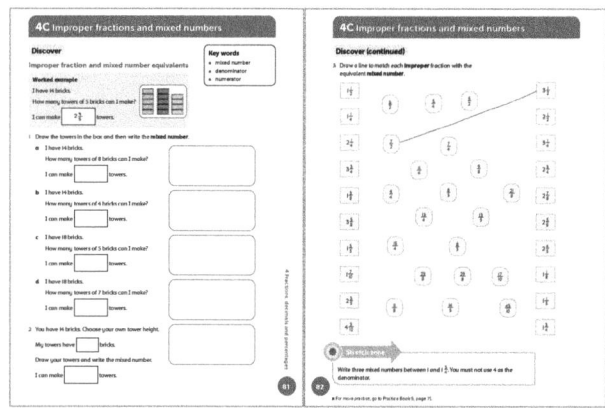

Unit 4 Fractions, decimals and percentages

Differentiation

Supporting: Work with students to model the use of cubes to make mixed numbers.

Consolidating: Ask students to describe their strategies for changing mixed numbers to improper fractions and vice versa.

Extending: Challenge students to describe their mental strategies for changing mixed numbers to improper fractions and vice versa.

Stretch zone: *Write three mixed numbers between 1 and $1\frac{3}{4}$. You must not use 4 as the denominator.*

Students should use their knowledge of ordering fractions between 0 and 1 to find the mixed numbers and equivalent fractions to find fractions $< 1\frac{3}{4}$ but without 4 as the denominator.

 Reflection time

Students should work together to agree on a definition for a mixed number. Ask pairs to discuss how they can change a mixed number into an improper fraction and vice versa. Record the clearest explanation on a poster to display as part of your working wall.

Practice Book: Students complete Practice Book page 71. They can do this directly after the Main activity, as homework, or as the focus of a separate mathematics session to help students consolidate their learning and build fluency.

Encourage students to continue to make towers using cubes or to sketch towers to represent each mixed number.

Differentiated outcomes	
All students	should represent simple mixed numbers by making cube towers.
Most students	will make more complex mixed-number towers.
Some students	may make a wide range of mixed-number towers and describe and use a mental strategy to convert mixed numbers to improper fractions and vice versa.

Answers

Student Book pages 81–82

1. **a** $1\frac{6}{8} = 1\frac{3}{4}$ **b** $3\frac{2}{4} = 3\frac{1}{2}$ **c** $3\frac{3}{5}$ **d** $2\frac{4}{7}$

2. Students choose their own tower height so the answers will vary.

3. $1\frac{1}{2} = \frac{3}{2}$ $2\frac{1}{2} = \frac{5}{2}$

   $1\frac{1}{4} = \frac{5}{4}$ $3\frac{1}{4} = \frac{13}{4}$

   $2\frac{1}{4} = \frac{9}{4}$ $2\frac{3}{4} = \frac{11}{4}$

   $3\frac{3}{4} = \frac{15}{4}$ $2\frac{7}{8} = \frac{23}{8}$

   $1\frac{3}{8} = \frac{11}{8}$ $2\frac{5}{8} = \frac{21}{8}$

   $3\frac{5}{8} = \frac{29}{8}$ $2\frac{4}{5} = \frac{14}{5}$

   $1\frac{3}{5} = \frac{8}{5}$ $1\frac{1}{8} = \frac{9}{8}$

   $1\frac{7}{10} = \frac{17}{10}$ $1\frac{1}{5} = \frac{6}{5}$

   $2\frac{3}{5} = \frac{13}{5}$ $1\frac{3}{4} = \frac{7}{4}$

   $4\frac{3}{10} = \frac{43}{10}$

Practice Book page 71

1. $2\frac{5}{8}$

2. $3\frac{1}{6}$

3. $4\frac{3}{5}$

4. $5\frac{2}{5}$

5. $3\frac{3}{4}$

Stretch zone: Check that students have drawn an appropriate diagram for $3\frac{2}{5}$.

4C Improper fractions and mixed numbers

Explore Student Book pages 83–84 • Practice Book page 72

Specific learning focus
- Compare mixed numbers with different denominators.
- Subtract mixed numbers from 10.

Global skills
- **Creative skills:** problem solving/exploring
- **Real-world skills:** presenting information/interpreting information
- **Interpersonal skills:** communication

Key vocabulary
- improper fraction, numerator, denominator

Resources
- mini whiteboards and markers

Language support
Model the key vocabulary carefully as students convert between mixed numbers and improper fractions. Use examples such as:
- $\frac{7}{2}$ is larger than $\frac{26}{8}$ because $\frac{7}{2}$ is $3\frac{1}{2}$ and $\frac{26}{8}$ is $3\frac{1}{4}$, and $\frac{1}{2}$ is greater than $\frac{1}{4}$.

Introductory activity

Ask pairs of students to use their whiteboards and draw diagrams to work out the answer to the question: *Which is larger: $3\frac{1}{2}$ or $\frac{13}{4}$?* Share responses and diagrams when pairs have solved the problem. Students may show that $3\frac{1}{2} = \frac{7}{2} = \frac{14}{4}$, which is clearly larger than $\frac{13}{4}$. They may also convert $\frac{13}{4}$ to a mixed number and then use their knowledge of fractions to say that $3\frac{1}{2}$ is larger because $\frac{1}{2}$ is larger than $\frac{1}{4}$.

Repeat for which is the larger of $\frac{19}{8}$ or $2\frac{1}{4}$. $\left(2\frac{1}{4} = \frac{9}{4} = \frac{18}{8}\right.$, which is smaller than $\frac{19}{8}$.)

Students answer the questions on page 83 of the Student Book (or you can introduce subtracting mixed numbers from 10 first, which is the focus of the activity on page 84 of the Student Book).

 Main activity

Read the problem on page 84 of the Student Book as a class. Point out that each row started as 10 cakes, and students need to decide how many are left by looking at the divisions on the cakes in the row. They should decide what fraction is needed to make up a whole cake, then add more cakes to bring the total to 10. Go through the worked example together to ensure that students are clear what is expected. Encourage students to draw diagrams such as the one in the Student Book to help them work out how much cake is left each time. Some students may also choose to draw number lines to help them. Model how to draw and label a number line for the worked example and count on from $4\frac{3}{4}$ to 10 in one step of $\frac{1}{4}$ to 5 and then a second step of 5 to 10.

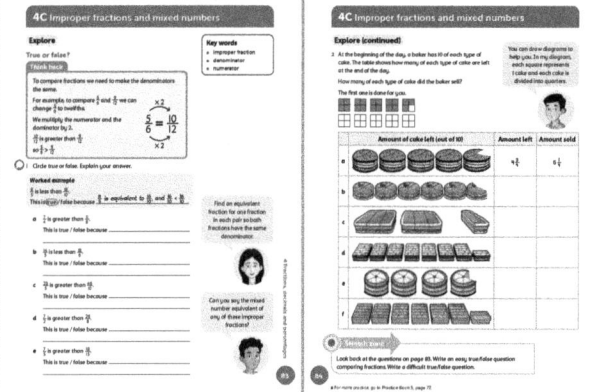

Differentiation

Supporting: Model how to use diagrams and number lines to help students to compare the sizes of fractions as well as work out how much cake is left.

Consolidating: Ask students to explain how they convert between mixed numbers and improper fractions to compare the sizes of the fractions.

Extending: Challenge students to use mental methods to help them to compare the sizes of the fractions.

Stretch zone: *Look back at the questions on page 83. Write an easy true/false question comparing fractions. Write a difficult true/false question.*

Students can make up their own statements for others in the class to solve.

 Reflection time

Select students to share their strategies for working on the activities in the Student Book. Can they describe how they used the denominators to help them compare the fractions? How did they find the missing amounts of cake on page 84?

Unit 4 Fractions, decimals and percentages

Ask students who completed the Stretch zone activity to share their questions with the class to answer. Did they find them easy or difficult to answer? Why?

Practice Book: Students complete Practice Book page 72. They can do this directly after the Main activity, as homework, or as the focus of a separate mathematics session to help students consolidate their learning and build fluency.

Work through a similar question as a class so students are clear on what the activity involves.

Differentiated outcomes	
All students	should use diagrams to help them to compare the sizes of the fractions.
Most students	will convert between mixed numbers and improper fractions to help them to compare the sizes of the fractions.
Some students	may use mental methods to help them to compare the sizes of the fractions.

Student Book pages 83–84

Possible examples of explanations:

1. a True because $\frac{7}{4} = \frac{14}{8} > \frac{11}{8}$
   b False because $\frac{14}{3} = \frac{28}{6} > \frac{18}{6}$
   c False because $\frac{23}{5} = \frac{46}{10}$
   d True because $\frac{7}{2} = \frac{28}{8} > \frac{26}{8}$
   e True because $\frac{7}{5} = \frac{21}{15} > \frac{18}{15}$

2. a (Provided) Left $4\frac{3}{4}$, Sold $5\frac{1}{4}$
   b Left $5\frac{3}{5}$, Sold $4\frac{2}{5}$
   c Left $2\frac{1}{3}$, Sold $7\frac{2}{3}$
   d Left $7\frac{2}{5}$, Sold $2\frac{3}{5}$
   e Left $3\frac{4}{5}$, Sold $6\frac{1}{5}$
   f Left $6\frac{2}{5}$, Sold $3\frac{3}{5}$

Practice Book page 72

Answers will vary because students choose the fraction that they will write as an improper fraction and as a mixed number. Check that they are drawn in approximately the correct positions on the number lines.

Stretch zone: Answers will vary. Check that students have selected a fraction in the correct range, written it as mixed number and improper fraction, and marked it correctly on the number line.

4D Adding and subtracting fractions

Discover Student Book page 85 • Practice Book page 73

Specific learning focus
- Add and subtract fractions.

Global skills
- **Creative skills:** investigating
- **Interpersonal skills:** communication/teamwork

Key vocabulary
- numerator, denominator, multiple

Resources
- fraction wall
- mini whiteboards and markers

Language support

Continue to reinforce how to pronounce fractions by emphasising them as you say any calculations, for example 'three **sixths** subtract one **sixth** equals 5 **sixths**'.

Introductory activity

Write $\frac{2}{5} + \frac{1}{5}$ on the board. Ask, *What do you notice about these two fractions?* (different numerators, the same denominator) *What will the denominator of the answer be? Why?* (a fifth because when the denominators of the two fractions you are adding are the same, the denominator of the answer will be the same) Reinforce by saying, *two **fifths** plus one **fifth** equals 3 **fifths**.* Use a bar model to further reinforce.

Repeat with other additions, including some that equal 1 whole, as well as subtractions.

Unit 4 Fractions, decimals and percentages

Main activity

Ask students to look at the worked example on page 85 of the Student Book. Ask them how this calculation is different from the ones they saw in the Introductory activity (it has different denominators). Use this as a further reminder that when adding (or subtracting) fractions, they should have the same denominators. Explain that they can use a fraction wall (as on page 85 of the Student Book) to find equivalent fractions so the denominators in the calculation are the same.

Students complete the questions on page 85 by finding suitable equivalent fractions for one of the fractions in the calculation. Once they have written their six number sentences, they should swap with a partner to solve them, writing their answers on their whiteboards. After they have solved each other's number sentences, they should compare answers as well as calculation strategies.

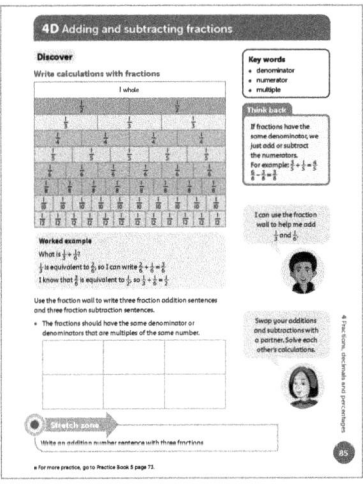

Differentiation

Supporting: Help students to find the equivalent fractions for the calculations using the fraction wall.

Consolidating: Ask students to explain how they know their calculations are correct.

Extending: Challenge students to add or subtract fractions with a variety of denominators.

Stretch zone: *Write an addition number sentence with three fractions.*

Students may be further challenged to use three different denominators.

 Reflection time

Ask several students to share their solutions. Ask students to work in pairs to find further fraction additions and subtractions, using the fraction wall, if necessary, to find the answers. Check their answers and understanding. Ask students whether they could add or subtract any two fractions by finding equivalent fractions. They may say that any fractions can be made into equivalents that have the same denominators so they can always be added or subtracted.

Practice Book: Students complete Practice Book page 73. They can do this directly after the Main activity, as homework, or as the focus of a separate mathematics session to help students consolidate their learning and build fluency.

Encourage students to continue to use the fraction wall to find equivalent fractions or to check their answers.

Differentiated outcomes	
All students	should use equivalent fractions to add or subtract fractions using a fraction wall for support.
Most students	will use equivalent fractions to add or subtract fractions with different denominators.
Some students	may use equivalent fractions to add more than two fractions with different denominators.

Answers

Student Book page 85

Students will have different answers because they choose their own addition and subtraction sentences with fractions.

Practice Book page 73

1 $\frac{5}{10} = \frac{1}{2}$

2 $\frac{8}{12} = \frac{2}{3}$

3 $\frac{9}{10}$

4 $\frac{1}{8}$

5 $\frac{4}{6} = \frac{2}{3}$

6 $\frac{5}{8}$

7 $\frac{6}{10} = \frac{3}{5}$

8 $\frac{7}{12}$

Stretch zone: The numerators increase by 1 each time and the denominators increase by 2 each time.

4D Adding and subtracting fractions

Explore Student Book pages 86–87 • Practice Book page 74

Specific learning focus
- Solve calculations with fractions

Global skills
- **Creative skills:** exploring
- **Interpersonal skills:** communication

Key vocabulary
- denominator, numerator, equivalent fraction, multiple

Resources
- mini whiteboards and markers

Language support
Support students by referring to fractions as being a fraction of the whole, and encourage them to talk about, for example, $\frac{5}{8}$ of the whole.

 Introductory activity

Write the following problem on the board.

Rachael has a bar of chocolate. She eats $\frac{1}{2}$ of the bar in the morning and eats another $\frac{3}{10}$ of the bar in the afternoon. How much of the bar has she eaten altogether?

Ask, *What equivalent fraction could you use to help you complete the calculation?* Give students some time to complete the calculation with a partner on their whiteboards before asking some pairs to share their calculation and explain how they did it. They can refer to the fraction wall on page 86 of their Student Book to help them solve the problem or use their knowledge of equivalent fractions.

Write on the board $\frac{1}{2} + \frac{3}{10}$ and ask students to do the calculation. Choose a student to model the answer on the board and discuss.

 Main activity

Write the following problem on the board.

Reva has $\frac{2}{3}$ of a bar of chocolate left from the day before. She now eats $\frac{1}{6}$ of the bar. How much of the bar has she got left now?

Give students some time to complete the calculation before asking a student to come out and show how they solved it on the board.

If none model it, show the class how to make an equivalent fraction by thinking about multiples, that is, 6 is a multiple of 3 because $2 \times 3 = 6$. They can multiply both the denominator and the numerator by 2 to make an equivalent fraction.

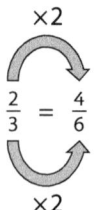

Now ask students to complete the calculations on pages 86–87 of the Student Book in pairs. Prompt them to consider equivalent fractions for each calculation.

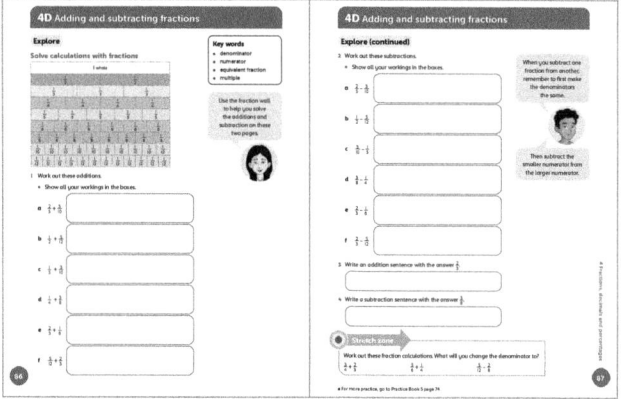

Differentiation

Supporting: Help students to find possible equivalent fractions using a fraction wall.

Consolidating: Ask students to explain how they found equivalent fractions for the answers to their calculations.

Extending: Challenge students to mentally solve the fraction calculations.

Stretch zone: *Work out these fraction calculations. What will you change the denominator to?*

$\frac{3}{4} + \frac{2}{3}$ $\frac{5}{6} + \frac{1}{4}$ $\frac{5}{12} - \frac{2}{8}$

Students may use their knowledge of mixed numbers to complete the additions.

Reflection time

Ask several students to share one of the fraction calculations they completed. Ask each student to talk through how they chose their equivalent fractions to complete each calculation. Repeat with another question and different pairs of students. If other students in the class used a different strategy to answer the question, invite them to share it.

Practice Book: Students complete Practice Book page 74. They can do this directly after the Main activity, as homework, or as the focus of a separate mathematics session to help students consolidate their learning and build fluency.

Unit 4 Fractions, decimals and percentages

Work through one question together to help students think about strategies for writing a calculation with different denominators. For example, to make an addition sentence with the answer $\frac{4}{6}$, start with two fractions with the same denominators such as $\frac{3}{6} + \frac{1}{6}$ and then convert $\frac{3}{6}$ to $\frac{1}{2}$ to make $\frac{1}{2} + \frac{1}{6} = \frac{4}{6}$.

Differentiated outcomes	
All students	should use equivalent fractions to add or subtract fractions using the fraction wall for support.
Most students	will use equivalent fractions to add or subtract fractions by thinking about multiples.
Some students	may be able to add or subtract some fractions with different denominators mentally, for example $\frac{1}{2} - \frac{5}{12}$.

Answers

Student Book pages 86–87

1. a $\frac{7}{10}$ d $\frac{5}{8}$
   b $\frac{11}{12}$ e $\frac{5}{6}$
   c $\frac{5}{10} = \frac{1}{2}$ f $\frac{13}{12} = 1\frac{1}{12}$

2. a $\frac{1}{10}$ d $\frac{1}{8}$
   b $\frac{1}{12}$ e $\frac{3}{6} = \frac{1}{2}$
   c $\frac{1}{10}$ f $\frac{3}{12} = \frac{1}{4}$

3 and **4** Students choose their own sentences.

Practice Book page 74

Answers will vary because students choose their own calculations.

Stretch zone: Check that students' calculations have the answer $\frac{2}{5}$. Examples could be: $\frac{1}{5} + \frac{2}{10}$, $\frac{3}{5} - \frac{2}{10}$ or $\frac{4}{5} - \frac{4}{10}$.

4E Multiplying fractions

Discover Student Book page 88 • Practice Book page 75

Specific learning focus
- Multiply a fraction by a whole number using a bar model or number line.

Global skills
- **Creative skills:** investigating

Key vocabulary
- denominator, numerator, multiplier, bar model

Resources
- grid paper

Language support
Continue to encourage students to count on aloud so they have plenty of practice saying aloud fractions that can be challenging to pronounce. Count together to model the pronunciation.

Introductory activity

Draw a rectangle on the board and divide it into four equal parts. Ask students to say how many parts there are (4) and what fraction of the rectangle each part represents ($\frac{1}{4}$). Label each part. Write the following calculations on the board:

$1 \times \frac{1}{4} =$ $2 \times \frac{1}{4} =$ $3 \times \frac{1}{4} =$

Ask students how many parts they should shade to show the first calculation (1 part). Shade 1 of the 4 quarters.

$\frac{1}{4}$	$\frac{1}{4}$	$\frac{1}{4}$	$\frac{1}{4}$

Now ask how many parts they should shade to show $2 \times \frac{1}{4}$ (2 parts). Can they tell you how much of the rectangle is shaded? They may say $\frac{2}{4}$, or they may recognise that this is $\frac{1}{2}$. Finally, shade another part to show $3 \times \frac{1}{4} = \frac{3}{4}$.

Main activity

Write on the board $5 \times \frac{1}{7}$. Refer students to the first worked example on page 88 of the Student Book. Ask, *How many equal parts has the rectangle, or bar, been divided into?* (7) *What fraction of the bar is each part?* ($\frac{1}{7}$) Ask students to count how many of the parts have been shaded and what fraction of the bar this shows. (5 and $\frac{5}{7}$)

Now draw a number line from 0 to 1, marked in sevenths. Start at 0 and make 5 jumps along the line, counting as you go: *one seventh, two sevenths, … five sevenths*. Explain to students that $5 \times \frac{1}{7}$ can be shown as a shaded bar, or as jumps along a number line.

Point out the second worked example, $10 \times \frac{1}{7}$. Ask, *What do you notice?* They should notice that the number line goes beyond 1 in this example, because $7 \times \frac{1}{7} = 1$ and this shows $\frac{10}{7}$. They can see that $10 \times \frac{1}{7} = \frac{10}{7}$, or $1\frac{3}{7}$.

Students should now complete the activities on page 88, using **bar models** or number lines to complete each multiplication, drawing their number lines on a separate sheets of grid paper.

As students work, move around the class reinforcing the language of whole and parts by asking questions about their bar models such as:

- *How many parts are there in the whole?*
- *What fraction is each part?*
- *How many parts are shaded?*

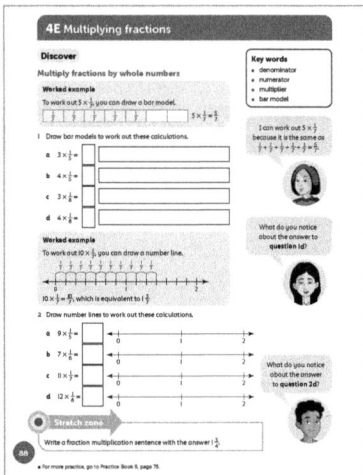

Differentiation

Supporting: Help students by modelling the use of the bar model and number line.

Consolidating: Ask students to explain how they use the bar model and number line to help them calculate.

Extending: Challenge students to make their own similar fraction multiplication problems.

Stretch zone: *Write a fraction multiplication sentence with the answer $1\frac{3}{4}$.*

Students will probably write $7 \times \frac{1}{4}$. Ask them to show this on a number line or as a bar model.

 Reflection time

Go through the calculations and ask students to describe how they completed each calculation. They should be encouraged to use the correct language, including 'equal parts' and 'equivalent' when describing their strategy. For example, for multiplying by sixths, the bar would be marked in 6 equal parts. If the total was 'seven sixths', this is equivalent to 'one and one sixth'.

Practice Book: Students complete Practice Book page 75. They can do this directly after the Main activity, as homework, or as the focus of a separate mathematics session to help students consolidate their learning and build fluency.

Tell students to draw a bar model for each calculation. Any answers that are improper fractions should also be recorded as mixed numbers.

Differentiated outcomes	
All students	should multiply unit fractions using a bar model or number line with support.
Most students	will multiply unit fractions using a bar model or number line.
Some students	may write their own multiplication problems to solve using a bar model or number line.

Answers

Student Book page 88

1 a $\frac{3}{5}$ b $\frac{4}{5}$ c $\frac{3}{8}$ d $\frac{4}{8} = \frac{1}{2}$

2 a $\frac{9}{5} = 1\frac{4}{5}$ b $\frac{7}{6} = 1\frac{1}{6}$ c $\frac{11}{7} = 1\frac{4}{7}$ d $\frac{12}{8} = 1\frac{4}{8} = 1\frac{1}{2}$

Check that students have completed the bars correctly.

Practice Book page 75

1 $\frac{4}{9}$

2 $\frac{7}{9}$

3 $\frac{4}{11}$

4 $\frac{8}{11}$

5 $\frac{10}{9} = 1\frac{1}{9}$

6 $\frac{11}{10} = 1\frac{1}{10}$

7 $\frac{9}{8} = 1\frac{1}{8}$

Check that students have drawn the bar models correctly to represent each calculation, shading the correct number of parts.

Stretch zone: Each answer is 1 whole and a unit fraction.

4E Multiplying fractions

Explore Student Book pages 89–91 • Practice Book page 76

Specific learning focus
- Multiply non-unit fractions using bar models and number lines.

Global skills
- **Creative skills:** problem-solving/exploring
- **Interpersonal skills:** communication

Key vocabulary
- denominator, numerator, multiplier, bar model

Resources
- counting stick marked 0–2 in fifths
- mini whiteboards and markers

Language support
Add a diagram to your working wall, showing how to partition mixed numbers to multiply by a whole number, labelling it and writing out the steps to describe the strategy.

Introductory activity

Ask the class to count together in steps of $\frac{1}{5}$, from 0 to 2 and back, firstly in fifths and then in mixed numbers, as you point to the fractions on a counting stick marked 0–2 in fifths:

$\frac{1}{5}, \frac{2}{5}, \frac{3}{5}, \frac{4}{5}, \frac{5}{5}, \frac{6}{5}, \frac{7}{5}, \frac{8}{5}, \frac{9}{5}, \frac{10}{5}$

$\frac{1}{5}, \frac{2}{5}, \frac{3}{5}, \frac{4}{5}, 1, 1\frac{1}{5}, 1\frac{2}{5}, 1\frac{3}{5}, 1\frac{4}{5}, 2$

Now repeat, counting in steps of $\frac{2}{5}$:

$\frac{2}{5}, \frac{4}{5}, \frac{6}{5}, \frac{8}{5}, \frac{10}{5}$

$\frac{2}{5}, \frac{4}{5}, 1\frac{1}{5}, 1\frac{3}{5}, 2$

Main activity

Write on the board $3 \times \frac{2}{7}$. Refer students to the first worked example on page 89 of the Student Book. Ask, *How many equal parts has the rectangle, or bar, been divided into?* (7) *How can we describe them as fractions of the whole bar?* (sevenths) *What fraction of the bar is each coloured part?* ($\frac{2}{7}$) Ask students to count how many of the coloured parts have been shaded and what fraction of the bar this shows. ($\frac{6}{7}$) Ask students to write this on their whiteboards as a repeated addition and a multiplication calculation. ($\frac{2}{7} + \frac{2}{7} + \frac{2}{7}$ and $3 \times \frac{2}{7} = \frac{6}{7}$)

Now draw a number line from 0 to 1, marked in sevenths. Ask, *How can we show this on a number line?* Agree that, starting at 0, they make 3 jumps of $\frac{2}{7}$ along the line. Choose a student to mark each jump on the number line as the class counts: *two-sevenths, four-sevenths, six-sevenths.*

Ask them to look at the second worked example on page 89 of the Student Book, $3 \times \frac{3}{8}$. Ask, *What do you notice?* They should notice that the number line goes beyond 1 in this example, because $3 \times \frac{3}{8} = \frac{9}{8}$, which is equivalent to $1\frac{1}{8}$.

Now ask them to look at the Think back example on page 90. Talk through the example to show how a mixed number can be partitioned to multiply it, in this case $4 \times 2\frac{1}{5}$ is the same as $4 \times 2 + 4 \times \frac{1}{5}$.

Students should now complete the activities on pages 89–91, using bar models or number lines to complete each multiplication.

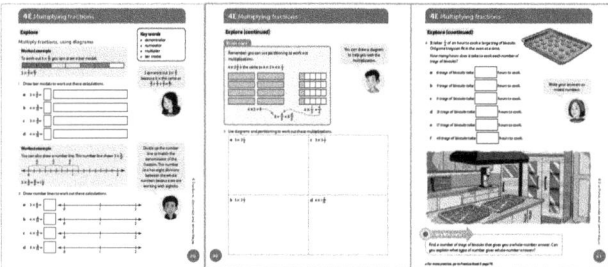

Differentiation
Supporting: Help students by modelling the use of the bar model and number line.

Consolidating: Ask students to explain their strategy for multiplying mixed numbers by whole numbers and show you, using a bar model or with number rods.

Extending: Challenge students to make their own similar fraction multiplication problems and represent these pictorially.

Stretch zone: *Find a number of trays of biscuits that gives you a whole-number answer. Can you explain what type of number gives whole-number answers?*

Students may find that trays with a multiple of 4 biscuits will take a whole number of hours to cook.

Reflection time
Go through the calculations and ask students to describe how they completed each calculation. Encourage them to use the correct language, including 'equal parts', 'equivalent' and 'partitioning', when describing their strategies.

Practice Book: Students complete Practice Book page 76. They can do this directly after the Main activity, as homework, or as the focus of a separate mathematics session to help students consolidate their learning and build fluency.

Unit 4 Fractions, decimals and percentages

Talk through the worked example on page 76 of the Practice Book, drawing a bar model to represent partitioning to calculate the answer. Explain that the examples use brackets to help you see how you are multiplying the whole and fractional parts separately first and then adding them. Encourage students to continue to present the calculations visually using diagrams, bar models or number lines, drawing them on a separate sheet of paper or their whiteboards, to help them with their calculations.

Differentiated outcomes	
All students	should multiply non-unit fractions using a bar model or number line with support.
Most students	will multiply non-unit fractions and mixed numbers by a whole number using a bar model or number line and record answers as non-unit fractions and mixed numbers, where appropriate.
Some students	may write their own multiplication problems including non-unit fractions and mixed numbers to solve using a bar model or number line.

Answers

Student Book pages 89–91

1. a $\frac{6}{9} = \frac{2}{3}$ b $\frac{12}{13}$ c $\frac{9}{11}$ d $\frac{12}{13}$
2. a $\frac{12}{9} = 1\frac{3}{9} = 1\frac{1}{3}$ b $\frac{16}{13} = 1\frac{3}{13}$ c $\frac{12}{11} = 1\frac{1}{11}$ d $\frac{15}{13} = 1\frac{2}{13}$
3. a $6\frac{3}{5}$ b $6\frac{3}{7}$ c $9\frac{3}{7}$ d $4\frac{8}{13}$
4. a $1\frac{1}{2}$ b $2\frac{1}{4}$ c $2\frac{3}{4}$ d $5\frac{1}{4}$
   e $4\frac{1}{4}$ f $11\frac{1}{4}$

Practice Book page 76

	Multiplication	Partition	Solve each part	Answer
1	$3 \times 2\frac{1}{7}$	$= (3 \times 2) + (3 \times \frac{1}{7})$	$= 6 + \frac{3}{7}$	$6\frac{3}{7}$
2	$3 \times 3\frac{1}{5}$	$= (3 \times 3) + (3 \times \frac{1}{5})$	$= 9 + \frac{3}{5}$	$9\frac{3}{5}$
3	$2 \times 5\frac{2}{7}$	$= (2 \times 5) + (2 \times \frac{2}{7})$	$= 10 + \frac{4}{7}$	$10\frac{4}{7}$
4	$2 \times 6\frac{3}{11}$	$= (2 \times 6) + (2 \times \frac{3}{11})$	$= 12 + \frac{6}{11}$	$12\frac{6}{11}$
5	$4 \times 5\frac{2}{9}$	$= (4 \times 5) + (4 \times \frac{2}{9})$	$= 20 + \frac{8}{9}$	$20\frac{8}{9}$

Stretch zone: $4 \times 3\frac{2}{5} = (4 \times 3) + (4 \times \frac{2}{5}) = 12 + \frac{8}{5} = 13\frac{3}{5}$

Students should notice that when they multiply the fractional part of the mixed number it gives an answer that is an improper fraction but the other calculations did not.

Unit 4 Fractions, decimals and percentages

4F Ordering fractions

Discover Student Book page 92 • Practice Book page 77

Specific learning focus
- Use equivalence and common denominators to help order fractions.

Global skills
- **Creative skills:** problem solving
- **Self-development skills:** reflecting on learning

Key vocabulary
- denominator, numerator, multiple, lowest common denominator, ascending order

Resources
- mini whiteboards and markers
- fraction wall

Language support
Ask students to explain, in their own words, what each of the key words mean. Explain any words they are unsure of, including examples.

Introductory activity

Write on the board $\frac{3}{7}, \frac{1}{7}, \frac{8}{7}, \frac{4}{7}, \frac{2}{7}$.

Ask students to write them in **ascending order** on their whiteboards. Remind them that this means from smallest to largest. Ask them to discuss with a partner how they know the order, then ask some students to share their reasoning. Students may say that that because the denominator is the same they can order the fractions by comparing the numerators; the larger the numerator, the larger the fraction.

Now write the fractions $\frac{1}{5}, \frac{1}{3}, \frac{1}{9}, \frac{1}{4}, \frac{1}{6}, \frac{1}{2}, \frac{1}{7}, \frac{1}{8}$ and ask students to order these from smallest to largest and explain how they did it. Students should notice that these are all unit fractions so they can compare the denominators to order the fractions; the larger the denominator, the smaller the fraction.

Main activity

Write on the board $\frac{3}{4}, \frac{7}{16}$ and $\frac{5}{8}$.

Ask students whether they can order these fractions. *How are these different from the fractions sets in the Introductory activity?* (Both the numerator and denominators are different.) *How can you compare the fractions? Can you use a **common denominator** to help you?* Ask them to look at the denominators of the fractions and see what they notice. (All the denominators are multiples of 4, and so all will divide into 16.) Ask students to convert each fraction into an equivalent fraction with denominator 16:

$$\frac{12}{16}, \frac{7}{16}, \frac{10}{16}.$$

Now they can arrange them in ascending order.

Ask students to look at the worked example on page 92 of the Student Book. Discuss each of the ordered lines of fractions and ask students what they notice. Ask them to think about how you could compare each group of three fractions as efficiently as possible. Leave some time for them to discuss this in pairs. They should notice that in the first set the numerators are all the same and are unit fractions. They need only compare the denominators and the fraction with the smallest denominator will be the largest fraction. For the second set, the denominators are all the same. They need only compare the numerators and the fraction with largest numerator will be the largest fraction. Use the third set to reinforce the idea of finding a common denominator and using equivalent fractions to help compare and order fractions.

Students now complete the activities on page 92 of the Student Book in pairs.

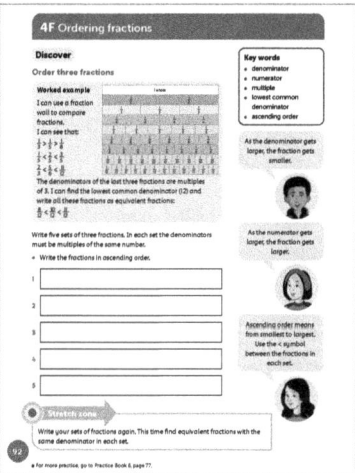

Differentiation

Supporting: Start students off with the set where they need only compare numerators and then denominators, reinforcing with the fraction wall. Next, show them a set where two fractions have the same denominator so they need only make one fraction equivalent in the first instance. Help students to find equivalent fractions so they can put fractions in order, modelling how to convert to an equivalent fraction alongside using the fraction wall to show equivalence.

Consolidating: Ask students to explain how to find the common denominator of a set of fractions, encouraging them to use key vocabulary.

Extending: Challenge students to order more than three fractions by finding common denominators.

Stretch zone: *Write your sets of fractions again. This time find equivalent fractions with the same denominator in each set.*

Students might be able to find different sets of equivalent fractions by using different common denominators.

Unit 4 Fractions, decimals and percentages

Reflection time

Choose students to describe how they found the common denominator for a set of fractions. For example, they might say they looked for a number that was a multiple of each of the denominators in the set. They should be able to describe how this helps them order the fractions. Ask some students to share their set of fractions for the class to see whether everyone can order them.

Practice Book: Students complete Practice Book page 77. They can do this directly after the Main activity, as homework, or as the focus of a separate mathematics session to help students consolidate their learning and build fluency.

Review what a factor is – a number, that when multiplied by another makes a given number. Ask, *What are all the factors of 12?* (1, 12, 2, 6, 3 and 4) Remind students of the relationship to multiples: this means that 12 is a multiple of 1, 2, 3, 4, 6 and 12.

Differentiated outcomes	
All students	should order a set of three fractions using common denominators with support.
Most students	will order a set of three fractions using common denominators.
Some students	may order sets of more than three fractions using common denominators.

Answers

Student Book page 92

Students choose their own sets of fractions, so answers will vary. Check that they have ordered each set in ascending order and have used the less-than symbol correctly each time.

Practice Book page 77

1 $\frac{1}{12}, \frac{2}{12}, \frac{4}{12}, \frac{8}{12}$

2 For example, $\frac{3}{4}, \frac{5}{6}, \frac{3}{10}, \frac{7}{8}$ – in order: $\frac{3}{10}, \frac{3}{4}, \frac{5}{6}, \frac{7}{8}$

3 For example, $\frac{2}{3}, \frac{1}{6}, \frac{4}{9}, \frac{5}{12}$ – in order: $\frac{1}{6}, \frac{5}{12}, \frac{4}{9}, \frac{2}{3}$

4 For example, $\frac{3}{4}, \frac{7}{8}, \frac{1}{12}, \frac{5}{16}$ – in order: $\frac{1}{12}, \frac{5}{16}, \frac{3}{4}, \frac{7}{8}$

Stretch zone: For example, $\frac{7}{5}, \frac{11}{8}, \frac{17}{10}, \frac{19}{12}$ – in order: $\frac{11}{8}, \frac{7}{5}, \frac{19}{12}, \frac{17}{10}$

4F Ordering fractions

Explore Student Book page 93 • Practice Book page 78

Specific learning focus
- Recognise equivalence between decimals and fractions and use this to help order fractions.

Global skills
- **Creative skills:** investigating
- **Interpersonal skills:** communication

Key vocabulary
- denominator, numerator, multiple

Resources
- mini whiteboards and markers

Language support
Add a 0–1 double number line with both tenths and decimal fractions written on it to your working wall. Encourage students to practise counting up and down it out loud.

 Introductory activity

Count up and back in tenths using a counting stick several times. Repeat, counting in decimal numbers, saying, *zero, zero point one, zero point two, … one.*

Next, try a mixture of both. For example, start with a fraction and then a decimal number and continue this pattern: *zero, one tenth, zero point two, three tenths, zero point four, … one.*

 **Main activity**

Write these fractions and decimal fractions on the board: $\frac{1}{2}$, 0.5, $\frac{1}{4}$, $\frac{1}{5}$, $\frac{1}{10}$, 0.9, $\frac{4}{5}$

Ask pairs to discuss how to order these fractions in order of size, smallest first. Take feedback from pairs on what they did.

Draw a number line on the board from zero to 1 with nine markers evenly spaced between them.

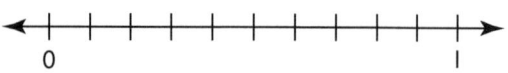

Invite a student to mark $\frac{1}{2}$ on the top of the number line. Ask, *What is a half in tenths? How do you write this as a decimal fraction?* Write 0.5 below $\frac{1}{2}$ on the number line. Invite other students to mark on the other fractions

Unit 4 Fractions, decimals and percentages

and decimal fractions and then to write their equivalent above or below the line. When marking $\frac{1}{4}$, discuss that this is halfway between 0.2 and 0.3. Ask, *What is the decimal fraction equivalent to $\frac{1}{4}$?* Introduce 0.25 as the equivalent decimal fraction to $\frac{1}{4}$.

Students now complete the activities on page 93 of the Student Book in pairs. They should copy the number line onto their whiteboards so that they can draft an answer to the first question before they write a solution in the Student Book.

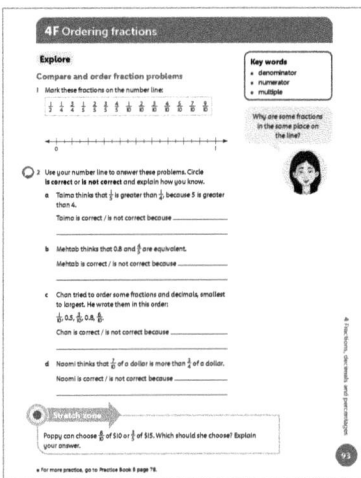

Differentiation

Supporting: Encourage students in using the number line to mark fractions.

Consolidating: For a specific question, ask students to explain how they can use their number line to help them compare and order the fractions.

Extending: Ask students to make up their own problems using fractions that are not on the number line.

Stretch zone: *Poppy can choose $\frac{8}{10}$ of $10 or $\frac{3}{5}$ of $15. Which should she choose? Explain your answer.*

Students should say that $\frac{8}{10}$ of $10 = $8, while $\frac{3}{5}$ of $15 = $9, so that is the larger amount.

Reflection time

Draw a 0–1 number line on the board and choose students to mark the fractions from question 1 on it. Ask students to think about the question in the speech bubble: *Why are some fractions in the same place on the line?* Take students' feedback. Can they tell you the decimal equivalences for some of the fractions on the number line?

Invite students to share the answers to their problems for question 2 and to explain how they used their number line to decide whether the students were correct or not. Does the class agree?

Practice Book: Students complete Practice Book page 78. They can do this directly after the Main activity, as homework, or as the focus of a separate mathematics session to help students consolidate their learning and build fluency.

Tell students to continue to mark the fractions and decimals on the number lines. Encourage them to find and mark the mid-point first to make placing the other fractions easier.

Differentiated outcomes	
All students	should use the number line to compare fractions and will remember simple equivalent fractions.
Most students	will remember equivalent fractions and use these to order and compare fractions.
Some students	may find fractions between those that are placed on the number line.

Answers

Student Book page 93

1 Check that students have marked the fractions correctly on the number line.

2 a not correct: $\frac{1}{5}$ is smaller than $\frac{1}{4}$.

   b correct: $0.8 = \frac{8}{10} = \frac{4}{5}$

   c not correct: $0.5 > \frac{3}{10}$ and $0.8 > \frac{6}{10}$

   d not correct: $\frac{7}{10} = 0.7$, but $\frac{3}{4} = 0.75$

Practice Book page 78

Check that the fractions and their decimal equivalents have been positioned correctly on the number lines.

1 0.6

2 0.8

3 $\frac{2}{5}$

4 $\frac{1}{10}$

5 0.7

Stretch zone: $\frac{4}{5} = \frac{8}{10} = 0.8 = 80\%$

Unit 4 Fractions, decimals and percentages

4G Thousandths

Discover Student Book pages 94–95 • Practice Book page 79

Specific learning focus
- Read and represent decimals to three decimal places.

Global skills
- **Creative skills:** exploring

Key vocabulary
- whole, tenths, hundredths, thousandths

Resources
- mini whiteboards and markers
- 1–9 digit cards
- Resource sheet 4: base-10 cards (cut the sheet into cards)
- place-value grids

Language support
Remind students how to read decimal numbers. Practise with examples, for instance show 32.672 and say it aloud as, *thirty-two point six seven two*.

 **Introductory activity**

Write 3.7 on the board and ask students what they know about this number. Agree that it is a mixed number with 3 ones and 7 tenths, and can be written as 3.7, $3\frac{7}{10}$ or $\frac{37}{10}$.

Now write 3.74 on the board. Ask pairs to discuss what they know about this number. Agree that it is a mixed number with 3 ones, 7 tenths and 4 hundredths. You can show students that this can be written as 3.74 or $\frac{374}{100}$. Draw the place-value columns on the board to show how the hundredths appear as a column to the right of the tenths:

```
    O . t h
    3 . 7 4
```

Now write 3.746 on the board and ask students what they think this number could represent. Extend the place-value columns to include thousandths and show 3.746 in the correct columns as 3 ones, 7 tenths, 4 hundredths and 6 thousandths. Ask students how this can be written as a fraction. ($\frac{3746}{10000}$)

Ask a student to come to the front of the class and select three digit cards. (There is no digit card for zero for this activity.) Students should make six different decimal numbers (also called decimal fractions) with these digits. For example, if they choose 4, 2 and 7, they can make:

0.247   0.427   0.274   0.472   0.742   0.724

Pairs should draw number lines from 0–1 on their whiteboards and place these decimal numbers on their number lines as accurately as they can.

Ask a pair to describe how they decided where to place the numbers. Remind them if necessary that they would first compare the digit in the tenths place, then the digit in the hundredths place and finally the digit in thousandths place.

Main activity

Tell students that when we are thinking about decimal numbers using base-10 equipment, the equipment represents different values. Show them a thousands-block and explain that we can use it to represent one whole. *How many thousands-blocks do you need to represent 9?* (9) *How many do you need to represent 6?* (6) Ask, *If the thousands-block represents one whole, what will the cubes, rods and hundreds-block represent? Can you say how to represent 3.746 using base-10 equipment?*

Give pairs time to investigate using base-10 equipment or, alternatively, using the base-10 cards copied and cut out from Resource sheet 4.4.

As pairs work, ask questions to support them, for example:

- *I want to build a thousands-block with hundreds-blocks. How many do I need? Can you show me?* (10)
- *Think of a 0–1 number line to show one whole. If I split it into ten equal parts, how would I mark each part as a fraction?* (tenths) *What do you think that means for the value of each of our ten parts of our one whole block?*

Discuss as a class, inviting students to explain their thinking, using the base-10 equipment to support their thinking. Choose a pair to represent 3.746 with base-10 equipment for the class.

Next, on the board, draw a ladder like the one on page 95 of the Student Book. Demonstrate the game in question 6 by playing it with the class. Invite students to pick digit cards, say the number they make, and give reasons for where they place it on the ladder.

Students should complete the activities on pages 94–95, and check answers with a partner.

Unit 4 Fractions, decimals and percentages

As pairs play the ladder game, ask questions to help them consolidate their understanding of decimal fractions, for example:

- What do the values of the digits in 1.632 represent? How do you know?
- Which is larger, 1.573 or 15.73? How do you know?
- When do you see numbers with three decimal places in real life?

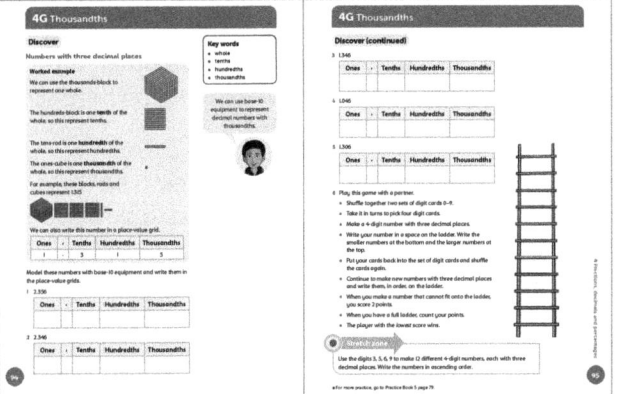

Differentiation

Supporting: Help students to make decimals to thousandths using base-10 equipment and place numbers on a number line or place-value grid to aid understanding.

Consolidating: Ask students to explain their thinking as they make and place their decimals on the ladder.

Extending: Suggest students use five cards and make numbers with tens, ones and three decimal places.

Stretch zone: *Use the digits 3, 5, 6, 9 to make 12 different 4-digit numbers, each with three decimal places. Write the numbers in ascending order.*

Check that students have made 12 numbers, for example 3.569, 5.369, 3.596, 5.396, and have ordered them from smallest to largest correctly. They can also represent them using base-10 equipment.

 Reflection time

Invite pairs of students to share the numbers they made with the class, explaining what the value of each digit is. Divide the class into two teams and play the ladder game together.

Practice Book: Students complete Practice Book page 79. They can do this directly after the Main activity, as homework, or as the focus of a separate mathematics session to help students consolidate their learning and build fluency.

Encourage students to use a dictionary to ensure that they are writing each number in words correctly.

Differentiated outcomes	
All students	should use base-10 equipment or place-value grids and counters to make numbers to three decimal places with support.
Most students	will use base-10 equipment or place-value grids and counters to make numbers to three decimal places.
Some students	may extend the game to making numbers consisting of five digits from tens to thousandths.

Answers

Student Book pages 94–95

Check that students have made the numbers correctly with base-10 equipment.

	Ones	.	Tenths	Hundredths	Thousandths
1	2	.	3	5	6
2	2	.	3	4	6
3	1	.	3	4	6
4	1	.	0	4	6
5	1	.	3	0	6

6 Answers will vary according to the decimal fractions made during the game. Check that that they have been positioned appropriately on the ladder.

Practice Book page 79

Students will make their own numbers using the digit cards. Check that they have written the numbers correctly and ordered them correctly in the tables.

Stretch zone: For example, 5.137

Unit 4 Fractions, decimals and percentages

4G Thousandths

Explore Student Book page 96 • Practice Book page 80

Specific learning focus
- Recognise and use decimals with up to three places in the context of measurement.

Global skills
- **Creative skills:** exploring
- **Real-world skills:** displaying/interpreting information

Key vocabulary
- seconds, tenths, hundredths, thousandths

Resources
- mini whiteboards and markers
- counting stick

Language support
Writing problems and defining good questions helps to support language development. Write a word bank on the board to help students to write questions including, for example:
- decimal, decimal fraction (*Which of these decimal fractions is greater, 0.371 or 0.092?*)

 Introductory activity

Hold up a counting stick and tell students that the ends of the counting stick are 0 and $\frac{1}{100}$. If there are 10 divisions along the stick, ask students how much each division represents. Agree that each division is $\frac{1}{1000}$, or 0.001.

Count together as a class in $\frac{1}{1000}$s from 0 to $\frac{1}{100}$ and back:

zero, one thousandth, two thousandths, three thousandths, four thousandths, five thousandths, six thousandths, seven thousandths, eight thousandths, nine thousandths, ten thousandths.

Then repeat the count in decimals:

zero, zero point zero zero one, zero point zero zero two ... , zero point zero zero nine, zero point zero one.

 **Main activity**

Introduce the topic of the Olympics and ask questions to lead a discussion, for example: *What can you tell me about the Olympics? What position are you in if you get gold, silver or bronze medals? If 29 competitors were placed higher than you, what position would you be placed? Do you know any sports included in the Olympics? How is the winner decided? What unit of measure would be used?*

Read together the introduction on page 96 of the Student Book and then ask students to look at the first table, which shows the 100-metre running times from the 2012 Olympic Games. Working in pairs, students should study the table and see who the fastest runner was and what their time was. Ask them to explain how they know they have found the fastest time. Can they see which athlete came last in the race? How do they know? Ask them to find out which athletes came second and third. Ask them to describe their strategy for placing the athletes in order. Pay particular attention to what students notice about the times in terms of tenths, hundredths and thousandths. Make a list of all the responses. Write the list on the board for the whole class to refer to.

Ask students to complete the blank table where they place all the athletes in order from fastest to slowest. They can then complete question 2 in a similar way, listing in order the high jump results.

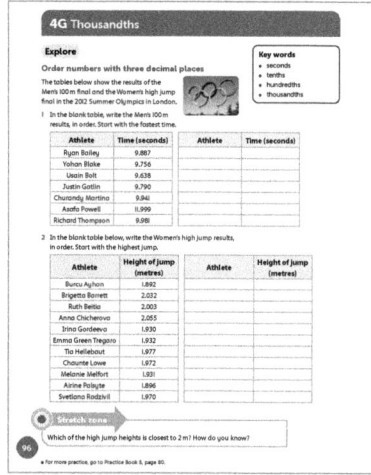

Differentiation
Supporting: Help students place the athletes' results in order by using a place-value grid.

Consolidating: Ask students to explain their strategies for ordering the times and the heights.

Extending: Challenge students to tell you a time or a height between two of the given times or heights.

Stretch zone: *Which of the high jump heights is closest to 2 m? How do you know?*

2.003 m is nearest to 2 m as it is only 3 thousandths of a metre away (or 3 mm).

Students could do practical activities that involve measurement and ordering, for example finding the time it takes students to write their name five times or the distance to various points in the classroom in metres and centimetres. Alternatively, use other data from athletics or other sporting events.

 Reflection time

Ask students to work in pairs to write a finishing time between Ryan Bailey and Yohan Blake. Ask them to show you their answers on their whiteboards. Then ask them to write a height that is higher than Ruth Beitia's jump,

Unit 4 Fractions, decimals and percentages

but lower than Airine Palsyte's jump. Now ask pairs to write their own questions for the rest of the class. Select five of these questions for the class to answer.

Practice Book: Students complete Practice Book page 80. They can do this directly after the Main activity, as homework, or as the focus of a separate mathematics session to help students consolidate their learning and build fluency.

Students look at race results, comparing and finding the difference between racing times. Discuss how they can do this, working through an example on how to find the difference between two reaction times. Show students how to count on along a number line to find the difference as well as use column subtraction. Repeat with a second example, this time inviting students to take you through the steps to find the difference.

Differentiated outcomes	
All students	should order numbers up to three decimal places with support.
Most students	will successfully order decimal numbers to three decimal places.
Some students	may give numbers that come between two decimals with three decimal places.

Answers

Student Book page 96

1 9.638; 9.756; 9.790; 9.887; 9.941; 9.981; 11.999

2 2.055; 2.032; 2.003; 1.977; 1.972; 1.970; 1.932; 1.931; 1.930; 1.896; 1.892

Practice Book page 80

1 No

2 No

3 0.028 seconds

4 0.007 seconds

5 0.005 seconds

Stretch zone: 9.741 seconds. Gatlin could have won if his reaction time had been 0.070 seconds or less.

4H Rounding decimals

Discover Student Book page 97 • Practice Book page 81

Specific learning focus
- Round decimals to the nearest tenth by reading hundredths.

Global skills
- **Creative skills:** problem solving
- **Real-world skills:** displaying information/interpreting information

Key vocabulary
- rounding, nearest tenth, nearest whole number

Resources
- mini whiteboards and markers
- 0–9 digit cards

Language support

Provide students with sentence frames to describe how they have rounded numbers, for example:

_____ rounded to the nearest whole number is _____.

_____ rounded to the nearest ten**th** is _____.

 **Introductory activity**

Write 1.875 on the board. Ask pairs to discuss everything they know about this number. Can they think of a sentence that would include this number? (They may suggest measurements – a length or a mass perhaps.) Draw this place-value grid on the board and together decide in which column to put each digit for 1.875:

Ones	.	Tenths	Hundredths	Thousandths
1	.	8	7	5
	.	3	4	6
	.	3	4	6
	.	0	4	6
	.	3	0	6

Ask, *What does this show?* Agree this shows that the 1 represents one whole and the 8 eight tenths. The 7 represents seven hundredths and the 5 five thousandths.

Ask students to write some numbers with three decimal places on their whiteboards. Then ask them to pick pairs of their numbers and compare them using the > and < symbols. Remind them that > means greater than and < means less than, if necessary. Invite students to write some of their examples on the board.

Unit 4 Fractions, decimals and percentages

Main activity

Ask a student to come to the front of the class and select four digit cards. They should make two different numbers with two decimal places, for example 23.51 and 99.46. Ask them to place these in order, explaining their decisions as they do this, which should be about looking at the tens digits.

Display the numbers on the place-value grid showing the tens, ones, tenths and hundredths. Ask students how they would **round** each number to the nearest 1 or whole number. They may say to look at the tenths digits and then round up if the tenths are 5 or more, otherwise round down. Then ask them to think about how they would round the number to the nearest tenth. They should explain their decisions as they do this. Prompt them if necessary to look at hundredths digits to round to tenths.

Ask students to look at the activities on page 97 of the Student Book. Pair students with similar understanding together for this activity and support those who need to draw number lines to support them with rounding.

While students are working, ask them to explain how they know which number to round their decimal to. Ask further questions such as:

- *Is 3.44 less than or greater than 3.414? How do you know?*
- *Which is greater, 2.57 or 2.507? How do you know?*
- *Which digit helps you round 4.37 to the nearest tenth?*
- *What is 9.96 rounded to the nearest tenth?*

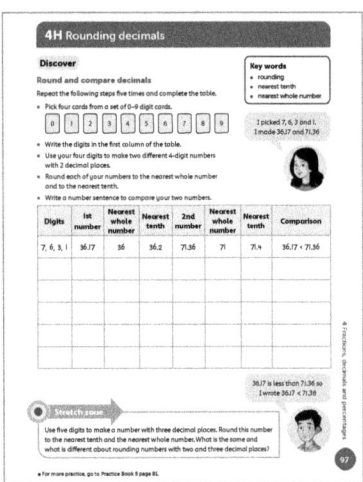

Differentiation

Supporting: Support students to round to the nearest tenth using numbers with two decimal places and a number line for support.

Consolidating: Ask students to explain their thinking as they round numbers to the nearest tenth.

Extending: Challenge students to describe how they would round three places of decimals to the nearest whole number.

Stretch zone: *Use five digits to make a number with three decimal places. Round this number to the nearest tenth as well as the nearest whole number. What is the same and what is different about rounding numbers with two and three decimal places?*

Check that students have rounded their 5-digit number to the nearest whole number and tenth successfully. In their own words, they should explain that when rounding any numbers with decimals to the nearest whole number or tenth, they both rely on the digit in the next column to the right. They differ by which column this is, the tenths or hundredths. Students can also explore whether rounding to the nearest tenth and whole number always gives different answers.

Reflection time

Invite individual students to share the numbers they made. As a class, work out what they are when rounded to the nearest tenth and whole number. Continue to record the numbers in the place-value grid as well as on a number line to help see how the numbers were rounded, as necessary. Invite other students to write number sentences for these numbers using < or >.

Practice Book: Students complete Practice Book page 81. They can do this directly after the Main activity, as homework, or as the focus of a separate mathematics session to help students consolidate their learning and build fluency.

Encourage students to continue to use number lines and place-value grids to help them round to the nearest tenth and whole number.

Differentiated outcomes	
All students	should round numbers with two places of decimals to the nearest tenth and whole number using number lines and place-value grids for support.
Most students	will round numbers with two decimals to the nearest whole or tenth.
Some students	may quickly round any decimals to the nearest whole.

Answers

Student Book page 97

Answers will vary according the decimal numbers made. Check that students have rounded the numbers correctly to the nearest whole number and tenth.

Practice Book page 81

Answers will vary because students are making their own numbers from the digit cards. Check that for their chosen digits, they have formed appropriate numbers and rounded them correctly.

Stretch zone: Check that students have labelled their number line effectively and placed each number on the line with a reasonable degree of accuracy.

Unit 4 Fractions, decimals and percentages

4H Rounding decimals

Explore Student Book page 98 • Practice Book page 82

Specific learning focus
- Round decimal measurements to the nearest unit.

Global skills
- **Creative skills:** exploring
- **Real-world skills:** presenting information/interpreting information

Key vocabulary
- rounding, nearest tenth, nearest whole number

Resources
- till receipts showing prices to two places of decimals (or facsimiles taken from the internet)
- catalogues or magazines containing measurements to two places of decimals
- 0–9 digit cards

Language support
Remind students of the abbreviations for metric units of measure and money, for example: ml = millilitre, kg = kilogram.

Introductory activity

Remind students of these relationships for money and measurement:

100¢ = $1

100 cm = 1 m

1000 ml = 1 litre

1000 g = 1 kg

In small groups, students should work together to make a poster that uses the measurements and prices from the catalogues, magazines or till receipts that you have given them. Encourage students to use place-value grids to illustrate the prices or measurements, for example: $36.42 and 7.462 m.

$36.42

$10	$1	decimal point	10¢	1¢
3	6	.	4	2

7.462 m

10 m	decimal point	10 cm	1 cm	1 mm
7	.	4	6	2

They can include in their posters the prices or measures expressed using a different unit of measure as well as saying what these would be when rounded to the nearest whole unit and tenth.

Each group should share their poster with the class once all groups have completed the activity.

 Main activity

Ask students to look at page 98 of the Student Book and read the instructions on how to complete the table together. In pairs, students choose four digit cards and make a measurement with two decimal places in either metres, kilograms or litres. They should then complete each row of the table using different digits and a variety of the three units. Pairs can work together to round their measurements to the nearest tenth or whole unit or they can complete their tables individually and then their partner can check their answers.

Ask questions to encourage students to explain their answers, for example:

- *Can you give an example of a measurement with three decimal places? Can you think of an example with two decimal places? Can you give me an example with one decimal place?*
- *Which of these is the largest measure: 2.1 kg, 2.01 kg, 2.001 kg, 2.11 kg? How do you know?*
- *What digit did you look at first to round your measurement to the nearest whole?*

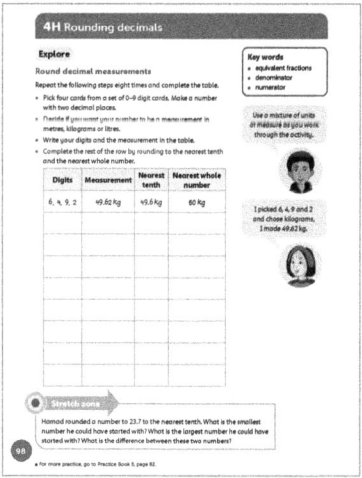

Differentiation

Supporting: Students can continue to use place-value grids and to draw number lines to help them round to the nearest whole and tenth.

Consolidating: Ask students to explain how they are rounding measures to the nearest whole and tenth.

Extending: Challenge students to use five cards, make make measures with up to three decimal places and round these to the nearest whole and tenth. Ask, *Is rounding measures to the nearest whole and tenth different from rounding numbers in no context?* Can they explain their answer?

Unit 4 Fractions, decimals and percentages

Stretch zone: *Hamad rounded a number to 23.7 to the nearest tenth. What is the smallest number he could have started with? What is the largest number he could have started with? What is the difference between these two numbers?*

Students should recognise that the smallest number would be 23.70 and the largest would be 23.74 and the difference is 0.04.

 Reflection time

Invite pairs to share their work, explaining how they made and rounded their numbers. Model the answers using number lines when appropriate. Have a discussion about when it is and isn't useful to round measurements and money and ask students to give examples.

Practice Book: Students complete Practice Book page 82. They can do this directly after the Main activity, as homework, or as the focus of a separate mathematics session to help students consolidate their learning and build fluency. This activity can be extended asking students to look at other past Olympic rankings. For example, ask, *Do the athlete's rankings change when their times or distances are rounded?*

Differentiated outcomes	
All students	should recognise how decimals are used in money and measurement and round to the nearest whole number and to the nearest tenth with support such as place-value grids and numbers lines.
Most students	will understand how place value to two decimal places is used in money and measurement and round to the nearest tenth and whole number.
Some students	may quickly round numbers mentally to the nearest whole and tenth.

Answers

Student Book page 98

Answers will vary according to the decimal numbers made. Check that students have rounded the amounts correctly to the nearest tenth and whole number.

Practice Book page 82

Athlete	Position	Time (seconds)	Time rounded to one decimal place	Time rounded to the nearest whole number
Elaine Thompson	1st	10.71	10.7	11
Tori Bowie	2nd	10.83	10.8	11
Shelly-Ann Fraser-Pryce	3rd	10.86	10.9	11
Marie-Josee Ta Lou	4th	10.86	10.9	11
Dafne Schippers	5th	10.90	10.9	11
Michelle-Lee Ahye	6th	10.92	10.9	11
English Gardner	7th	10.94	10.9	11
Christania Williams	8th	11.80	11.8	12

3 Rounding to one decimal place does not distinguish athletes within a tenth of a second interval.

Stretch zone: In close races such as 100 m, times to three decimal places may be needed to separate athletes.

Unit 4 Fractions, decimals and percentages

4I Percentages

Discover 1
Student Book page 99 • Practice Book page 83

Specific learning focus
- Understand percentage as the number of parts in every 100.

Global skills
- **Creative skills:** investigating
- **Real-world skills:** displaying information/interpreting information

Key vocabulary
- hundredth, percentage, out of 100

Resources
- blank 100-square grids

Language support
Display the posters students make as part of the Main activity, with the key vocabulary highlighted, for them to refer to in later lessons.

Introductory activity

Write a **percentage** on the board. What can the students tell you about it? Where have they seen it before? Jot down their suggestions. Ensure that examples include, for example, test scores and price discounts, popularity, improvement or increase.

Establish that a percentage is a special fraction, and that 1% means 1 out of 100. Explain that this means that 100% is the whole amount and that amount can be anything.

Display a blank 100-square grid. Say, *This represents 100%. So, what does one square represents as a percentage?* (1%) Tell students to think about what they know about fractions on a 100-square. *What does one square represent as a fraction out of 100?* ($\frac{1}{100}$) *What do this tell us?* (that 1% = $\frac{1}{100}$)

 Main activity

Give out blank 100- square grids. Students should work on completing the activities on page 99 of the Student Book, first reading the introduction to percentages together as a class. Pairs of students should continue with this activity, making as many different fraction and percentage equivalents as they can and illustrating them by shading the 100-square grids, for example: $\frac{4}{5}$ of 100 = 80, so $\frac{4}{5}$ = 80%.

Students can then make a poster showing some of the different percentages and equivalents they found.

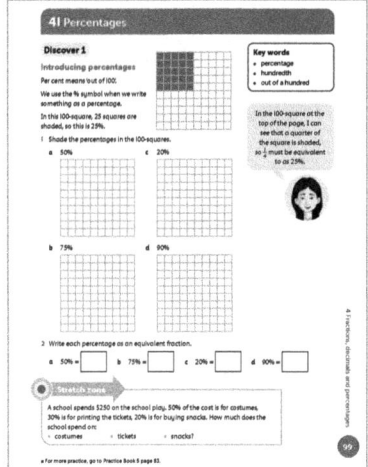

Differentiation

Supporting: Help students to find quick ways of counting the squares in the 100-square by drawing on their knowledge of fractions.

Consolidating: Ask students to explain their thinking when finding equivalent fractions and representing them as percentages.

Extending: Challenge students to find fraction and decimal equivalents for each percentage from 20% to 30%.

Stretch zone: *A school spends $250 on the school play. 50% of the cost is for costumes, 30% is for printing the tickets, 20% is for buying snacks. How much does the school spend on costumes, tickets and snacks?*

Students should work out that the school spent $125 on costumes, $75 on tickets and $50 on tickets by, for example, dividing 250 by 10 to find 10% ($25) and then using this to work out 20% and 30% by multiplying by 2 and 3.

Reflection time

Invite small groups to share their posters with the rest of the class. They should talk about the posters, carefully describing how they worked out the different representations.

Practice Book: Students complete Practice Book page 83. They can do this directly after the Main activity, as homework, or as the focus of a separate mathematics session to help students consolidate their learning and build fluency.

Look at the worked example together. Discuss how the diagram shows $\frac{42}{100}$, 0.42 and 42%. Agree that because there are 100 squares altogether, one square represents $\frac{1}{100}$, 0.01 and 1%. 42 squares are shaded so that is $\frac{42}{100}$, 42 × 0.01 = 0.42 and 42 × 1% = 42%. As the focus of the lesson was on percentage and fraction equivalences, you may prefer to have students focus on these only.

Unit 4 Fractions, decimals and percentages

Differentiated outcomes	
All students	should shade a 100-square to show percentages and find equivalent fractions with support
Most students	will shade a 100-square to show percentages and find equivalent fractions independently.
Some students	may quickly find a range of equivalent percentages and fractions and begin to identify some decimal equivalences.

Answers

Student Book page 99

1 Check that the correct amount of the 100-square has been shaded in each case.

2 a $\frac{1}{2}$ b $\frac{3}{4}$ c $\frac{1}{5}$ d $\frac{9}{10}$

Practice Book page 83

	Fraction	Decimal	Percentage
1	$\frac{2}{5}$	0.4	40%
2	$\frac{3}{5}$	0.6	60%
3	$\frac{11}{20}$	0.55	55%
4	$\frac{33}{100}$	0.33	33%
5	$\frac{99}{100}$	0.99	99%

Stretch zone: $\frac{1}{10} = 0.1 = 10\%$; $\frac{3}{8} = 0.375 = 37.5\%$, for example.

41 Percentages

Discover 2 — Student Book page 100 • Practice Book page 84

Specific learning focus
- Find percentages of amounts of money.

Global skills
- **Creative skills:** problem solving
- **Real-world skills:** displaying information/financial literacy
- **Interpersonal skills:** communication/teamwork

Key vocabulary
- percentage, hundredth, out of a hundred

Resources
- mini whiteboards and markers
- flipchart paper

Language support
Refer students to the first speech bubble on page 100 of the Student Book to help them explain their thinking.

 Introductory activity

Remind students of equivalent fractions and percentages, in particular:

$\frac{1}{2} = 50\%$

$\frac{1}{4} = 25\%$

$\frac{1}{10} = 10\%$

Split the class into five groups. Make these mixed-attainment groups so that students can share their understanding of percentages. Write $250 on the board and ask students to write on their whiteboards as many fractions and percentages as they can of this amount of money, for example: $\frac{1}{2}$ of $250 = $125, 50% of £$250 = $125. Share all the different fractions that groups find.

Encourage students to use percentages they know to find new percentages. For example, if they know 10% is $25, then 20% is $50 (double) and 5% is $12.50 (half).

 Main activity

Using five sheets of flipchart paper, write a different amount of money in the middle of each sheet:

$80 $120 $300 $500 $1000

The class remains divided into five groups. Give each group one of the flipchart sheets. Allow groups three minutes to write down as many percentages of their value as they can. After three minutes, move each flipchart sheet one group clockwise and the new group should add to the list. Encourage students to use percentages that have already been found to find new percentages.

Unit 4 Fractions, decimals and percentages

Ask questions to encourage students to explain their answers, for example:

- Which fraction will give you 25% of an amount? ($\frac{1}{4}$) When you want to find $\frac{1}{4}$ of amount, what operation do you use? (division) What do you divide by? (4)
- Which of these is the largest: 10% of $50 or 20% of $40? How do you know?
- You know that half of $80 is $40. You know that half is the same as 50%. How does this help you work out 25%?
- I know that 1% of $100 = $1. How can I use this information to find 1% of multiples of $100?

Students then complete the activity on page 100 of the Student Book, adding new percentages of $260 to the spider diagram.

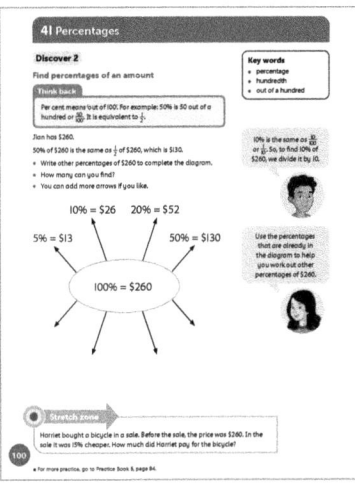

Differentiation

Supporting: Help students to calculate simple percentages such as 50% and 10%. Ask them to recall your strategies and then support them to use these to work out other simple percentages.

Consolidating: Ask students to explain to you or another student how they calculated new percentages of a specific amount.

Extending: Challenge students to use percentages of amounts to find equivalent fractions of amounts.

Stretch zone: Harriet bought a bicycle in a sale. Before the sale, the price was $260. In the sale it was 15% cheaper. How much did Harriet pay for the bicycle?

Students will likely add the $13 and $26 (5% and 10% of $260) to find 15% of $260 ($39). They subtract $39 from $260 to get the final price that Harriet paid: $221. They can then write their own problems about price changes.

 Reflection time

Look together at the flipchart sheets compiled earlier and discuss the percentages that students have found for the different amounts. Ask students to explain how they found the amounts and how they might have used previously found percentages to find new ones.

Practice Book: Students complete Practice Book page 84. They can do this directly after the Main activity, as homework, or as the focus of a separate mathematics session to help students consolidate their learning and build fluency.

Some students may wish to shade 100-squares to help them work out equivalences. As with the last lesson, you may prefer that students continue to focus on percentages and fraction equivalences.

Differentiated outcomes	
All students	will find percentages of amounts with support.
Most students	should use known percentages of amounts to derive additional percentages of amounts.
Some students	may use percentages to work out price changes in sales, for example.

Answers

Student Book page 100

Students find their own percentages of $260, so answers will vary.

Practice Book page 84

	Fraction	Decimal	Percentage
1	$\frac{1}{2}$	0.5	50%
2	$\frac{1}{4}$	0.25	25%
3	$\frac{3}{4}$	0.75	75%
4	$\frac{7}{10}$	0.7	70%
5	$\frac{4}{5}$	0.8	80%
6	$\frac{8}{10}$	0.8	80%
7	$\frac{1}{5}$	0.2	20%
8	$\frac{24}{100}$	0.24	24%

Stretch zone: Tam gets the higher score.

Unit 4 Fractions, decimals and percentages

4I Percentages

Explore Student Book pages 101–102 • Practice Book page 85

Specific learning focus
- Solve problems using percentages of amounts.

Global skills
- **Creative skills:** problem solving
- **Real-world skills:** displaying information/financial literacy

Key vocabulary
- hundredth, percentage, per cent, out of 100

Resources
- flipchart paper and coloured pens
- timers (optional)

Language support

Students may be unfamiliar with some of the word-problem contexts. Provide clear descriptions of any new vocabulary or support with a visual image. However, encourage them to begin to infer what a word might mean. For example: 'A carnival must be a big event because I know 680 people are at it. Half of the people are children so it is an event that children can attend.'

Introductory activity

Write the following problem on the board.

A library has 100 new books. 60% of the new books are fiction and the rest are non-fiction. How many of each type of book are there?

Draw a bar divided into 10 equal parts and ask, *What percentage does each part represent?* (10) *How many parts should be shaded to represent 60%?* (6) *How many books does each equal part represent?* (10) *So how many books are fiction and how many are non-fiction?* (60 and 40). Label the bar model as you add information.

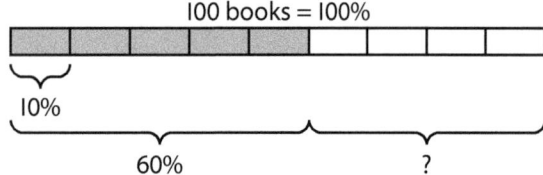

As you discuss the problem, keep emphasising that '**per cent**' means out of 100. Ask, *If there are 100 items, why is this easy to work out?* Students should say that if there are 100 items, then any number of them is a percentage because 'per cent' means out of 100.

 **Main activity**

Write the following problem on the board.

A car showroom has 40 new cars. 35% of the cars are from Japan, and the rest are from Europe. How many cars are from Europe?

First, discuss the context, then draw a bar model. Ask, *What do we need to find out? What is the key information? How can we represent the problem using a bar model?*

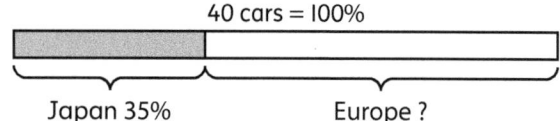

What easy percentage could you use to work out how many cars are from Europe? Agree 10% because you can use this to find how many cars would be 5% then use the 10% and 5% figures to calculate how many cars make 35% or 65%.

You could divide the bar into 10 equal parts and ask students how many cars each part represents. Prompt them to use sentences such as 40 cars = 100% so 4 cars = 10% or 40 cars divided into 10 equal groups equals 4 cars.

Leave pairs some time to work through the remaining calculations. As a class, discuss how pairs worked out the answer. For example, they might have calculated that 14 cars were from Japan (3 × 10% of 40 and 1 × 5% of 40 = 12 + 2) so 26 cars are from Europe (40 – 14 = 26). Alternatively, they might have worked out what 65% of 40 cars is to reach the same answer. Amend the bar model to show all the new information.

Students should now complete the questions on page 101 of the Student Book. Encourage students to draw a bar model to help them, if necessary, and work out how much each 10% represents.

As students work, model phrases and ask questions to help with the problems, for example:

- *100% is …*
- *How can you work out 5% of an amount? Do you know any other ways to work this out?*
- *100% is 30. What other percentages of 30 do you know? How did you work that out?*

Before looking at page 102 of the Student Book, show students a blank 100-square. Say, *If this is one whole, what is one square?* Give students some time to consider their answer in pairs. Share answers and explanation as a class and agree: 0.01 because, for example, each square is 1 of 100 small squares and we write $\frac{1}{100}$ as 0.01.

Ask them to stay in their pairs to complete the first table. If appropriate, they could time each other to see how quickly they can complete the table. They then join with another pair to complete the second table, discussing their answers as they work. Students may find it helpful to continue to shade 100-squares to find equivalences.

128 Unit 4 Fractions, decimals and percentages

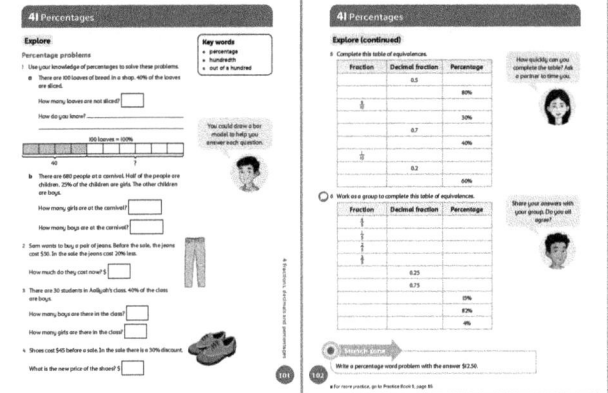

Differentiation

Supporting: Help students to calculate simple percentages such as 50% and 10%. Ask them to explain their strategies.

Consolidating: Ask students to explain how they used one percentage to calculate a new percentage.

Extending: Challenge students to write their own percentage problems for a partner to solve.

Stretch zone: *Write a percentage word problem with the answer $12.50.*

Encourage students to make up several problems where $12.50 could be a percentage of different amounts.

Reflection time

Invite students to share their strategies for solving the word problems. They can explain how they used a bar model to help them, or which known percentages they used to find new ones.

Discuss as a class the answers to the fraction/decimal/percentage tables and ask students to explain how they found the missing numbers.

Practice Book: Students complete Practice Book page 85. They can do this directly after the Main activity, as homework, or as the focus of a separate mathematics session to help students consolidate their learning and build fluency.

Some students may find it challenging to record all equivalences on one number line. Alternatively, they could sketch multiple number lines on a separate sheet of paper, aligning them so the intervals match, and record percentages, decimals, fractions and tenths on separate number lines.

Differentiated outcomes	
All students	should recognise and calculate simple percentages of amounts with support.
Most students	will recognise and calculate a range of percentages that are multiples of 5%.
Some students	may write their own percentage word problems.

Answers

Student Book pages 101–102

1. **a** 60; 60% of 100 is 60

   b 85 girls, 255 boys

2. $40

3. 12 boys, 18 girls

4. $31.50

5.

Fraction	Decimal fraction	Percentage
$\frac{1}{2}$	0.5	50%
$\frac{4}{5}$	0.8	80%
$\frac{9}{10}$	0.9	90%
$\frac{3}{10}$	0.3	30%
$\frac{7}{10}$	0.7	70%
$\frac{2}{5}$	0.4	40%
$\frac{1}{10}$	0.1	10%
$\frac{1}{5}$	0.2	20%
$\frac{3}{5}$	0.6	60%

6.

Fraction	Decimal fraction	Percentage
$\frac{4}{5}$	0.8	80%
$\frac{1}{5}$	0.2	20%
$\frac{2}{5}$	0.4	40%
$\frac{3}{5}$	0.6	60%
$\frac{1}{4}$	0.25	25%
$\frac{3}{4}$	0.75	75%
$\frac{3}{20}$	0.15	15%
$\frac{41}{50}$	0.82	82%
$\frac{1}{25}$	0.04	4%

Practice Book page 85

Check that the fractions and their decimal and percentage equivalents have been positioned correctly on the number lines.

1. **a** $\frac{4}{10}$, 0.4, 40%; $\frac{7}{10}$, 0.7, 70%; $\frac{2}{10}$, 0.2, 20%; $\frac{9}{10}$, 0.9, 90%

   b $\frac{2}{10}$, 0.2, 20%; $\frac{1}{10}$, 0.1, 10%; $\frac{8}{10}$, 0.8, 80%; $\frac{3}{10}$, 0.3, 30%

2. **a** For example, $\frac{4}{5}$, 0.8, 80%

   b For example, $\frac{3}{10}$, 0.3, 30%

Stretch zone: For example, 60%, 0.65, $\frac{65}{100}$ (= $\frac{13}{20}$), and 56%, 0.56, $\frac{14}{25}$

Unit 4 Fractions, decimals and percentages

4J Proportion

Discover Student Book page 103 • Practice Book page 86

Specific learning focus
- Use fractions to describe and estimate a simple proportion of an amount.

Global skills
- **Creative skills:** problem solving
- **Real-world skills:** presenting information

Key vocabulary
- numerator, denominator, proportion

Resources
- coloured counters or cubes for each pair (at least 12 counters in four different colours)
- paper

Language support
Provide students with sentence frames to use when describing proportions of amounts, for example:

The proportion of ___ that are ___ is $\frac{\square}{\square}$ or ___%.

___ % of ___ is/are ___.

$\frac{\square}{\square}$ of ___ is/are ___.

___ out of ___ is/are ___.

Introductory activity

Give pairs of students coloured counters. They should have at least 12 counters in at least four different colours. Ask them to create an array using two colours so that one-quarter of the counters are red.

Ask, *What fraction of your array is red counters?* Agree that $\frac{1}{4}$ of the array is red counters.

Explain that we can also say: *The **proportion** of counters that are red is $\frac{1}{4}$.* Explain that proportion tells us how much there is of something in the whole. We can describe proportions using fractions or percentages. For example, if $\frac{1}{4}$ of a set of counters are red, then 1 out of every 4 counters is red. This matches how the fraction is written, with the numerator showing how many items, out of the total number, which is the denominator.

Ask students to create a different array where $\frac{1}{4}$ of the counters are red. Ask, *What proportion of your array is red?* ($\frac{1}{4}$) Repeat this once more, so that they have created three different arrays with $\frac{1}{4}$ of the counters red. Ask students to record the proportion of red in their arrays. For example, they may have 1 out 4 red counters, or 2 out of 8 red counters.

130 **Unit 4** Fractions, decimals and percentages

Main activity

Ask students, in pairs, to use their counters to create arrays where the proportion of counters are these fractions: $\frac{1}{2}, \frac{3}{5}, \frac{3}{4}, \frac{3}{10}, \frac{4}{5}$.

Students should draw these arrays on a sheet of paper and write the proportions as fractions and percentages.

Ask pairs to complete the activities on page 103 of the Student Book. They should draw images to help them solve the problems or continue to use counters.

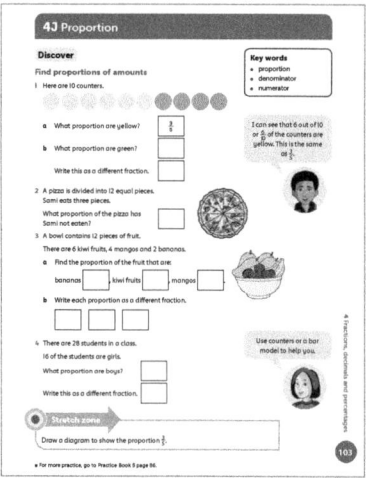

Differentiation

Supporting: Help students to use counters to model proportions.

Consolidating: Encourage students to use drawings to model proportions.

Extending: Challenge students to use mental methods to solve problems and then check by drawing a diagram or using counters.

Stretch zone: *Draw a diagram to show the proportion $\frac{3}{5}$.*

Check that students' diagrams show $\frac{3}{5}$. Diagrams may include arrays or bar models but do not need to be limited to these representations.

Students can pose their own proportion problems to share with their family at home.

Reflection time

Share solutions to each problem and ask one pair to model their solution. Share the images that students have used to solve problems. Model the problem from question 4 practically, using students from the class. Repeat this for other proportions including:

- 25 students, 40% are girls
- 20 students, 75% are boys.

Practice Book: Students complete Practice Book page 86. They can do this directly after the Main activity, as homework, or as the focus of a separate mathematics session to help students consolidate their learning and build fluency.

Encourage students to use a dictionary to look up any vocabulary they are unfamiliar with.

Differentiated outcomes	
All students	should model simple proportions with counters with support.
Most students	will solve proportion word problems using drawings to help them.
Some students	may use mental methods to solve the problems, relating proportion to division.

Answers

Student Book page 103

1. **a** (Provided) $\frac{3}{5}$ **b** $\frac{2}{5}$ **c** $\frac{4}{10}$
2. $\frac{3}{4}$
3. **a** bananas $\frac{1}{6}$, kiwi fruits $\frac{1}{2}$, mangos $\frac{1}{3}$
   b $\frac{2}{12}, \frac{5}{10}, \frac{3}{9}$
4. boys $\frac{12}{28} = \frac{3}{7}$

Practice Book page 86

1. $\frac{2}{5}$ are women, $\frac{3}{5}$ are men; 40% are women, 60% are men.
2. $\frac{9}{10}$ are goats, $\frac{1}{10}$ are sheep; 90% are goats, 10% are sheep.
3. $\frac{1}{5}$ are orange, $\frac{4}{5}$ are green; 20% are orange, 80% are green.
4. $\frac{4}{5}$ are lemon, $\frac{1}{5}$ are chocolate; 80% are lemon, 20% are chocolate.

Stretch zone: Check that students have drawn a bar model for 75%.

Unit 4 Fractions, decimals and percentages

4J Proportion

Explore Student Book page 104 • Practice Book page 87

Specific learning focus
- Solve word problems using proportions and fractions.

Global skills
- **Creative skills:** problem solving
- **Real-world skills:** interpreting information
- **Interpersonal skills:** communication/teamwork

Key vocabulary
- numerator, denominator, proportion, fraction, equivalent fraction

Resources
- mini whiteboards and markers
- coloured counters or cubes in three different colours

Language support
Students may be unfamiliar with some of vocabulary in the word problems. Provide clear descriptions of any new vocabulary or support with a visual image. You may suggest that they look up the meaning of certain vocabulary in a dictionary as well.

 **Introductory activity**

Ask students to work in pairs to create a new problem similar to those on page 103 of the Student Book. Pairs exchange problems and solve them. Choose three problems to work on as a whole class. Model solving the problems with counters or diagrams.

 Main activity

Give pairs of students a pile of three different-coloured counters. Ask them to work out the proportions of each colour in their pile, and to write the proportions as fractions on their whiteboards. Encourage them to check their answers by adding the numerators to ensure that they add up to make the denominator.

Set the following problem.

I have a packet of sweets. $\frac{1}{2}$ of them are red, $\frac{1}{6}$ are green and $\frac{1}{3}$ are yellow. There are between 15 and 20 sweets altogether. How many of each colour do I have?

Ask students to work with a partner to solve this problem. Encourage them to use diagrams to help them. Take feedback. Establish that for these proportions there must be a number that can be divided equally between 2, 3 and 6. $18 \div 2 = 9$, $18 \div 3 = 6$ and $18 \div 6 = 3$, so there must be 18 sweets: 9 red, 3 green and 6 yellow. 18 is a multiple of 2, 3 and 6.

Explain the proportion problems on page 104 of the Student Book. Emphasise that, for each question, students need to find the number that is a multiple of the denominators of the fractions. Emphasise that you expect students to use counters or drawings to help them solve the problems.

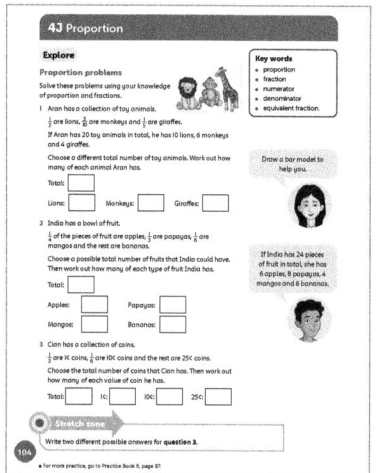

Differentiation

Supporting: Help students to use counters to model proportions.

Consolidating: Encourage students to use diagrams to model proportions.

Extending: Challenge students to use mental methods to solve problems and to model the problems as part of checking their answers.

Stretch zone: *Write two different possible answers for question 3.*

With 12 coins, 4 are 1¢ coins, 2 are 10¢ coins and 6 are 25¢ coins. With 24 coins, 8 are 1¢, 4 are 10¢ and 12 are 25¢.

 Reflection time

Invite pairs to share how they worked out the problems and to give their possible solutions. Establish that there isn't enough information to find exact answers.

Practice Book: Students complete Practice Book page 87. They can do this directly after the Main activity, as homework, or as the focus of a separate mathematics session to help students consolidate their learning and build fluency.

Encourage students to continue to draw diagrams to help them solve the problems.

Unit 4 Fractions, decimals and percentages

Differentiated outcomes	
All students	should model simple proportions with counters with support.
Most students	will find possible solutions to problems using drawings to help them.
Some students	may use mental methods to solve the problems, relating proportion to division.

Answers

Student Book page 104

Examples of possible answers:

1 There are 30 animals in total. There are 15 lions, 9 monkeys and 6 giraffes.

2 There are 12 pieces of fruit in total. There are 3 apples, 4 papayas, 2 mangos and 3 bananas.

3 Cian has 18 coins in total. There are 6 cents, 3 10¢ coins and 9 25¢ coins.

Practice Book page 87

1 4 aeroplanes

2 6 grapes

3 10 green

Stretch zone: Check that students have written an appropriate problem about 16 cherries.

4K Ratio

Discover Student Book page 105 • Practice Book page 88

Specific learning focus
- Use ratio to solve problems.

Global skills
- **Creative skills:** problem solving
- **Real-world skills:** interpreting information

Key vocabulary
- ratio, proportion

Resources
- coloured counters
- mini whiteboards and markers
- large sheets of paper

Language support

Provide students with sentence frames to use when describing simple ratios:

There are _____ _____ for every _____ _____.

The ratio of _____ to _____ is _____ : _____.

 Introductory activity

Ask pairs to recreate the arrays from 4J Discover Introductory activity, where one-quarter of the counters are red. Build your own array where one-quarter are red and three-quarters are yellow.

Tell students that another way of comparing amounts is called **ratio**. Explain that proportion compares one part to the whole. Ratio compares part to part. For example, in the array where $\frac{1}{4}$ of the counters are red, we can say that the ratio of red to yellow is 1 : 3 ('one to three'). This means that for every red counter, there are 3 yellow counters.

Using the same counters, ask pairs of students to create arrays to show ratios of:

1 : 4   1 : 5   2 : 3

 Main activity

Ask pairs to discuss the link between the ratio and the proportion. They should notice that the denominator for the fraction is the sum of all the 'parts' in the ratio. For example, an array in the ratio 1 : 4 contains $\frac{1}{5}$ of one colour and $\frac{4}{5}$ of the other colour.

Look together at the worked example on page 105 of the Student Book. Ask students to model the example using counters of different colours to represent the different ingredients. Discuss the text in the speech bubble and ask students to compare three ingredients and record it as a ratio.

They should then answer the questions about the recipe together, working on one large sheet of paper. (For the Stretch zone task, students should use their whiteboards.)

Unit 4 Fractions, decimals and percentages

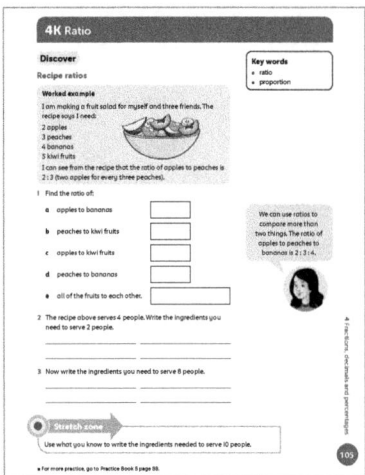

Look at the worked example and discuss the answer. Ask, *Would it matter if they had shaded different squares? Why not? Can you show me?* Students can draw their own examples of bars to show the ratio 3 grey squares for every 1 black square. Agree that as long as 3 squares are shaded for every 1 black square shaded, the drawing would still show the ratio 3 : 1.

Differentiated outcomes	
All students	should calculate simple ratios using counters with support.
Most students	will calculate simple ratios using counters.
Some students	may use mental methods to solve the problems.

Answers

Student Book page 105

1 **a** 2 : 4 **b** 3 : 5 **c** 2 : 5 **d** 3 : 4 **e** 2 : 3 : 4 : 5

2 2 people: 1 apple, $1\frac{1}{2}$ peaches, 2 bananas, $2\frac{1}{2}$ kiwi fruits

3 8 people: 4 apples, 6 peaches, 8 bananas, 10 kiwi fruits

Practice Book page 88

1 2 grey, 3 black

2 6 grey, 2 black

3 8 grey, 2 black

4 4 grey, 16 black

5 12 grey, 6 black

Stretch zone: Check that students have drawn a grid to show 5 : 2 (it should have a multiple of 7 squares).

Differentiation

Supporting: Help students to use counters to model ratios.

Consolidating: Encourage students to use drawings to model ratios.

Extending: Challenge students to use mental methods to solve ratio problems.

Stretch zone: *Use what you know to write the ingredients needed to serve 10 people.*

For 10 people, the recipe is 5 apples, $7\frac{1}{2}$ peaches, 10 bananas, $12\frac{1}{2}$ kiwi fruits.

 Reflection time

Invite groups to share their new recipes and to explain how they calculated their answers. For question 1e, point out that the ratio is not simply comparing two of the fruits with each other, but all four fruits. As there are 2 apples, 3 peaches, 4 bananas and 5 kiwi fruits, the ratio for all the fruits is 2 : 3 : 4 : 5

Ask, *How can you use what you know to write a recipe for 6, 15 and 16 people?*

Practice Book: Students complete Practice Book page 88. They can do this directly after the Main activity, as homework, or as the focus of a separate mathematics session to help students consolidate their learning and build fluency.

Unit 4 Fractions, decimals and percentages

4K Ratio

Explore Student Book page 106 • Practice Book page 89

Specific learning focus
- Use ratio to solve word problems.

Global skills
- **Creative skills:** problem solving
- **Real-world skills:** interpreting information

Key vocabulary
- ratio, proportion

Resources
- coloured counters
- mini whiteboards and markers
- large sheets of paper

Language support
Remind students of how to record ratios as, for example, 2 : 5. Explain that you say this ratio as *two to five*.

Introductory activity

Review what students learned about ratio in 4K Discover.

Ask questions to check that students understand the concept of proportion and ratio, for example:

- *How are ratio and proportion the same? How are they different?*
- *What does a ratio of 3 : 8 mean?*

Remind them that: proportion compares part of a whole to the whole; ratio compares the relative sizes of two or more parts of the whole. They are different, but they both compare amounts.

Demonstrate how you can show both proportion and ratio, using two blue counters and three red counters. Say: *There are two blue counters for every three red* so the ratio is 2 : 3. Tell students that the proportion compares the number of counters of a particular colour to the whole number of counters. The proportion of blue counters is $\frac{2}{5}$ or 40%. The proportion of red counters is $\frac{3}{5}$ or 60%. When you combine these proportions, they make $\frac{5}{5}$ or the whole 100%.

Give pairs three blue counters and five red counters and ask them to find the ratio of blue to red counters and the proportion of blue and red counters. Agree that the ratio is 3 : 5 and the proportions are $\frac{3}{8}$ and $\frac{5}{8}$.

Main activity

Give students piles of two different-coloured counters. Tell them to count their counters and write the ratios and proportions of the different colours on their whiteboards.

Set the following problem.

I have 10 packets of crisps. $\frac{3}{5}$ of the packets are cheese and onion flavour. The other packets are plain flavour. What is the ratio of the two flavours?

Take feedback. Establish that $\frac{3}{5}$ of 10 is 6, so there are 6 packets of cheese and onion flavoured crisps. The proportion of plain crisps is $\frac{2}{5}$. $\frac{2}{5}$ of 10 is 4, so there are 4 packets of plain crisps. The ratio is 4 : 6. Ask students to tell you another way of writing this ratio (2 : 3). Repeat this with different numbers and proportions, using fractions that students are familiar with.

Ask students to solve the problems on page 106 of the Student Book. Students can work in pairs and use counters or bar models to help them solve each problem.

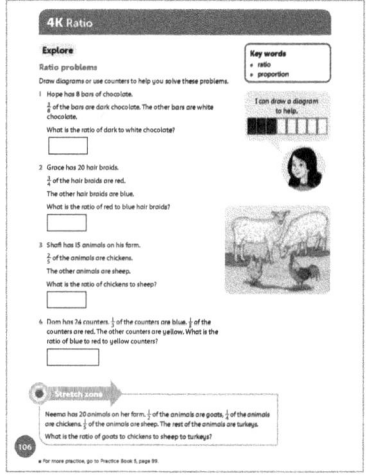

Differentiation

Supporting: Help students to use counters to model ratios.

Consolidating: Encourage students to use diagrams to model ratios.

Extending: Encourage students to use mental methods to solve ratio problems and explain their method to you.

Stretch zone: *Neema has 20 animals on her farm. $\frac{1}{2}$ of the animals are goats, $\frac{1}{4}$ of the animals are chickens. $\frac{1}{5}$ of the animals are sheep. The rest of the animals are turkeys. What is the ratio of goats to chickens to sheep to turkeys?*

10 : 5 : 4 : 1

Students can make up similar problems for other students to solve.

 ### Reflection time

Invite groups to share their solutions to the problems and to explain how they calculated their answers. Ask pairs to create their own problems and swap them with another pair. Select three problems to solve as a whole class.

Unit 4 Fractions, decimals and percentages

Practice Book: Students complete Practice Book page 89. They can do this directly after the Main activity, as homework, or as the focus of a separate mathematics session to help students consolidate their learning and build fluency.

Review millilitre and litre equivalences. Ask questions such as, *How many millilitres are there in 2 litres? What fraction of a litre is 500 ml? I pour 150 ml of juice each into 5 glasses. How much juice did I pour altogether?*

Differentiated outcomes	
All students	should solve simple ratio problems using counters with support.
Most students	will solve ratio problems by modelling with counters or diagrams.
Some students	may use mental methods to solve ratio problems.

Answers

Student Book page 106

1. 3 : 5
2. 3 : 1
3. 2 : 3
4. 8 : 12 : 4

Practice Book page 89

8 litres of water

Stretch zone: 6 bottles of squash (needs 5.5 but you can't buy half a bottle) 11 litres of water

4 Fractions, decimals and percentages

Connect Student Book page 107

Big idea

I can write equivalent fractions, decimals and percentages. I can use my knowledge of equivalents to solve problems, and to calculate with fractions.

Global skills

- **Creative skills:** investigating
- **Interpersonal skills:** teamwork
- **Self-development skills**: reflecting on learning

Key vocabulary

- equal shares, fair, fraction, proportion

Resources

- counters or cubes
- large sheets of paper

Language support

Students may need additional support to understand the rules of the fair shares game. Go through these slowly and clearly, modelling actions where possible.

 **Introductory activity**

Write 'Fractions' in the middle of the board. Draw five arrows from the word. Ask students to tell you the types of fractions they have been learning about (fractions, decimal fractions, percentage). Put these at the end of the arrows and add 'proportion' and 'ratio'. Next, ask them to form small groups to discuss: *What do you remember about each of these?* Take feedback. Make notes of what they say in the appropriate places on the diagram. Ask students to give examples, for example: 'Improper fractions are when the numerator is larger than the denominator. $\frac{11}{3}$ and $\frac{24}{5}$ are improper fractions.'

 Main activity

Ask questions to check that students understand the concept of fair shares, for example:

I have a packet of 10 biscuits. My friend and I decide to share it equally so we both have a fair share. I give her 4 biscuits. Is that a fair share? If not, what would be a fair share?

Go through the instructions on page 107 of the Student Book together.

Students should work together in mixed-attainment groups to work out how to share the chocolate bars as fairly as possible. They can work together using counters or cubes or real chocolate bars to model the problem and use large sheets of paper to record their solutions, showing the share each student gets, which they can then copy into their Student Books. Emphasise that the students must choose a chair one at a time and then not change places.

Unit 4 Fractions, decimals and percentages

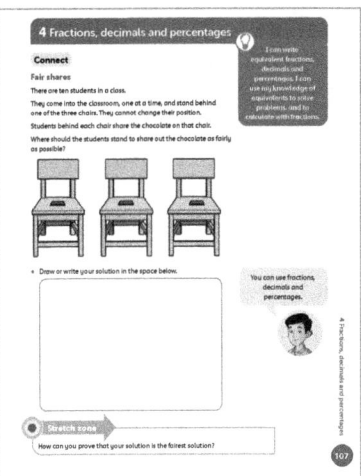

Differentiation

Supporting: Pair less-confident students with more-confident students within the group.

Consolidating: Ask students to explain how they worked out how to divide the chocolate fairly.

Extending: Ask students to write their solution in full sentences.

Stretch zone: *How can you prove that your solution is the fairest solution?*

Students may be able to describe how they worked out the share of chocolate as each student arrived at the chairs. They may show that their solution is the fairest possible by recording the shares at each stage in a systematic way.

Reflection time

Ask groups to share the results of their investigations. Groups should have found that 4 students stand behind 1 chair and they get $\frac{1}{4}$ of a chocolate bar each. 3 students each stand behind the other 2 chairs and they will get $\frac{1}{3}$ of a chocolate bar each. Ask them to explain how they calculated the fractions in each case, and to share any drawings they used to help them.

Differentiated outcomes	
All students	should work out the shares of chocolate for each group of students with support.
Most students	will work out each student's share of chocolate.
Some students	may record systematically to prove the fairest solution.

Answers

Student Book page 107

4 students at one chair, 3 at each of the others. The group of 4 will each get $\frac{1}{4}$ of a bar each, the groups of 3 will each get $\frac{1}{3}$ of a bar each.

---

4 Fractions, decimals and percentages

Review Student Book page 108 • Practice Book page 90

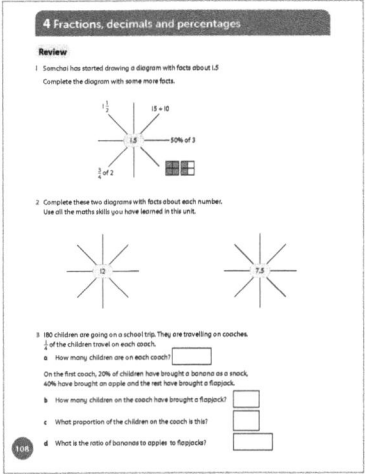

Global skills:
- **Creative skills:** problem solving
- **Real-world skills** interpreting information
- **Interpersonal skills:** communication/teamwork
- **Self-development skills:** reflecting on learning

Student Book

With young students, assessment activities are most effective when carried out as an everyday classroom activity. Students should be able to explore fractions in different contexts and extend their previous knowledge of calculating to multiply fractions. They should develop strategies for finding fractions of amounts and be able to use their knowledge of equivalent fractions to order and compare fractions, sometimes with the help of fraction walls or number lines. They should also be able to extend decimal fractions to three decimal places, and calculate percentages, and ratios and proportions.

As they work on the Student Book activity on page 108, they should be able to draw upon their work across the unit to extend the spider diagrams to include more facts, diagrams, drawings and number lines if appropriate.

If students are struggling to come up with ideas, remind them that fractions, decimals and percentages often come in different representations. Can they recall different drawings or equipment they used? Point out that they could link the numbers to other units in mathematics, such as adding and subtracting, multiplying and dividing, along with fractions and percentages.

Unit 4 Fractions, decimals and percentages

Watch for any students who may be struggling to solve the word problems due to poor understanding of the vocabulary. Provide definitions or show them images to explain as necessary.

Answers

Student Book page 108

1. Check that students have added some facts about 1.5 to the diagram.

2. Check that students have added correct facts about each number.

3. **a** 45 **b** 18 **c** $\frac{2}{5}$ **d** 1 : 2 : 2

Practice Book

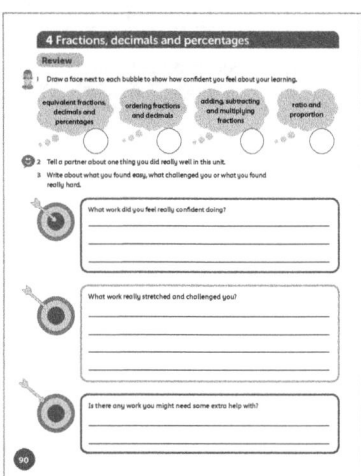

With students in the upper primary years, it is appropriate to complete this Practice Book review as a whole-class discussion. You may choose to keep a record of the class discussion or a copy of the Review page for your own records. The Review provides an opportunity for students to reflect on their learning from the unit, to discuss any areas of mathematics that they feel went particularly well, and any areas that they feel less confident about.

Ensure that students have a copy of the Student Book to support them as they discuss and answer the questions in the Practice Book.

Allow students plenty of time for discussion before asking them to complete the Practice Book page individually, and then, if appropriate, to share their responses with the rest of the class. If students complete this self-assessment at home, encourage them to discuss this with adults.

Make a note of areas that students still feel unsure about. As fractions, decimals and percentages are found in many everyday contexts, revisit these ideas regularly. Build fraction work into other aspects of everyday practice, for example when sharing things or finding a part of an amount or quantity.

Additional material

There are additional end-of-unit assessment available on the *Oxford Owl for School* website.

5 Length, mass, capacity and volume

Overview

Big idea

The Big idea for this unit is that measure is always only accurate to a chosen degree of accuracy. We can never measure a line exactly. The measurement is always to the nearest mm, cm, m and so on. Students need to decide an appropriate degree of accuracy to use. The same principles apply to measuring mass and capacity, whereby every measure is to an appropriate degree of accuracy.

By this stage, students are developing a good understanding of length, mass and capacity. They should know that an object in any orientation has the same length. They should also understand that, because of its density, a small object can sometimes be heavier than a larger object.

Students need to use their understanding of place value and their knowledge of decimals to help them convert between units of measure. In this unit, students develop their knowledge of scales. This will help them to use measures to an appropriate degree of accuracy.

Students should be aware of the use of imperials units and how to convert them, and metric measures, and their awareness of measuring space extends to 3D in this unit with experience of comparing, estimating and measuring volumes.

Look out for

- **Students who do not start measuring from zero on a ruler.** Model for them how to do this correctly using the zero mark and not the end of the ruler.
- **Students who frequently forget equivalences between units.** For example, they might forget how many millilitres make one litre. Measure is a topic that is not frequently taught. Rehearse facts relating to units of measure regularly and refer to them when covering fractions, decimals and mental and written calculation.

Possible misconceptions

- **Students may confuse the words 'mass' and 'weight'.** You can use the definitions in the following table and use the example of a person in space to explain the difference:

Measure	Definition	Equivalent measure in space
mass	a measurement of how much matter is in an object.	the same as it is on Earth because the amount of matter making it up is still the same
weight	a measurement of how hard gravity is pulling on that object	0, because there is no pull of gravity

- **Students may confuse the words 'capacity' and 'volume'.** These may or may not be the same. You can use the following definitions to explain the difference.

Capacity	Definition
capacity	the amount a container can hold
volume	the amount of a substance inside the container

It is important to use the correct vocabulary with students. Ask them to use sentences to describe mass and volume. You can ask them to point to the capacity of a bottle and the volume of liquid inside it to support the definitions.

- **Students may find judging the capacity of containers challenging and often believe that tall, thin containers hold more than shorter, wider ones.** Provide plenty of opportunities to compare containers. Students need lots of practice to improve their skills and this will help when they start estimating and measuring volume.

Key vocabulary

- length, height, width, kilometre (km), metre (m), centimetre (cm), millimetre (mm), kilo-, milli-, centi-, ruler, metre stick, tape measure, estimate
- mass, weight, heavy/light, weigh, kilogram (kg), gram (g), scales
- capacity, volume, litre (l), millilitre (ml), container, measuring jug/cylinder
- cubic centimetre (cm^3), cubed, dimensions
- units of measure, imperial unit, metric unit, convert between units, scale
- base-10 system, conversion, approximate

Coverage in lessons

Learning objective	E	5A	5B	5C	5D	5E	5F	5G	5H	C	R
Convert between different units of metric measure (for example kilometre and metre; centimetre and metre; centimetre and millimetre; gram and kilogram; litre and millilitre).	✓	✓	✓	✓	✓	✓		✓	✓	✓	✓
Understand and use equivalences between metric units and common imperial units such as inches, pounds and pints.							✓		✓	✓	✓
Estimate volume, for example using 1 cm³ blocks to build cuboids (including cubes) and capacity, for example using water.								✓			
Use all four operations to solve problems involving measure (for example length, mass, volume, money) using decimal notation, including scaling.		✓	✓		✓	✓	✓	✓	✓	✓	✓

5 Length, mass, capacity and volume

Engage Student Book page 109

Big question
- How do I convert between different units of measure?

Global skills
- **Creative skills:** exploring
- **Self-development skills:** reflecting on learning

Key vocabulary
- length, height, width, kilometre (km), metre (m), centimetre (cm), millimetre (mm)
- mass, weight, heavy/light, weigh, kilogram (kg), gram (g), capacity, volume, litre (l), millilitre (ml), imperial measures, metric units

Resources
- mini whiteboards and markers
- coloured pencils
- large sheets of paper
- internet access or magazines and newspapers containing images related to measure

Language support
Working in mixed-attainment groups will support the language development of less-confident speakers of English. Ensure that students record all new and key vocabulary systematically in their glossaries as it is introduced across the unit.

Introductory activity

Ask small mixed-attainment groups of students to write on their whiteboards as many different metric units as they can remember that we use to measure length. Some students may also know the names of imperial measures, such as miles, feet and inches. Ask students to write next to the unit what they might use to measure length – for example a ruler, a tape measure, a trundle wheel. Finally, ask students to write down any equivalences they know, for example:

10 mm = 1 cm

100 cm = 1 m

1000 m = 1 km

Record them on the board, reading each amount aloud to model the correct pronunciation of each.

Repeat this activity for mass, capacity and volume.

Main activity

Ask students to work in pairs to discuss and answer at the questions in the speech bubbles and match the units and pictures on page 109 of the Student Book.

Pairs should then join up to make small groups of four, if possible, to create measurement posters. Give them magazines and newspapers or internet access so that they can add images to the posters of, for example, objects that you measure in particular units.

As students work, circulate among the groups, asking questions to assess their understanding of measure. Ask, for example:

- *I want to measure the width of my garden. What measurement would be most practical to use? What unit of measure might I choose to use?*
- *How far would you estimate your home is from the school?*
- *What unit of measure would you use to measure the amount of water in a fish tank?*
- *Name a unit of measure used for measuring masses? Can you name a larger/smaller one?*

Differentiation

Supporting: Give students a chart of common equivalences. To reinforce and embed these equivalences, ask questions such as, *How many millilitres are there in one metre?*

Consolidating: Ask an individual student to tell you everything they know about a common metric unit of measure.

Extending: Choose several of the images included in their measurements poster. Can they estimate, for example, what the height of the objects shown would be in real life?

Reflection time

Display page 109 of the Student Book, on an IWB if possible. Discuss the questions in the speech bubbles with students.

Ask groups to share their posters to prompt discussion about learning. These posters should then be displayed for reference throughout the unit.

Answers

Student Book page 109

millilitre, ml, measuring jug

gram, g, kitchen scales, ant,

metre, m, ruler, tree

kilogram, kg, bathroom scales, elephant

centimetre, cm, tape measure, shoes, ruler

kilometre, km, road sign, road

litre, l, measuring jug, large water bottle

millimetre, mm, ruler, ant

5A Units of measure

Discover
Student Book pages 110–111 • Practice Book page 91

 **Introductory activity**

Give each pair of students a tape measure. Ask them to measure four different items in the classroom. These could include books, pencils, desks, or even the length of their hand. For each item they measure, they should estimate first.

Students read the measurement from the tape measure to a suitable degree of accuracy, and record it on their whiteboards, then write it using a different **unit of measure**. For example, a book is 18 cm wide, this can also be 180 mm; an object that is 35 mm tall could also be recorded as 3.5 cm. Remind and review as necessary. For example, emphasise the movement of the digits as the numbers getting 10, 100, 1000 times bigger or smaller as the units change by recording equivalences in lists, both from smallest unit to greatest and vice versa.

35 mm	2.1 m
3.5 cm	210 cm
0.035 m	2100 mm

Note that this will be discussed in more detail in the next lesson.

Students should record each of the measurements on page 110 of the Student Book.

Specific learning focus
- Measure the length and mass of objects in metric units and convert from m to cm and from kg to g and vice versa.

Global skills
- **Creative skills:** investigating
- **Real-world skills:** research/presenting information
- **Interpersonal skills:** communication/teamwork

Key vocabulary
- units of measure, kilo-, centi-, milli-

Resources
- metre sticks, 30-cm rulers, tape measures
- mini whiteboards and markers
- weighing scales (large and small), measuring jugs (some with capacities greater than 1 litre)
- access to the internet or other reference resources

Language support
Ask students to refer to their measurements posters for support. Continue to model the key vocabulary for them and encourage them to read aloud the measurements they are making.

 **Main activity**

Look together at the activity on page 111 of the Student Book. Explain that students are going to continue to measure in pairs, but the focus is on mass and capacity. Ask them to tell you how they will measure each object. Explain any practical details of how to do this measuring. For example, they should use the scales on a flat surface with nothing else touching them and pour jugs of water carefully and fill them almost to the top but not completely.

Unit 5 Length, mass, capacity and volume

You may want to run this part of the activity in tandem with the Introductory activity to make classroom management easier, particularly if you have a limited amount of measuring equipment available. You can split the class into thirds, having one third focus on measuring length, while another third focuses on mass using the scales and the final third measures capacity of containers.

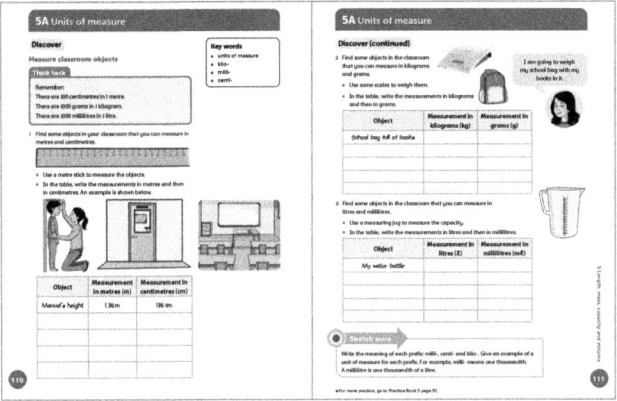

Differentiation

Supporting: Help students to read their measurements and record them accurately.

Consolidating: Ask students to describe one of their measurements and tell you more about it. For example, *Is it taller than it is wide? Is it less or more than half a metre? By how much?*

Extending: Challenge students to describe their measurements in different ways, for example in part units as a decimal or as a mixed number.

Stretch zone: *Write the meaning of each prefix:* **milli**-, **centi**- *and* **kilo**-. *Give an example of a unit of measure for each prefix. For example, milli- means one thousandth. A millilitre is one thousandth of a litre.*

Provide students with resources or internet access so they can research the meaning of each prefix. You might also ask them to research the prefix 'deci-'.

 Reflection time

Ask students to describe or show the objects they measured and their measurements. Ask, *Were some objects harder or easier to measure? Why do you think that was?*

Practice Book: Students complete Practice Book page 91. They can do this directly after the Main activity, as homework, or as the focus of a separate mathematics session to help students consolidate their learning and build fluency.

Students will need access to measuring tools to find both the length and mass of six objects.

Differentiated outcomes	
All students	should measure length, mass and capacity of everyday classroom objects with an increasing degree of accuracy.
Most students	will choose an appropriate unit of measure and measure accurately to the nearest whole unit.
Some students	may record some lengths, masses and capacities in more than one way using, for example, decimals or mixed numbers.

Answers

Student Book pages 110–111
Answers will vary because students choose different things to measure in the classroom. Check that students have written equivalent amounts correctly.

Practice Book page 91
Answers will vary because students choose different things to measure in the classroom or at home. Check that students have written equivalent amounts correctly.

Unit 5 Length, mass, capacity and volume

5A Units of measure

Explore Student Book page 112 • Practice Book page 92

Specific learning focus
- Convert metric units to different formats.

Global skills
- **Creative skills:** problem solving
- **Real-world skills:** interpreting information
- **Interpersonal skills:** communication/teamwork

Key vocabulary
- units of measure, kilo-, centi-, milli-

Resources
- mini whiteboards and pens
- squared paper
- blank place-value grids
- measuring jugs with a capacity of at least 2 litres with millilitre intervals marked and other containers with varying capacities

Language support
Create a display showing conversion of units by multiplying or dividing by 10, 100, 1000 and so on, with worked examples in length, mass and capacity.

Introductory activity

Display 3.4 m on the board and say *three point four metres*. Show this using a place-value grid.

Ask students what this means. (It is three whole metres and point four of a metre.) Explain to students that moving the digits one place to the left will make the measure 10 times larger, so 3.4 m becomes 34 m, for example. Link this clearly to multiplying and dividing. Ask students to convert 3.4 m into cm and mm by moving the digits to the right.

Repeat, using measures for mass and capacity, for example 6.2 kg and 1.8 l.

Main activity

Students should work in pairs on the activity on page 112 of the Student Book. Pair a student with a good understanding of measurement with a less-confident student.

Ask questions to support the use of key vocabulary, for example:

- *How many grams are there in 6 kilograms?*
- *How can we write 5 l 125 ml in a different way?*

- *Is 1.2 m longer or shorter than 125 cm? How do you know?*
- *Is 2.75 kg heavier or lighter than 2 kg 700 g? How do you know?*

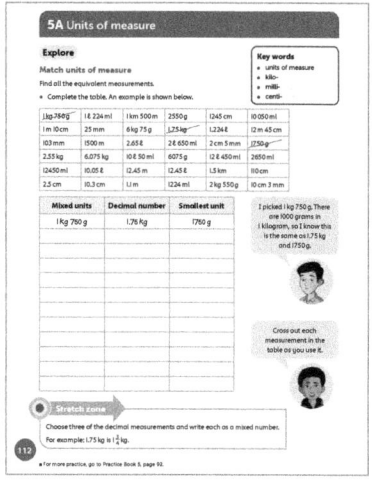

Differentiation

Supporting: Use a place-value grid to help students to record measures, then talk through moving the digits to make them larger or smaller by a factor of 10, 100 or 1000.

Consolidating: Ask students to explain how they converted between units.

Extending: Challenge students to express measures in more than two different ways.

Stretch zone: *Choose three of the decimal measurements and write each as a mixed number.*

For example: 1.75 kg is $1\frac{3}{4}$ kg.

Students may find it useful to use a place-value grid to support them with these conversions. Check that they have recorded each decimal measurement as a mixed number accurately, for example $2\frac{55}{100}$ and $6\frac{75}{1000}$. You may want to encourage students to record them in their simplest form as well, for example $2\frac{11}{20}$ and $6\frac{3}{40}$.

 Reflection time

Take a measurement and say, for example, *What object, or group of objects, in this classroom do you think may have a mass 2.55 kg?* Students, in small groups or pairs, then go around the room trying to find objects with a total mass of 2.5 kg, drawing on their experience of weighing in the last lesson. Each group or pair can look at a different measurement from the table on page 112 of the Student Book. Discuss their findings, as a class.

Practice Book: Students complete Practice Book page 92. They can do this directly after the Main activity, as homework, or as the focus of a separate mathematics session to help students consolidate their learning and build fluency.

This activity focuses on capacity. Students will need a measuring jug with a capacity of at least 2 litres with millilitre intervals marked on it as well as six other containers with varying capacities. Using a measuring jug, or an image of a measuring jug scale, show students how to round up or down to the nearest 100 ml.

Differentiated outcomes	
All students	should multiply and divide to convert to larger or smaller units with support.
Most students	will explain how to multiply and divide to convert to larger or smaller units.
Some students	may have a sense of the size of each of the measurements and be able to record them in different ways, including as a mixed number.

Answers

Student Book page 112

1 l 224 ml	1.224 l	1224 ml
1 km 500 m	1500 m	1.5 km
2550 g	2.55 kg	2 kg 550 g
10050 ml	10 l 50 ml	10.05 litres
1245 cm	12 m 45 cm	12.45 m
1 m 10 cm	110 cm	1.1 m
25 mm	2 cm 5 mm	2.5 cm
6 kg 75 g	6.075 kg	6075 g
103 mm	10.3 cm	10 cm 3 mm
2.65 l	2 l 650 ml	2650 ml
12 l 450 ml	12 450 ml	12.45 l

Practice Book page 92

Answers will vary because students choose their own containers to measure. Check that the conversion between litres and millimetres is correct.

5B Measuring length

Discover Student Book page 113 • Practice Book page 93

Specific learning focus
- Read, choose, use and record standard units to estimate and measure length to a suitable degree of accuracy.

Global skills
- **Creative skills:** exploring
- **Self-development skills:** reflecting on learning

Key vocabulary
- estimate, metre, centimetre (cm), millimetre (mm)

Resources
- mini whiteboards and markers
- A4 paper
- pieces of string about 20 cm long
- modelling clay
- rulers or tape measures

Language support

Model the sentence structure for showing measure equivalences, for example:
- 2 metres is equivalent to 200 centimetres.
- There are 200 cm in 2 m.
- The difference between my estimate and the actual length of my snake was 2.5 cm.

Introductory activity

Split students into pairs and ask one student in each pair to draw a curved line on their whiteboard. They should both estimate the length of the line. Give each pair a piece of string. Ask them to place the string along their line. Then ask them to measure their string to the nearest half **centimetre**. Ask them to record their answer in centimetres (for example 7.6 cm). Ask students to draw another curved line for their partner to measure. Repeat this a few times. Take feedback. Ask, *Did your estimates improve? Why do you think that is?*

Main activity

Ask students to work in mixed-attainment groups of four on the activity on page 113 of the Student Book. Read through the instructions together, making sure that students understand that they must roll the modelling clay into 'snakes' and then estimate their lengths. They order them from shortest to longest and record them in their tables. They then measure the actual lengths and find the difference between their estimates and the actual lengths. Encourage students to make snakes in a variety of lengths and use both **millimetres** and centimetres to estimate and measure their lengths.

As groups work, ask questions as you circulate to extend the activity as well as assess their understanding, for example:

- *How did you come up with your estimate for the length of your snake?*
- *What calculation strategy did you use to find the difference between your estimate and the actual length of your snake?*
- *Which snake is longer? How much longer?*
- *What is the difference in length between your shortest and longest snake? Can you record that in two ways?*

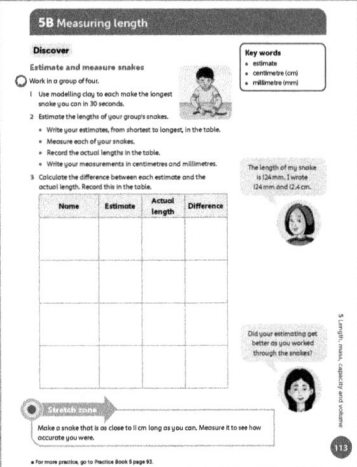

Differentiation

Supporting: Provide students with tape measures and model measuring lengths of snakes to the nearest centimetre.

Consolidating: Ask students to explain their estimates for the lengths of snakes to the nearest centimetre and millimetre.

Extending: Ask students to describe each length in at least three ways. What would be the length of two or more of their snakes, if they combined them? Encourage them to calculate the answer and then measure to check.

Stretch zone: *Make a snake that is as close to 11 cm long as you can. Measure it to see how accurate you were.*

Ask students to describe to you their strategy for making a snake 11 cm long. For example, did they think about how long the other snakes they made were and how they compared to 11 cm? Did they imagine how long 10 cm was on a ruler and try to make a snake a little bit longer?

 Reflection time

Take the length of one snake from each group and write it on the board. Ask students to order these lengths from shortest to longest. Take suggestions on how else to record the measurements as well.

Ask students to work in pairs and to make a snake that is exactly 5 cm long. Measure all the snakes to see which pair's snake is the closest to this length.

Practice Book: Students complete Practice Book page 93. They can do this directly after the Main activity, as homework, or as the focus of a separate mathematics session to help students consolidate their learning and build fluency.

Review how to measure to the nearest 10 centimetres as needed. You may want to extend this activity by asking students to work in groups to see how far they can jump, measure the distance and round it to the nearest 10 cm. They can compare their own jumps, their jumps to those of other students in their group, or to the distances listed in the Practice Book.

Differentiated outcomes	
All students	should estimate and measure lengths to the nearest centimetre with support.
Most students	will estimate and measure lengths to the nearest centimetre and millimetre.
Some students	may convert confidently between millimetres and centimetres and calculate differences confidently.

Answers

Student Book page 113

Answers will vary because students make their snakes of diffrent lengths. While students are working, observe their measuring and note who has good measuring skills and who requires more practice.

Practice Book page 93

	Name	Jump in m	Jump in cm	Jump to nearest 10 cm
1	Tanvi	1.72 m	172 cm	170 cm
2	Camila	1.8 m	180 cm	180 cm
3	Carla	1.95 m	195 cm	200 cm
4	Grace	2.41 m	241 cm	240 cm
5	Jada	2.43 m	243 cm	240 cm
6	Zainab	2.52 m	252 cm	250 cm
7	Harper	2.75 m	275 cm	280 cm

Stretch zone:

Lili	1640 mm
Tanvi	1720 mm
Camila	1800 mm
Carla	1950 mm
Grace	2410 mm
Jada	2430 mm
Zainab	2520 mm
Harper	2750 mm

5B Measuring length

Explore
Student Book page 114 • Practice Book page 94

Specific learning focus
- Interpret a scale to measure distances on a map.

Global skills
- **Creative skills:** investigating
- **Real-world skills:** interpreting information
- **Interpersonal skills:** teamwork/communication

Key vocabulary
- scale, kilometre (km), metre (m)

Resources
- mini whiteboards and markers
- metre sticks, rulers
- tape measures at least 3 m long
- copies of a map of the local area
- string

Language support
Give students a list of key words or an annotated simple street map like the one on page 114 of the Student Book. This will encourage them to use the vocabulary to describe movement around and position on a street map.

Introductory activity
Show the class a **metre** stick. In small groups, ask students to use the metre stick to estimate the length, depth or height of an item in the classroom, such as the width of their table or desk, or the height of the door. Ask them to write their estimates on their whiteboards. Invite students, particularly those you wish to assess, to measure these items to check. Ask students to calculate how close their estimates were. Who was the closest?

Main activity
Ask students to work in groups to estimate and then measure five more things in the classroom using metre sticks or rulers. Tell them to record their estimates and measurements in at least two different ways. Then ask them to order the lengths they found from shortest to longest. Take feedback from each group.

Ask each group to select one of the items they have measured that was between 1 m and 2 m long. Explain that they are going to draw the length or width of the object to **scale**. Say that they are going to use a scale of 1 cm = 10 cm. Explain that this means that for every 10 cm long an object is, they will draw a 1 cm length.

Say, *For example: I am 170 cm or 1.7 m tall. So, to draw me to this scale you would need to draw a person that is 17 cm tall.*

Work through a number of examples, starting with, say, 20 cm = 2 cm, then 30 cm = 3 cm, 40 cm, 50 cm, 55 cm, 56 cm and then see whether students notice a pattern. Ask, *Can you see how we can calculate this? What operation can you use?* Say that to scale up they multiply by 10 and to scale down they divide by 10.

After groups have completed their scale drawings, they can share these with the class.

Next, introduce the activity on page 114 of the Student Book. Discuss the features of the map, introducing any new vocabulary, for example 'intersection', 'block', 'ferry terminal'. Ask students questions to encourage them to think about position, turns and following direction. For example: *I am travelling north on Nathan Road and I am going to turn onto Hillwood Road. Where will I turn?* Draw students' attention to the scale. Ask, *How is the same and different from the scale we used?* (Both give a scale using 1 cm but here 1 cm = 1 km (**kilometres**) instead of 10 cm.)

Students work in pairs so they can estimate, measure and compare their lengths. Suggest they use string to measure the distances on the map in the first instance and then lay it straight on a ruler to find the distance.

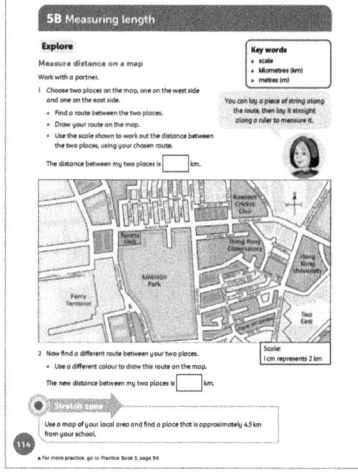

Differentiation
Supporting: Help students to make a 'ready reckoner' to support them to work out the actual distance between two places as well as see the emerging pattern. For example, they might use: 1 cm = 1 km, 1.5 cm = 1.5 km, 2 cm = 2 km or 2.5 cm = 2 km.

Consolidating: Ask students to share their thinking with you as they make the conversions.

Extending: Challenge students to use other scales to represent objects around the class.

Stretch zone: *Use a map of your local area and find a place that is approximately 4.5 km from your school.*

Students look at a map of the local area. If necessary, suggest they use the scale to work out how far 4.5 km would be on the map, find the school and then find a location 4.5 km away.

Reflection time

Invite students to share their routes from the activity. Write down the lengths of six of these routes on the board. Finish the lesson by asking students to order different lengths in ascending order.

Practice Book: Students complete Practice Book page 94. They can do this directly after the Main activity, as homework, or as the focus of a separate mathematics session to help students consolidate their learning and build fluency.

Students will need a tape measure that is at least 3 m long to complete the activity.

Differentiated outcomes	
All students	should measure accurately in centimetres and millimetres with support.
Most students	will use the scale accurately.
Some students	may use a range of their own scales to make scale drawings.

Answers

Student Book page 114

Students choose which places they will find routes for, so answers will vary. Check that the routes and distances are correct for the chosen places.

Practice Book page 94

Answers will vary because students choose their own objects to measure. Check that the objects have been written in the table in order of length, starting with the shortest, and that the conversion between metres and centimetres and the rounding is correct.

Stretch zone: Check that students have calculated the difference between the longest and shortest object correctly.

5C Centimetres and millimetres

Discover Student Book page 115 • Practice Book page 95

Specific learning focus
- Draw and measure lines to the nearest centimetre and millimetre.

Global skills
- **Creative skills:** exploring

Key vocabulary
- centimetre (cm), millimetre (mm)

Resources
- mini whiteboards and markers
- rulers
- A4 paper
- place-value grids

Language support
Support students with phrases and questions, such as:
- *What is the length of … ?*
- *The length of _____ is … .*
- *How can you write 1.3 cm in another way?*

 **Introductory activity**

Give each student a ruler and a sheet of A4 paper. First, ask them to measure the length and width of the paper. They should write these measurements in millimetres, mixed units and then in centimetres and millimetres using decimal notation as appropriate (for example 21 cm × 29.7 cm). Ask students to recall equivalents and suggest that they use a place-value grid where appropriate. Next, ask them to measure some items from their pencil cases such as a pencil sharpener, eraser and pencil. They should write the measurements to the nearest millimetre. Then ask them to draw lines of these lengths on their paper and label them with the name of the item and their length. Again, ask students to write these lengths in millimetres, centimetres and millimetres, and centimetres. They should work in pairs so that they check one another's measurements.

 **Main activity**

Look together at the activity on page 115 of the Student Book and explain what students should do. They will continue to practise measuring in centimetres and millimetres. They work in the same pairs and take it in turns to measure a line and to check the measurement. They can record the lines they draw for question 2 on a separate sheet of paper if they would prefer to have more space for drawing their lines. Continue to provide place-value grids for students to support working out equivalences.

Unit 5 Length, mass, capacity and volume

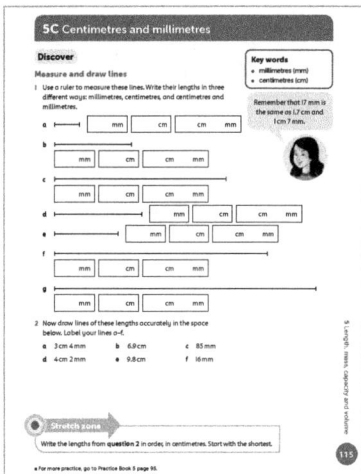

Differentiation

Supporting: Help students to measure in centimetres and millimetres. Ask them to tell you the measurement.

Consolidating: Ask students to explain how they knew how to write the measurements in millimetres, and in centimetres and millimetres.

Extending: Challenge students to write the lengths as metres.

Stretch zone: *Write the lengths from question 2 in order, in centimetres. Start with the shortest.*

You could also ask students to make up lengths in millimetres for a partner to draw and label in three different ways as in the activity on page 115 of the Student Book. They must measure to check to see whether they were accurate and that the lengths have been written correctly.

Reflection time

Ask each pair of students to try to draw three lines of these lengths on their whiteboards by estimating:

7 cm   9.5 cm   12 cm

They should compare their lines with another pair's lines and measure them with a ruler to see who was most accurate.

Practice Book: Students complete Practice Book page 95. They can do this directly after the Main activity, as homework, or as the focus of a separate mathematics session to help students consolidate their learning and build fluency.

Remind students that they may find it easier to convert some of the measurements to a different unit of measure before they draw the lines. Confirm that all lengths will need to be in the same unit of measure before they can total them to find the answer for the Stretch zone question.

Differentiated outcomes	
All students	should measure accurately in centimetres and millimetres with support.
Most students	will record measurements in different ways and order measurements.
Some students	may convert measurements to other units of length such as metres.

Answers

Student Book page 115

1. **a** 15 mm, 1 cm 5 mm, 1.5 cm
   b 46 mm, 4 cm 6 mm, 4.6 cm
   c 103 mm, 10 cm 3 mm, 10.3 cm
   d 52 mm, 5 cm 2 mm, 5.2 cm
   e 38 mm, 3 cm 8 mm, 3.8 cm
   f 128 mm, 12 cm 8 mm, 12.8 cm
   g 157 mm, 15 cm 7 mm, 15.7 cm

2. Check that students have drawn lines accurately to each of the lengths and labelled them correctly.

Stretch zone: Order: 1.6 cm, 3.4 cm, 4.2 cm, 6.9 cm, 8.5 cm, 9.8 cm

Practice Book page 95

Check the accuracy of the lines students have drawn.

Stretch zone: Check that this is a correct total of the lengths of the drawn lines. (35.2 cm)

5C Centimetres and millimetres

Explore Student Book page 116 • Practice Book page 96

Specific learning focus
- Estimate and measure lengths in centimetres and millimetres and find the difference between estimated and actual lengths.

Global skills
- **Creative skills:** exploring/investigating
- **Real-world skills:** presenting information/interpreting information
- **Interpersonal skills:** communication/teamwork

Key vocabulary
- estimate, centimetre (cm), millimetre (mm)

Resources
- mini whiteboards and markers
- rulers
- paper

Language support
Students may struggle with the nouns/adjectives in this section. You could write typical questions on the board, for example:
- How long is _____? (for length)
- How tall is _____? (for height)
- How wide is _____? (for width)
- How heavy is _____? (for weight)

Support their understanding with visual examples.

Introductory activity

Explain that you want to find out the width of your thumb and compare it to theirs. What unit of measure do they think they should use? How wide do they estimate their thumb will be? Ask students to record their estimates on their whiteboards to work in pairs to use their rulers to measure the width of each of their thumbs. Ask, *How many millimetres wide is your thumb?* Write the measurements on the board, such as 8 mm, 10 mm, 11 mm, 15 mm. Ask, *How can we write these lengths using decimals?* For example, you could write them as 0.8 cm, 1 cm, 1.1 cm, 1.5 cm.

Take feedback by asking: *Were estimates reasonable? Is there a big difference in the width of your thumb and your partner's? How much of a difference is there?*

 Main activity

Explain the activity on page 116 of the Student Book. Pairs should work together to measure and compare the length of their hands, and then use these measurements to estimate and measure other objects in the classroom.

As pairs work, ask questions to encourage them to share their thinking and measuring strategies, for example:

When you are estimating the length of something using your hands, what do you need to remember? (You need to line your wrist up with one end of your object and then line up your wrist on your other hand with the tip of your first hand's middle finger. You continue to do this for however many hands long your object is.)

My hand is 18 cm long and my rucksack is about 3 hand-lengths wide. What calculation could I do to make a reasonable estimate? (multiply 18 by 3) *What about to measure the height of my mug which is about half a hand-length tall?* (divide by 2)

If you know the length of your hand, when might it be useful to use this information to measure? (when you need an estimated length of an object that is not too long and you don't have a ruler or metre stick)

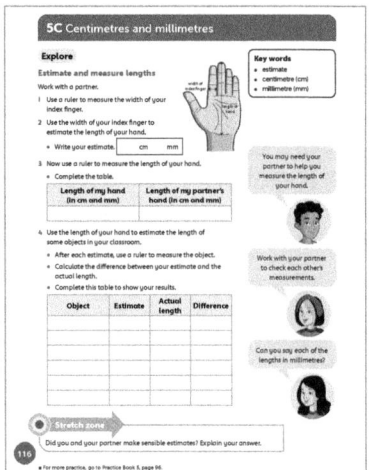

Differentiation

Supporting: Help students to measure the lengths of their hands and then model how to use their hands to measure objects to get an approximate length, width or height.

Consolidating: Ask students to explain to others how to use their hand lengths to measure lengths and convert to centimetres.

Extending: Challenge students to use their hands measured to the nearest millimetre as a measure for classroom objects.

Stretch zone: *Did you and your partner make sensible estimates? Explain your answer.*

Listen to students' explanations. How accurate do they feel an estimate needs to be to be sensible? Do they think they could improve their estimates next time? How?

150 Unit 5 Length, mass, capacity and volume

Reflection time

Take feedback from the activity on page 116 of the Student Book. Ask students to share their estimates and measurements for their items. Ask, for example, *How good was your estimate? Did your estimates get better with practice? Were you better at estimating in millimetres or centimetres? Can you say your measurements in millimetres as well as centimetres?*

Practice Book: Students complete Practice Book page 96. They can do this directly after the Main activity, as homework, or as the focus of a separate mathematics session to help students consolidate their learning and build fluency.

This activity is best done in pairs so you may prefer to do this during class time, possibly outdoors to provide more space and greater measuring opportunities.

Differentiated outcomes	
All students	should estimate and measure items in hand lengths with support.
Most students	will convert measurements from hand lengths to centimetres accurately.
Some students	may estimate and convert measurements from hand lengths to centimetres mentally.

Answers

Student Book page 116

Students estimate and measure their hands and other lengths in the classroom, so answers will vary. Check that the estimates and measures seem reasonable and that the differences have been calculated correctly.

Practice Book page 96

Answers will vary because students choose their own objects to measure. Check that the objects have been written in the table correctly and that the Egyptian measures used are accurate.

Stretch zone: Check that students have checked the Egyptian body ratios correctly.

5D Measuring mass

Discover Student Book page 117 • Practice Book page 97

Specific learning focus
- Read, choose, use and record standard metric units to estimate and measure mass to a suitable degree of accuracy.

Global skills
- **Creative skills:** investigating
- **Real-world skills:** research/presenting information
- **Interpersonal skills:** communication/teamwork

Key vocabulary
- estimate, mass, kilogram (kg), gram (g)

Resources
- weighing scales
- four or five items of food to weigh (pre-weighed if necessary); try to include some food items that are, for example, smaller but with a larger mass and larger with a smaller mass

Language support

Repeat the phrases and questions we commonly use to talk about mass, for example:
- *How heavy is … ?*
- *What unit of measure do you use to weigh a … ?*
- *How many grams are there in a … ?*
- *We measure a … in … .*
- *A … weighs/has a mass of … .*

 **Introductory activity**

Show the class four or five items of packaged food. List the weights of food on the board without telling students which weight refers to which package of food. Ask students to predict which weight matches which item. Invite several students to come to the front of the class to hold the packages. They should use this experience to try to order the items from lightest to heaviest.

Agree on an order and then use weighing scales to check the answers. *Were your predictions accurate? What did you notice after you picked some of the packets up?* Depending on the objects you choose and the predictions they made, they may notice that the larger the packet does not necessarily mean the heavier the **mass**.

Unit 5 Length, mass, capacity and volume

Main activity

Explain to students what they need to do in the activity on page 117 of the Student Book. They should work together, ideally in mixed-attainment groups of four students, to estimate the mass of different items in the classroom, then use a set of scales to weigh them.

They complete the activity by ordering the masses from lightest to heaviest. Encourage them to use known masses to help them estimate. They could use the items they weighed in the Introductory activity or you could suggest they choose another item to weigh and use this to help them make reasonable estimates. They could, for example, hold one object in one hand and then hold a second object in their other hand to compare and predict the mass of the second object. As they work, check that they are measuring accurately with the scales. Ensure that they use the same unit of measure when they compare their estimate with the actual mass.

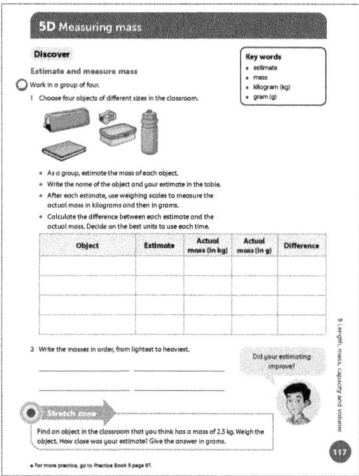

Differentiation

Supporting: Model aloud your thinking for making a reasonable mass estimate. Next, help students to read the masses from the scales accurately in two ways.

Consolidating: Ask students to explain how they reached one of their estimates.

Extending: Ask students to record masses in at least three different ways: grams, kilograms and grams, and kilograms.

Stretch zone: *Find an object in the classroom that you think has a mass of 2.5 kg. Weigh the object. How close was your estimate? Give the answer in grams.*

Students could also look for combinations of objects they think will have a mass of 2.5 kg.

Reflection time

Take feedback from the group activity from different students within each group. Ask questions such as: *Were your estimates close? Did you choose the best unit to use? Did your estimating improve? Can you explain why? Did you use the same unit each time to record your differences or did it change? How did you decide which was the best unit to use?*

Practice Book: Students complete Practice Book page 97. They can do this directly after the Main activity, as homework, or as the focus of a separate mathematics session to help students consolidate their learning and build fluency.

You could extend the activity by asking students to research one of the animals, focusing on facts involving measure and recording the measurements in more than one way.

Differentiated outcomes	
All students	should read scales accurately and record masses with support.
Most students	will estimate and record masses accurately in kilograms and grams.
Some students	may improve their estimates as they weigh more objects, as well as calculate some difference in mass mentally.

Answers

Student Book page 117

Students choose the items for which they estimate then measure mass, so answers will vary. Check that the estimates and measures seem reasonable and that the differences have been calculated correctly.

Practice Book page 97

	Animal	Mass in g	Mass in kg	Mass to nearest kg
1	skunk	750 g	0.75 kg	1 kg
2	manul	2600 g	2.6 kg	3 kg
3	porcupine	5250 g	5.25 kg	5 kg
4	meerkat	650 g	0.65 kg	1 kg
5	muskrat	1950 g	1.95 kg	2 kg
6	fox	4250 g	4.25 kg	4 kg
7	beaver	7750 g	7.75 kg	8 kg

Stretch zone: Check that students have found suitable information comparing the mass of different animals.

5D Measuring mass

Explore Student Book page 118 • Practice Book page 98

Specific learning focus
- Read, choose, use and record standard units to estimate and measure mass to a suitable degree of accuracy.

Global skills
- **Creative skills:** exploring
- **Interpersonal skills:** communication/teamwork

Key vocabulary
- kilogram (kg), gram (g)

Resources
- mini whiteboards and markers
- scales, some that measure masses of up to 5 kg
- 1 kg bags of rice
- modelling clay (at least 1 kg per group)

Language support
Repeat important phrases and questions, such as:
- *How heavy is … ?*
- *How many grams are there in … ?*
- *How many kilograms are equivalent to … g?*
- *What weight is equivalent to 1.5 kg? Can you think of another?*
- *We can measure the mass of something using … .*

Encourage students to use these while in discussion with their partners.

Introductory activity

Review the work covered in 5D Discover. Explain that mass is a measurement of how much matter is in an object, weight is a measurement of how hard gravity is pulling on that object.

Choose an object for students to estimate its mass, choosing an object for which estimating the mass would be difficult. Discuss the difficulty of estimating masses when there is not a known mass to compare with. Show a book to the class. Ask eight students (one at a time) to hold the **kilogram** bag of rice and tell them that it has a mass of 1 kg. Then ask them to hold the book. Finally, ask them to estimate the mass of the book. Write their estimates on the board, recording them in the format or unit of measure that they are given to you (ideally there will be a variety of formats). Weigh the book to check its mass. As a class, decide who gave the closest estimate, converting formats and units of measure to make comparison easier as needed.

Main activity

Give each group of four students a 1 kg bag of rice and a set of scales. Ask them to choose two items from the classroom. Tell each student in each group to hold the rice, then to hold each item. They estimate the weight of each item and record their estimate on their whiteboard. Then ask the groups to weigh their items, compare the masses with their estimates and decide who made the closest estimate.

Students should work in mixed-attainment small groups for the activity on page 118 of the Student Book. Each group needs modelling clay and a set of scales. Discuss with students how to weigh out the clay to make each fruit by converting the masses from kilograms to **grams**. They should use the scales to weigh out the correct number of grams for each fruit before making a model of the correct shape. For question 2, ask students to think about the unmarked increments on the scales and what they represent. Ask, *What is the maximum mass that can be weighed on each scale?* They should work collaboratively on the first part of the activity and then compare their answers to the last question, correcting any errors and explaining where they went wrong.

Observe students while they make estimates, make the fruits from modelling clay and then weigh them. Ask them to explain their answers to the questions in the speech bubbles in the Student Book: *When do you need to estimate in this activity? How did you get the correct mass?*

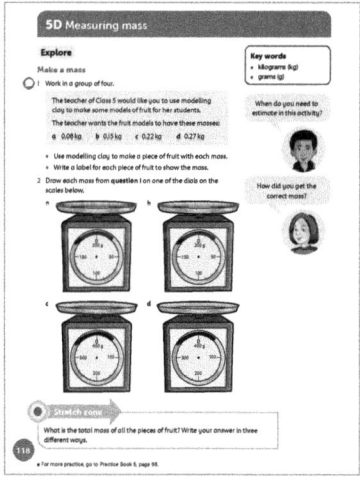

Differentiation

Supporting: Help students to convert from kilograms to grams, emphasising how the digits move to the left 3 times when they multiply by 1000.

Consolidating: Ask students to explain how they found the masses of the fruits in grams.

Extending: Challenge students to make their estimate in grams as accurate as possible.

Stretch zone: *What is the total mass of all the pieces of fruit? Write your answer in three different ways.*

The total mass of the fruits is: 720 g = 0 kg and 720 g = 0.72 kg. Ask students to explain to you how they added the masses. For example, did they convert all the masses to grams first or did they add the kilograms and then convert them to grams?

Unit 5 Length, mass, capacity and volume

You could ask students to make different-sized balls of modelling clay and estimate and measure their mass.

 Reflection time

Take feedback from the activity. Invite students in each group to share what they did and to explain when and why they made estimates and how they got the correct mass. Students are likely to say that they estimated how much clay to put on the scale when trying to find the exact amount of clay they would need for each fruit. They would then take away or add clay, to get the correct mass, estimating how much each time.

Practice Book: Students complete Practice Book page 98. They can do this directly after the Main activity, as homework, or as the focus of a separate mathematics session to help students consolidate their learning and build fluency.

Discuss rounding to the nearest kg before students begin this activity. They will need weighing scales that measure masses up to 5 kilograms to complete this activity.

Differentiated outcomes	
All students	should read digital scales accurately and mark masses on scale faces with support.
Most students	will estimate and record masses accurately in grams and convert masses from kilograms to grams.
Some students	may convert masses between grams and kilograms and vice versa mentally and improve their estimates as they weigh more pieces of clay.

Answers

Student Book page 118

1 Check that each group's fruits have the correct mass.

2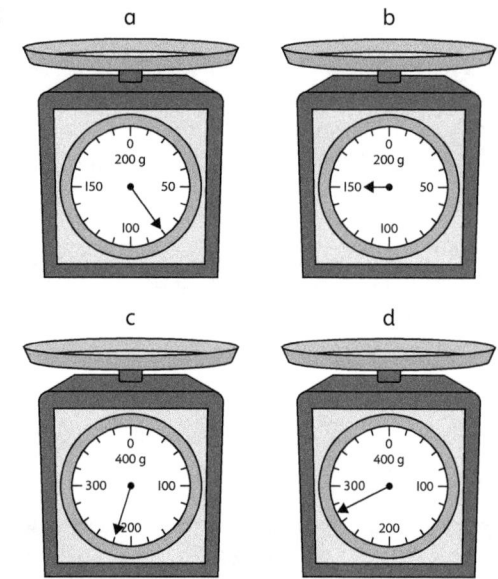

Practice Book page 98

Answers will vary as students choose their own objects to measure. Check that the objects have been written in the table correctly and that the conversion between grams and kilograms and the rounding is correct.

Stretch zone: Check that students have calculated the difference between the heaviest and lightest object correctly.

5E Measuring capacity

Discover Student Book page 119 • Practice Book page 99

Specific learning focus
- Read, choose, use and record standard units to estimate and measure capacity to a suitable degree of accuracy.

Global skills
- **Creative skills:** investigating
- **Interpersonal skills:** teamwork

Key vocabulary
- litre (l), millilitre (ml), capacity, volume

Resources
- mini whiteboards and markers
- measuring jugs
- range of packaged liquids such as a single-serving box of juice, a small bottle of shampoo, a can of drink

Language support

Add to your working wall a labelled illustration of a measuring jug. Surround it with speech bubbles with key questions and statements relating to measuring capacity. For example: 'The capacity of this jug is … .' 'We measure capacity in millilitres or litres.' 'There are 1000 millilitres in 1 litre. The abbreviation for litres is l and for millilitres is ml'.

 **Introductory activity**

Show the class four or five items of packaged liquids. List the volumes of liquids on the board without telling students which one refers to which package. Explain that the **volume** is a measure of how much liquid there is, which is not the same as **capacity**, or how much the container could hold. Show students a container that has 400 millilitres of liquid so they have something to compare with. Pairs of students should write on their whiteboards the packages and the volumes that they think match each package. One student can then come to the front and read the labels to tell the class which package contains which amount of liquid.

Unit 5 Length, mass, capacity and volume

 **Main activity**

Give each group of four a measuring jug and check that all students in the class know how to read the scale. Remind students of the method.

- First, you pour an amount of liquid into the jug.
- You look to see the level.
- You read the number closest to the level of liquid. When there is no number, you need to work out what the unmarked divisions are. This will depend on the scale on the measuring jug.

Look at page 119 of the Student Book together and discuss what students need to do. They should work together, ideally in mixed-attainment groups of four students, to estimate the capacity of different containers, then use the measuring jugs to find the actual capacities in **litres** and **millilitres**. They then calculate the difference between the estimated capacity of each container and its actual capacity. They complete the activity by ordering the capacities from smallest to greatest.

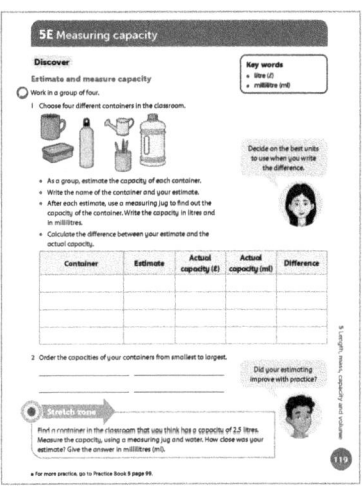

Differentiation

Supporting: Help students to read the scale on the measuring jug.

Consolidating: Ask students to explain how they estimated capacities before they measured.

Extending: Ask students to say capacities in millilitres for a partner to write in three different ways: millilitres, litres and millilitres, and litres.

Stretch zone: *Find a container in the classroom that you think has a capacity of 2.5 litres. Measure the capacity, using a measuring jug and water. How close was your estimate? Give the answer in millilitres (ml).*

You could also ask students to find a combination of containers that they think would have a total capacity of 2.5 litres.

 Reflection time

Take feedback from the group activity. Ask questions such as: *Were your estimates close? Did your estimates improve with practice? What can make estimating the capacity of a container challenging? Did your tallest container have the greatest capacity? Can you explain why/why not? Did you use the same unit each time to record your differences or did it change? How did you decide which was the best unit to use?*

Practice Book: Students complete Practice Book page 99. They can do this directly after the Main activity, as homework, or as the focus of a separate mathematics session to help students consolidate their learning and build fluency.

Discuss rounding to the nearest 100 ml before students begin this activity.

Differentiated outcomes	
All students	should read scales accurately and record capacities with support.
Most students	will estimate and record capacities accurately.
Some students	may improve their estimates as they continue to measure capacities and work out differences in capacity mentally.

Answers

Student Book page 119

Answers will vary depending on the containers students choose. Check that the estimates and measures seem reasonable and that the differences have been calculated correctly.

Practice Book page 99

	Container	Capacity in litres	Capacity in millilitres	Capacity to nearest 100 ml
1	small cup	0.125 l	125 ml	100 ml
2	glass	0.75 l	750 ml	800 ml
3	large cup	0.3 l	300 ml	300 ml
4	milk pan	1.75 l	1750 ml	1800 ml
5	soup pan	2.25 l	2250 ml	2300 ml
6	soup bowl	0.65 l	650 ml	700 ml
7	stew pot	3.5 l	3500 ml	3500 ml

Stretch zone: $4\frac{2}{3}$ glasses will fill the stew pot, so you would have to pour water from the glass 5 times to fill the pot.

Unit 5 Length, mass, capacity and volume

5E Measuring capacity

Explore Student Book pages 120–121 • Practice Book page 100

Specific learning focus
- Read, choose, use and record standard units to estimate and measure capacity to a suitable degree of accuracy.

Global skills
- **Creative skills:** exploring
- **Interpersonal skills:** communication/teamwork

Key vocabulary
- capacity, litre (l), millilitre (ml), volume

Resources
- mini whiteboards and markers
- measuring jugs or cylinders
- a glass
- 1-litre bottle of water
- range of different-sized and shaped containers that can be filled with water
- 0–9 digit cards

Language support
Support students with important question and phrase forms, for example:
- *What is the capacity of … ?*
- *There are … millilitres in … .*
- *In real life people use … to measure … .*

 Introductory activity

Remind students of the activity they did in the previous lesson, which included estimating capacities. Show a glass and a 1-litre bottle of water. Explain that the bottle holds 1 litre. Ask students to think about how the two capacities might differ.

Then ask them to estimate the capacity of the glass and write their estimates on their whiteboards. Pour the water from the bottle into the glass. Then measure the amount into a measuring jug. Choose a student to come to the front of the class to read the scale and another to check that they are correct. Compare students' estimates with the actual amount. Repeat this with different-sized and shaped containers in the classroom.

 **Main activity**

Students should work in pairs. To introduce the activity, choose a student to come to the front of the class. They should pick three digit cards and make the largest number they can with these digits. Explain that this becomes a volume they have to estimate. Then they will measure to see how close their estimate was. For example, if they choose 4, 1 and 8, they could make 841 ml.

They should estimate 841 ml of water by adding water into an unmarked container, then pour it into a marked measuring cylinder to check their estimate.

As pairs work, ask questions to encourage them to explain and show their understanding, for example: *What did you think about first when you were estimating the volume of your water? How can looking at the measuring cylinder before you estimate help you to make a reasonable estimate? Did your estimate improve? Why do you think that was?*

Pairs should make at least three different volumes, estimating and measuring one volume and then the next.

Explain that they are going to do a very similar activity but this time thinking about capacity. Ask, *How do you think it will be different?* Some students may realise that they may need to estimate the capacity of containers and then measure them.

Read together the instructions on page 120 of the Student Book. Students should then complete the activities on pages 120–121 in their pairs, marking capacities on measuring cylinders, totalling and ordering the capacities.

As pairs work, ask questions relating to measuring and capacity, for example:

- *The larger the capacity, the taller the jug. Is that true? Why or why not?*
- *I find the difference between two capacities that are measured in millilitres so I must record the difference in litres. Is that true? Why or why not?*

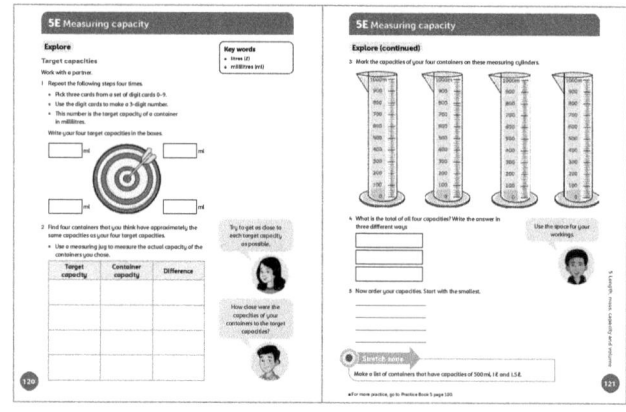

156 Unit 5 Length, mass, capacity and volume

Differentiation

Supporting: Help students to read the scale on the measuring container.

Consolidating: Ask students to explain how they estimated capacities before they measure.

Extending: Ask students to make capacities using four digit cards, then write them in litres and millilitres.

Stretch zone: *Make a list of containers that have capacities of 500 ml, 1 l and 1.5 l.*

Students could find different containers such as labelled drink bottles and check the accuracy of the labelling.

 Reflection time

Take feedback from the Main activity. Invite pairs to explain what they did. Ask, *Were you able to find containers with capacities close to those of your target capacities? How close were your estimates? Did they improve with practice?*

Practice Book: Students complete Practice Book page 100. They can do this directly after the Main activity, as homework, or as the focus of a separate mathematics session to help students consolidate their learning and build fluency.

Ask students to look at each volume and then the scale on the measuring containers. They should notice that the volume is in litres and the scale is in millilitres. Agree that they will need to convert the volumes to millilitres before they mark the amounts on the containers.

If possible, set aside some time for students to explain their answers to the Stretch zone question in pairs or small groups.

Differentiated outcomes	
All students	should read scales accurately and record capacities with support.
Most students	will estimate and record capacities accurately.
Some students	may write estimates and capacity measures in different ways.

Answers

Student Book pages 120–121

Answers will vary because students use digit cards to make capacities they will work with. Check that their estimates and measures seem reasonable and that the differences and totals have been calculated correctly.

Practice Book page 100

Answers will vary because students choose their own volumes to mark on the measuring cylinders.

Stretch zone: Again, answers will vary because students choose their own volumes to mark on the measuring cylinders.

5F Imperial units

Discover Student Book page 122 • Practice Book page 101

Specific learning focus
- Select and use standard units and convert between imperial and metric units.

Global skills
- **Creative skills:** investigating
- **Real-world skills:** research/presenting information
- **Interpersonal skills:** communication/teamwork

Key vocabulary
- base-10 system, imperial units, metric units, conversion, approximate

Resources
- scissors
- large sheets of paper
- internet access or reference books relating to imperial measures
- rulers with both inches and centimetres marked (paper versions can be downloaded from the internet)
- tape measures

Language support

Encourage students to read the results of their research aloud to you. Focus particularly on the units. You may need to model the language for students so that they hear the correct pronunciation.

 Introductory activity

Write on the board the following equivalences.

1000 mm = 1 m   1000 g = 1 kg   1000 ml = 1 litre

Ask students to discuss what is the same and what is different about them. Discuss the fact they all are all **metric units** and they all use the **base-10 system**, meaning that each successive unit is 10 times larger than the one before, which makes it easy to calculate with.

Unit 5 Length, mass, capacity and volume

Explain that metric is the system most commonly used in the world, but that a few countries, such as the US, use the imperial system and the UK uses a mixture of metric and **imperial units**. Imperial units do not use the base-10 system, which can make calculations harder. Write on the board the following imperial equivalences.

12 inches = 1 foot

3 feet = 1 yard

1 pound = 16 ounces

1 pint = 20 fluid ounces

Ask students to calculate some imperial equivalences. Ask, for example, *How many inches are there in 16 feet?*

 Main activity

Give pairs of students a ruler with both inches and centimetres marked on it. Ask them to work out how many centimetres are approximately equal to 1 inch. (2.5) Ask them to measure several objects and record their lengths in both inches and centimetres. Ask them to consider which calculation they could do to find out how long an object was in centimetres if they knew its length in inches, for example 6 inches. They may suggest multiplying 6 by 2.5 or counting up in 6 steps of 2.5. Ask them to check their answers using the ruler.

Next, ask students to work in small groups to read the text on page 122 of the Student Book and research a set of imperial units. Tell them that they will use their research to create a poster.

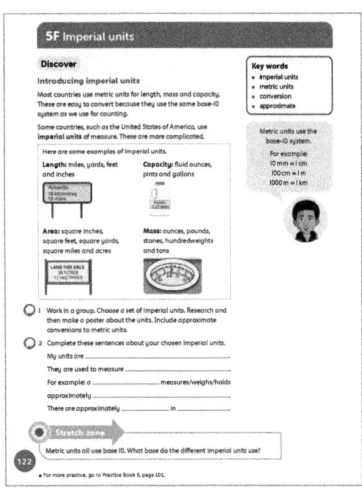

Differentiation

Supporting: Ensure that students have a clear and appropriate role within their group.

Consolidating: Ask students to share examples of things they might measure using different imperial units. What do they think they might measure, based on what they know so far about imperial units?

Extending: Students could research imperial and metric equivalences and include some examples of conversions in their group's poster.

Stretch zone: *Metric units all use base 10. What base do the different imperial units use?*

Encourage students to think about the equivalences from the Introductory activity. Can they see a pattern? For example, feet and inches are in base 12 because there are 12 inches to every foot.

 Reflection time

Ask each group to present their research into imperial units, sharing their examples of objects and their measures and, if any groups did so, explaining how they converted to metric units.

Practice Book: Students complete Practice Book page 101. They can do this directly after the Main activity, as homework, or as the focus of a separate mathematics session to help students consolidate their learning and build fluency.

Students will need a measuring tape or ruler as well as another person to help them to measure their actual height.

Differentiated outcomes	
All students	should think of some objects measured in imperial units.
Most students	will think of objects to represent most of the imperial units.
Some students	will think of a range of objects to represent all the imperial units.

Answers

Student Book page 122

1 Students choose their own sets or imperial units and make a poster about them, so answers will vary.

2 Answers will vary because students give their own definitions and examples of the imperial units and relationships between the imperial and metric units. Check that students have correctly understood the imperial units.

Practice Book page 101

1 Shanghai Tower 632 m and 695.2 yards

2 River Nile 4180 km and 2612.5 miles

3 Smallest adult 55 cm and 22 inches

4 Distance from London to Dubai 5500 km and 3437.5 miles

5 Monitor lizard 160 cm and 64 inches

Stretch zone: Answers will vary depending on each student's height. Check that they have converted their height to feet and inches correctly.

5F Imperial units

Explore
Student Book page 123 • Practice Book page 102

Specific learning focus
- Convert metric units to imperial units.

Global skills
- **Creative skills:** problem solving
- **Real-world skills:** interpreting information

Key vocabulary
- imperial units, metric units, conversion, approximate

Resources
- mini whiteboards and markers
- a selection of items for weighing
- measuring scales

Language support
Model the language for students so that they hear the correct pronunciation of the names of the imperial units. Verbalise the conversions, for example:
- a book has a mass of 2 kg. 1 kg = 2.2 pounds, so the book has a mass of 4.4 pounds.

Introductory activity

Write on the board the conversion 1 kg = 2.2 pounds.

Use a counting stick and count up and back in multiples of 2.2 with students:

2.2, 4.4, 6.6, 8.8, 11, 13.2, 15. 4, 17, 6, 19, 8, 22.

Ask students to work in pairs to say how many pounds is the same as each of these masses:

10 kg   20 kg   50 kg   5 kg   200 kg   5000 kg

Now repeat using the conversion 8 km = 5 miles. Discuss how students could convert from km to miles. (divide by 8 and multiply by 5). Ask them to say how many miles is the same as each of these distances:

16 km   80 km   40 km   160 km   20 km   320 km

Ask, *How did you work each one out? Which known facts did you use to work out new ones?*

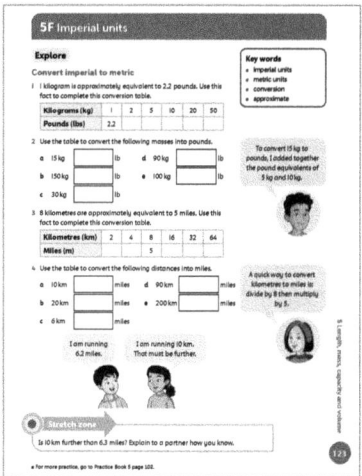

 Main activity

Students complete the questions on page 123 of the Student Book in pairs. As they work, ask how they are using the conversion rates. Encourage them to use **conversions** they have already done to work out other conversions. For example, they can find how many miles is equal to 6 km by adding the mile equivalents for 2 km and 4 km.

Students should work in pairs to find objects in the classroom to weigh. Alternatively, give them a selection of objects to weigh. They will need five items, starting from **approximately** 500 grams in mass. First they should estimate their masses in grams or kilograms and write their estimates on their whiteboards. Then they use scales to weigh the items in grams and kilograms and record the masses on their whiteboards.

When they have finished estimating and weighing, students should use the conversion rate of 1 kg = 2.2 pounds to convert each of the masses into pounds. Encourage them to use, or extend, the table in their Student Book to work out the masses in pounds.

Differentiation

Supporting: Help students to use multiplication to convert metric units into imperial units.

Consolidating: Ask students to describe their strategy for converting between units.

Extending: Challenge students to convert from imperial units back to metric units.

Stretch zone: *Is 10 km further than 6.3 miles? Explain to a partner how you know.*

Students should be able to explain to a partner that it is not because 10 km = 6.25 miles. Students can then make up similar questions for a partner, based on using the conversion rates.

Unit 5 Length, mass, capacity and volume

Reflection time

Pairs can share their findings with the class, presenting one of their objects. The rest of the class could estimate the mass. Then the pair could reveal the mass in kilograms and explain their conversion calculation and any strategy they used for converting.

Practice Book: Students complete Practice Book page 102. They can do this directly after the Main activity, as homework, or as the focus of a separate mathematics session to help students consolidate their learning and build fluency.

Tell students to estimate their conversions first by rounding to the nearest whole unit. Then they should use a calculator to check their estimates.

Differentiated outcomes	
All students	should convert kilograms to pounds and kilometres to miles with support.
Most students	will convert kilograms to pounds and kilometres to miles.
Some students	may convert from imperial units back to metric units.

Answers

Student Book page 123

1

Kilograms (kg)	1	2	5	10	20	50
Pounds (lbs)	2.2	4.4	11	22	44	110

2 **a** 33 lb **b** 330 lb **c** 66 lb **d** 198 lb **e** 220 lb

3

Kilometres (km)	2	4	8	16	32	64
Miles	1.25	2.5	5	10	20	40

4 **a** 6.25 miles **b** 12.5 miles **c** 3.75 miles
 d 56.25 miles **e** 125 miles

Practice Book page 102

	Object	Metric	Imperial
1	mass of elephant	2300 kg	5060 lbs
2	mass of human	635 kg	1397 lbs
3	mass of cat	2.9 kg	6.38 lbs
4	capacity of glass	450 ml	0.81 pints
5	capacity of can	330 ml	0.594 pints

Stretch zone: Metric units are in base 10, imperial units are in various bases. Base 10 is easier for calculations between units.

5G Volume

Discover Student Book page 124 • Practice Book page 103

Specific learning focus
- Measure volume in cubic centimetres.

Global skills
- **Creative skills:** investigating
- **Self-development skills:** reflecting on learning

Key vocabulary
- volume, capacity, cubic centimetres (cm³), cubed, dimensions

Resources
- 36 1-cm³ interlocking cubes (or similar) per pair
- mini whiteboards and markers

Language support

Support students with important question and phrase forms, for example:
- *What is the volume of … ?*
- *There are … cubic centimetres in … .*
- *In real life, people use … to measure … .*

Introductory activity

Ask students to take a handful of cubes and make a shape by linking them all together. They should then compare shapes with a partner. Ask, *Who has the larger shape? How do you know?* Discuss with the class that they are learning about volume, which is how much 3D space is taken up by a shape: the larger shapes are the ones that are made from more cubes. Ask students to count how many cubes they used for their shape.

Once they have found whose shape was larger (in other words, the shape with greater volume), ask pairs to make two shapes that have the same volume. Talk about the volume of a cube of dimensions 1 cm × 1 cm × 1 cm, connecting this to the volume being length × width × height so we say 1 cm³ pronounced 'one centimetre **cubed**'. To find a volume, you are counting how many of those 1 cm³ there are in the shape. Ask students to say what the volume of their shape is in **cubic centimetres (cm³).**

 Main activity

Give each pair of students 12 interlocking cubes each measuring 1 cm³. Ask them to recall the properties of a cuboid (6 rectangular faces, 12 edges, 8 vertices) and then ask them to use all 12 cubes to make a cuboid. When they have done this, ask them to write on their whiteboards the length, width and height of their cuboid in centimetres. Ask whether they can make a different cuboid using all 12 cubes. What are its **dimensions**? Can they make any more that are different?

160 Unit 5 Length, mass, capacity and volume

Length	Width	Height
12	1	1
6	2	1
4	3	1
3	2	2

Go through all the possibilities as a class and discuss how students knew they had found them all. Talk about any duplicates and the fact that these could be different orientations of the same shape.

Give pairs 24 additional interlocking cubes to complete the activity on page 124 of the Student Book.

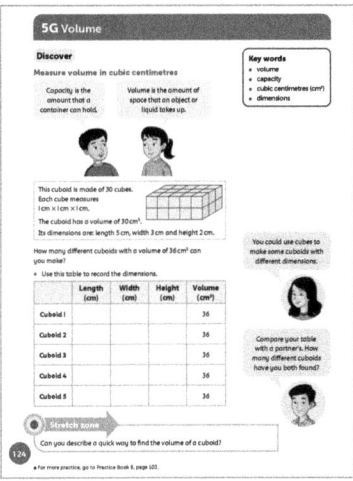

Differentiation

Supporting: Help students to build, measure and record the dimensions of cuboids they make, working systematically and starting with, for example, a 1 × 1 × 24 cuboid.

Consolidating: Ask students to explain how they calculated the volume of their shapes.

Extending: Challenge students to find all different cuboids with a volume of 36 cm³.

Stretch zone: *Can you describe a quick way to find the volume of a cuboid?*

Students may notice that the volume is found by multiplying the three dimensions.

 Reflection time

Ask students to share the dimensions of their 36 cm³ cuboids. Compare the dimensions and discuss whether the cuboids are different. For example, is a cuboid of 2 cm × 3 cm × 6 cm the same as a cuboid of 3 cm × 6 cm × 2 cm?

Practice Book: Students complete Practice Book page 103. They can do this directly after the Main activity, as homework, or as the focus of a separate mathematics session to help students consolidate their learning and build fluency.

Encourage students to continue to work with 24 interlocking cubes to complete this activity.

Differentiated outcomes	
All students	should make cuboids of a given volume with support.
Most students	will find and make five possible cuboids of a given volume.
Some students	may calculate the volume of a cuboid using its dimensions.

Answers

Student Book page 124
Possible answers include:

	Length (cm)	Width (cm)	Height (cm)	Volume (cm³)
Cuboid 1	1	1	36	36
Cuboid 2	1	2	18	36
Cuboid 3	1	3	12	36
Cuboid 4	1	4	9	36
Cuboid 5	1	6	6	36

Practice Book page 103
Possible answers include:

	Length (cm)	Width (cm)	Height (cm)	Volume (cm³)
Cuboid 1	1	1	24	24
Cuboid 2	1	2	12	24
Cuboid 3	1	3	8	24
Cuboid 4	1	4	6	24

Stretch zone: There are six different cuboids for 24 cm³, those in the table above, as well as 2 × 2 × 6 and 2 × 3 × 4.

Unit 5 Length, mass, capacity and volume

5G Volume

Explore Student Book pages 125–126 • Practice Book page 104

Specific learning focus
- Find volumes of 3D shapes and draw them on isometric paper.

Global skills
- **Creative skills:** exploring
- **Real-world skills:** presenting information

Key vocabulary
- volume, capacity, cubic centimetres (cm³)

Resources
- 12 1-cm³ interlocking cubes (or similar) per individual
- isometric paper

Language support
Reinforce the distinction between capacity and volume, with capacity being the amount a container can hold, and the volume being the amount of a substance inside the container.

Introductory activity

Give each student a set of 12 interlocking cubes. Ask students to make a 3D shape using all of the cubes. Explain that the shape does not have to be a cuboid. When everyone has finished, compare the shapes. Ask, *What is the same and what is different about the shapes you have made?* Students should agree that the shapes may look different, but they all have the same volume because everyone used 12 cubes to make them, so each shape has a volume of 12 cm³.

Main activity

Make a shape using cubes and display it. Ask students to see whether they can estimate its volume in cubic centimetres. Then ask them to copy the shape using their own cubes. Ask, *How can you know whether you have made the same shape? Can you work out the volume of your shape?*

Now introduce students to isometric paper as a way of drawing 3D shapes, by looking at the shape from an angle so they can see the three different dimensions. Model on the board how to draw a single cube on isometric dots, and then extend this to more complex shapes using additional cubes.

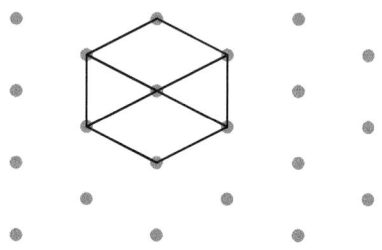

Students should then complete the activities on pages 125–126 of the Student Book. You may need to monitor as they draw to ensure that they get the perspective correct for their shapes. Remind them to keep the shapes in the same position as they draw.

For question 5, students move from counting cubes and reading volumes on a scale to calculating using dimensions. Ask students to imagine the swimming pool to contain large cubes of 1 m × 1 m × 1 m and ask how many there would be in the pool. Using length × width × height will give the capacity of the pool, and then they can find $\frac{3}{4}$ of this as the volume of water.

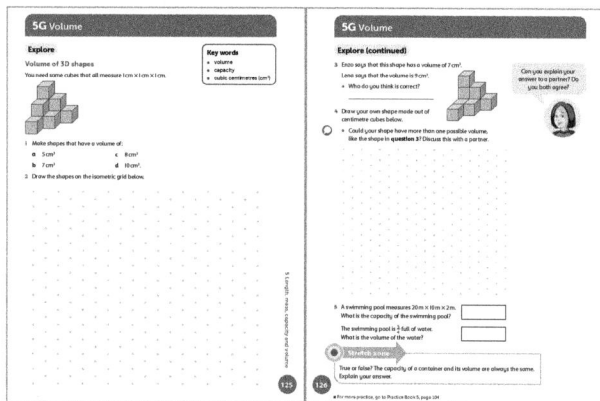

Differentiation

Supporting: Help students to count the cubes on complex shapes to find the volume.

Consolidating: Ask students to work out the volume of various shapes made from cubes.

Extending: Challenge students to draw complex shapes made from cubes and calculate their volumes.

Stretch zone: *True or false? The capacity of a container and its volume are always the same. Explain your answer.*

Students should be encouraged to give examples to show that this is false.

 Reflection time

Share together the drawings made on the isometric paper. Ask students to explain how they drew their shapes. Ask some students to tell the class the volume of their shapes and say how they can tell from the drawing.

Practice Book: Students complete Practice Book page 104. They can do this directly after the Main activity, as homework, or as the focus of a separate mathematics session to help students consolidate their learning and build fluency.

Students should continue to use interlocking cubes to help them, if possible.

Differentiated outcomes	
All students	should make shapes from cubes and find the volume with support.
Most students	will calculate the volume of shapes made from cubes.
Some students	may work out the volume of shapes made from cubes by looking at an isometric drawing of the shape.

5H Problem solving with measures

Discover Student Book page 127 • Practice Book page 105

Specific learning focus
- Solve word problems about length, mass, capacity and volume.

Global skills
- **Creative skills:** problem solving
- **Real-world skills:** interpreting information/presenting information

Key vocabulary
- convert between units, scale, conversion

Resources
- large sheets of paper
- card or balsa wood or similar, adhesive or sticky tape
- tape measures

Language support
Encourage students to read the results of their research aloud to you. Focus particularly on the units. You may need to model the language for students so that they hear the correct pronunciation and model this in the word problems.

Answers

Student Book pages 125–126

1. Check that students have made the shapes correctly.
2. Check that students' drawings are accurate.
3. Lena is probably correct because there could be some hidden cubes that help join it together.
4. Check that students' the drawings are accurate.
5. $400\,m^3$; $300\,m^3$

Practice Book page 104

Check that students have drawn shapes accurately.

Stretch zone: Check that students have given a good estimate of the volume of a box.

 Introductory activity

Ask students to review their learning about the different units of measure they have covered in this unit. Write each of the following in the centre of four separate large sheets of paper: 'Length', 'Mass', 'Capacity' and 'Volume'. Assign a group of students to each area of measurement and ask them to write down everything they can think of about that subject on the paper. For example, it could be the metric units used to measure it, or conversion to smaller or larger units, or something about imperial units. After a short time, move the sheets of paper around to each group in turn to see whether they can add anything.

 **Main activity**

Display the sheets of paper and discuss each in turn, asking students to explain what they know about each property and how they can use the units to measure and solve problems. Ask them to clarify, for example, *How do you convert cm into m?, How many ml are there in a litre?, Approximately how many miles are equivalent to 16 km?*

Recall with students that they have measured distances on maps using a scale, for example 1 cm = 10 m. Explain that scales can be written as ratios, as long as both figures are in the same units. (1 cm = 10 m would be written as 1:1000 as there are 1000 cm in 10 m.)

Students then work in pairs on the activity on page 127 of the Student Book. They should use the conversions and **scales** provided to complete the table, using calculators so they can focus on the concept of the conversion and not be burdened with the multiplication with decimal places.

For question 2, provide students with sufficient materials to make their model towers. They may need card and sticky tape or balsa wood and adhesive to build a tower of over 80 cm. This part of the activity may need to be completed over more than one lesson.

Unit 5 Length, mass, capacity and volume

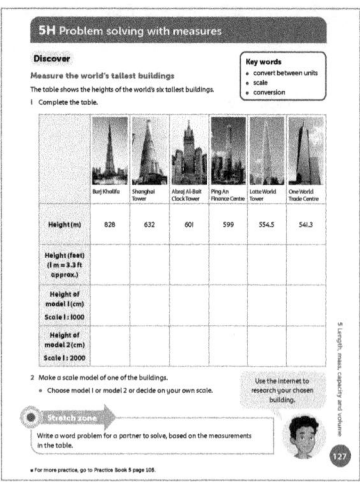

Stretch zone: *Write a word problem for a partner to solve, based on the measurements in the table.*

Encourage students to write a two-step problem. They should also be able to solve their problem.

Reflection time

Discuss together the methods used to convert the measures from metres into feet and centimetres. Can students describe what calculations they did and why?

Practice Book: Students complete Practice Book page 105. They can do this directly after the Main activity, as homework, or as the focus of a separate mathematics session to help students consolidate their learning and build fluency.

Students will need a tape measure and an adult to support them with their measuring, if possible, to complete this activity.

Differentiated outcomes	
All students	should complete the conversions with support.
Most students	will complete the conversions using a calculator.
Some students	may write their own conversion problems.

Differentiation

Supporting: Help students to convert the measures using a scale.

Consolidating: Ask students to explain their methods for completing each row of the table.

Extending: Challenge students to research the heights of other tall buildings and use the scale to convert their heights to centimetres.

Answers

Student Book page 127

1

	Burj Khalifa	Shanghai Tower	Abraj Al-Bait Clock Tower	Ping An Finance Centre	Lotte World Tower	One World Trade Centre
Height (m)	828	632	601	599	554.5	541.3
Height (feet)	2732.4	2085.6	1983.3	1976.7	1829.85	1786.29
Height (cm) of model 1 at 1:1000 scale	82.8	63.2	60.1	59.9	55.45	54.13
Height (cm) of model 2 at 1:2000 scale	41.4	31.6	30.05	29.95	27.725	27.065

2 Check that students have made their model to the correct height for the scale they chose.

Practice Book page 105

Answers will vary depending on the scale and units chosen, and on the size of the student. Check that they have measured and converted accurately.

Stretch zone: Check that students have measured and calculated the ratio correctly for their own body.

5H Problem solving with measures

Explore Student Book page 128 • Practice Book page 106

Specific learning focus
- Solve word problems using mixed measures and converting between units.

Global skills
- **Creative skills:** problem solving
- **Real-world skills:** interpreting information
- **Interpersonal skills:** communication/teamwork

Key vocabulary
- convert between units, scale

Resources
- mini whiteboards and markers

Language support
Encourage students to read the word problems aloud. Explain any unfamiliar vocabulary as necessary.

Introductory activity

Ask pairs of students to write on a whiteboard their answers to puzzles you set, for example:

Can you write:
- *a length between 95 cm and 1.04 m*
- *a mass that is half of 1.24 kg in grams*
- *a capacity that is double 595 ml*
- *a volume one quarter as much as 396 cm³?*

Pairs should take it in turns to give their answers. Repeat each answer as a whole class, making sure that students say the number and unit in full.

Main activity

Students complete the word problems on page 128 of the Student Book. They should be reminded to check the units that are given and give answers in the required units. Students work in pairs and check each other's answers. They should estimate first and use inverse operations to check their answers.

Ask questions to help students tackle the word problems, for example:
- *How many centimetres are there in a metre?*
- *How do you convert metres to centimetres?*
- *How many millilitres is the same as 3.5 litres? How do you know?*
- *What do you need to remember when adding and subtracting amounts?* (They need to be in the same unit of measure.)
- *What do you need to do first to solve this problem? What operation will you use?*

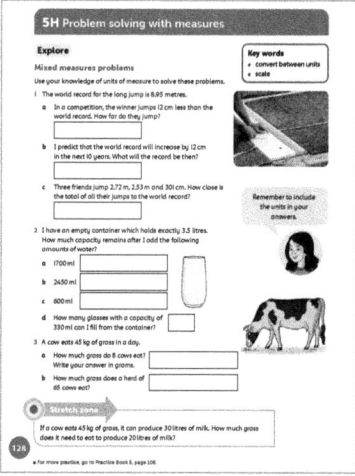

Differentiation

Supporting: Read questions together and support students to identify the key information and identify which operation(s) they will need to use to solve the problems. Help them to convert the units in each question to be the same.

Consolidating: Ask students to explain how they solved the word problems.

Extending: Challenge students to write their own word problems for a partner to solve.

Stretch zone: *If a cow eats 45 kg of grass, it can produce 30 litres of milk. How much grass does it need to eat to produce 20 litres of milk?*

Students should calculate that the amount of grass is 30 kg, which can be worked out using ratios.

Reflection time

Ask pairs of students to share with the class their strategies for solving the word problems. Ask them to explain how they converted the units to the ones required. Did they estimate their answers first? How close where their estimates to the actual answers?

Unit 5 Length, mass, capacity and volume

Practice Book: Students complete Practice Book page 106. They can do this directly after the Main activity, as homework, or as the focus of a separate mathematics session to help students consolidate their learning and build fluency. Encourage students to show their workings.

Differentiated outcomes	
All students	should complete the calculations with support.
Most students	will complete the calculations independently.
Some students	will estimate and calculate the answers independently, in some cases using mental methods.

Answers

Student Book page 128
1   **a** 8.83 m   **b** 9.07 m   **c** 0.69 m
2   **a** 1800 ml   **b** 1050 ml   **c** 2900 ml   **d** 10 glasses
3   **a** 360 000 g   **b** 2925 kg

Practice Book page 106
1   458 ml
2   147 ml
3   235 ml
4   3 soda cans
5   3 plant pots

Stretch zone: 32 spoonfuls

5 Length, mass, capacity and volume

Connect Student Book page 129

Big idea
- I can use my knowledge of metric and imperial units to convert between different units of measure.

Global skills
- **Creative skills:** problem solving
- **Real-world skills:** interpreting information/presenting information
- **Interpersonal skills:** communication/teamwork
- **Self-development skills:** reflecting on learning

Key vocabulary
- mass, weight, heavy/light, weigh, kilogram (kg), gram (g), capacity, litre (l), millilitre (ml)

Resources
- pancake recipe

Language support
Revisit the posters that you made at the beginning of the unit. Check that all the key vocabulary is understood and ask whether any new vocabulary needs to be added. Students can take photographs of these posters and keep them for reference.

 Introductory activity

Ask students if they know a recipe for making pancakes. Discuss what ingredients it may involve (for example milk, flour, eggs). *What units would you use to measure them? How much do you think you would need of each if you were making enough for four people?* Use an example of a pancake recipe to compare with students' answers.

 **Main activity**

Arrange students into mixed-attainment groups of four. Encourage them to try to interpret and begin the activity without any teacher input.

As students complete the task on page 129 of the Student Book. Ask questions, for example:

- *What do you notice about the ingredients in the instructions?*
- *What do you notice about the scales on the measuring equipment?*
- *What is different about the instructions and the scale on the weighing scales?*
- *What do you think Ibrahim should do?*
- *How can you change the masses in the instructions to help Ibrahim weigh the ingredients on his scales?*
- *How can you change the capacities in the instructions to help Ibrahim measure the ingredients in his jug?*

When groups have finished the task, ask them to rewrite both sets of instructions for different numbers of cakes, for example: 2 cakes, 10 cakes, 5 cakes and 15 cakes. Encourage students to use their knowledge of doubling and multiplying to do this (for example doubling for 2, multiplying by 10 for 10, halving that for 5, adding 10 and 5 for 15).

Unit 5 Length, mass, capacity and volume

Take feedback from the activity. Ask groups to show the measurements for the instructions that will enable Ibrahim to make his cake. Invite students to share their new instructions.

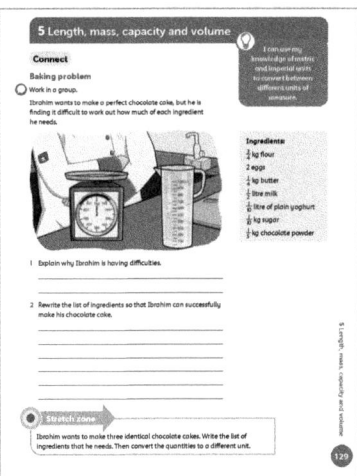

Differentiation

Supporting: Give students blank double number lines and show them how to use these as appropriate to help them convert amounts.

Consolidating: Ask students to support a less-confident student by explaining how to convert some of the ingredients.

Extending: Ask students to rewrite the recipe in several ways using metric units of measure.

Stretch zone: *Ibrahim wants to make three identical chocolate cakes. Write the list of ingredients that he needs. Then convert the quantities to a different unit.*

Students can make up their own problems similar to Ibrahim's that involve length, mass and capacity.

Reflection time

Ask students to talk about the different areas of measurement they have been studying in this unit. Which areas did they find easiest? Which were most difficult? Why?

Differentiated outcomes	
All students	should participate in group work and be able record the recipe using different units with support.
Most students	will convert ingredient amounts to a different unit of metric measure.
Some students	may be able to describe ingredient amounts in several different ways.

Answers

Student Book page 129

1. The list uses fractions. The marks on the scales are in whole grams and millilitres.

2. Students should rewrite Ibrahim's instructions so that he can successfully make his chocolate cake.

Unit 5 Length, mass, capacity and volume

5 Length, mass, capacity and volume

Review Student Book page 130 • Practice Book page 107

Global skills

- **Real-world skills:** presenting information/interpreting information
- **Interpersonal skills:** communication
- **Self-development skills:** reflecting on learning

Student Book

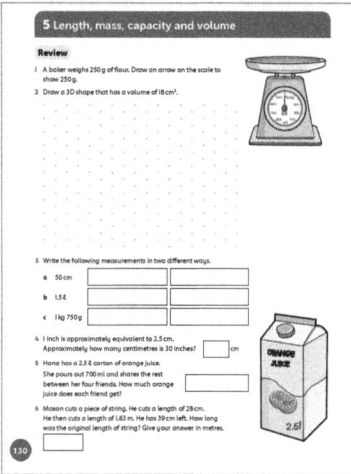

With young students, assessment activities are most effective when carried out as an everyday classroom activity. Students should be able to explore measures in different contexts and extend their previous knowledge to include conversions between metric units and from metric to imperial units.

As they work on the Student Book activity on page 130, they should be able to use their knowledge of measures and units. If some students struggle to draw a shape on isometric paper with a volume of 18 cm³, you may prefer to give them interlocking cubes and ask them to build it.

Point out how measures link to other units in mathematics, such as adding and subtracting, multiplying and dividing.

Answers

Student Book page 130

1. Check that the arrow is marked correctly halfway between the 200 g and 300 g marks.
2. Check that the drawing is of a 3D shape of the correct volume.
3. **a** 0.5 m, 500 mm **b** 1500 ml, 1 1500 ml, 1 1/2 l **c** 1.75 kg, 1750 g
4. 75 cm
5. 450 ml
6. 2.3 m

Practice Book

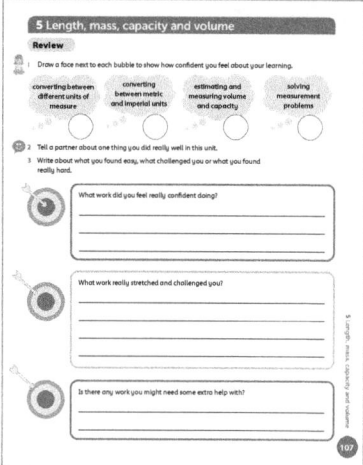

With students in the upper primary years, it is appropriate to complete this as a whole-class discussion. You may choose to keep a record of the class discussion or a copy of the Review page for your own records. Use the Student Book to briefly remind students of the areas of mathematics that they have worked on in this unit.

Ensure that students have a copy of the Student Book to support them as they discuss and answer the questions in the Practice Book.

Allow students plenty of time for discussion before asking them to share their responses with the rest of the class. If students complete this assessment at home, encourage them to discuss this with adults.

Make a note of areas that students still feel unsure about. As measures are found in many everyday contexts, revisit these ideas regularly. Build estimating and measuring into other aspects of everyday practice, for example when sharing things or finding a part of an amount or quantity.

Additional material

There are additional end-of-unit assessment available on the *Oxford Owl for School* website.

Unit 5 Length, mass, capacity and volume

6 Area and perimeter

Overview

Big idea

The Big idea for this unit is the development of students' knowledge and understanding of perimeter and area. Perimeter is the distance around the outside of an area (shape). Students will find perimeters of regular and irregular polygons by measuring and developing the formula. Area is the amount of space covered by a shape. Students will develop the use of appropriate formulae.

The relationship between perimeter and area is not a simple one and this is partly because the two measures have different dimensions. Perimeter is a one-dimensional measure of length, and hence is measured in centimetres or metres, or other equivalent measures in metric or imperial units. Area, on the other hand, is a measure of two-dimensional space. In starting to compare and measure areas, students are introduced to square units, such as square centimetres, written cm^2, or other equivalents such as m^2.

The relationship between two measures in different dimensions can cause intuitive ideas to be false, and students may assume that if the perimeter increases, then so will the area.

Look out for

- **Students who continue to need to count squares to find areas because they cannot see the relationship with multiplication.** Provide opportunities for them to look at different arrays when finding areas, using square units for rectangles and squares.

- **Students who confuse the terms 'perimeter' and 'area'.** Using the idea of fences to represent perimeter and fields to represent area can be a useful analogy. Help students to visualise the concepts. Tell them to sit on the perimeter of the carpet area when you want them to work in a circle, or to sit in the area of the carpet when you want them to fill the space.

Possible misconceptions

- **Students may think that different shapes have different areas.** This is not necessarily so. Give them pieces of paper that are the same size to cut up and rearrange so that they can appreciate that different shapes can have the same area.

- **Students may think that there is a relationship between the area and perimeter of shapes.** In fact, there is no relationship. For any given perimeter, there is a range of different areas. For any given area, there can be a range of perimeters. Students need to explore this early on in their experiences of these concepts. You could give students four small pieces of paper that are the same size. Tell them to place these side by side in different ways. This enables them to see that the perimeters are different simply by counting the edges of the pieces of paper (but also to see that the areas are the same).

Key vocabulary

- perimeter, estimate
- millimetres (mm), centimetres (cm), metres (m)
- composite shape, formula, rectilinear, regular, irregular, scale
- area, surface, square units, square centimetre (cm^2), square metre (m^2)

Coverage in lessons

Learning objective	E	6A	6B	6C	C	R
Measure and calculate the perimeter of composite rectilinear shapes in centimetres and metres.	✓	✓		✓	✓	✓
Calculate and compare the area of rectangles (including squares), and including using standard units, square centimetres (cm^2) and square metres (m^2) and estimate the area of irregular shapes.	✓		✓	✓	✓	✓

6 Area and perimeter

Engage Student Book page 131

Big question
- How do I calculate areas and perimeters of rectilinear shapes?

Global skills
- **Creative skills:** investigating
- **Self-development skills:** reflecting on learning

Key vocabulary
- perimeter, millimetres, centimetres, metres, area, surface, square units, square centimetres (cm^2), square metres (m^2)

Resources
- mini whiteboards and markers
- cm-squared paper, rulers, scissors

Language support
Check students' pronunciation and understanding of the vocabulary, for example:
- perimeter: 'all the way around the outside'
- area: 'the amount of surface'.

Create a poster using the responses to the Engage activity that highlights the words 'area' and 'perimeter'.

 **Introductory activity**

Ask students to think about their classroom. Say that the classroom is going to be refurbished with a new carpet. The carpet will be fixed using a length of edging. Ask students to discuss in pairs how they might work out how much carpet and edging is needed. They can use their whiteboards to write down ideas.

After a short time, take some feedback from pairs, prompting them with questions such as, *How can we know how much carpet is needed? How do we calculate the amount of edging required?* Students might suggest counting how many equal carpet squares would cover the floor, or estimating a length in metres for the edging using the lengths and width of the classroom.

 **Main activity**

Look together at page 131 of the Student Book. If you have access to an IWB, display the page.

Introduce the Big idea of the unit to students. Read the text and then have students work in mixed-attainment groups of four.

Give each group sheets of cm-squared paper. Explain that they each need to make 6 squares with sides of 10 cm. Tell them to use their squares to make the same patterns as in the Student Book and to investigate whether shapes with the same number of squares have the same area and same perimeter. Encourage students to discuss in their groups how to organise the activity before they begin. Explain that they will need to share their thinking with the rest of the class later.

Once they have made the shapes they can draw them on squared paper, recording their dimensions as well as their areas and perimeters. As students work, circulate among the groups asking questions to assess their understanding of measure, for example:

- *This side of your shape is made up of three squares. How long does that make this side? How do you know?*
- *How can you calculate the perimeter of shape? What about its area?*
- *Which of these two shapes has the greater perimeter? Why do you think that is?*
- *How does knowing the area of one of your squares help you work out the area for an entire shape?*

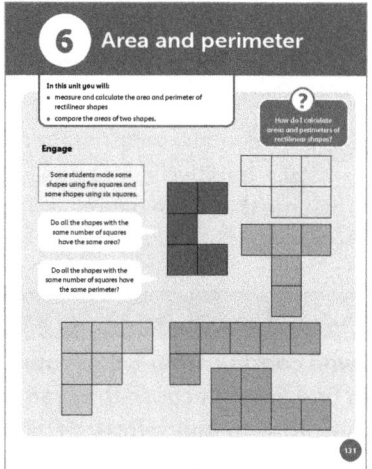

Differentiation
Supporting: Help students to find the area of one square by counting smaller squares and help them to see how they can use this to find the total area of their shape.

Consolidating: Ask students to explain the findings of the investigation and the fact that the areas are the same while the perimeters can vary.

Extending: Challenge students to find a systematic method to find perimeters and areas.

 Reflection time

Invite groups to share what they found out. Establish that the areas of their shapes are either 500 cm^2 or 600 cm^2, depending on whether they were made with 5 or 6 squares, and this does not change no matter how they are arranged. The perimeters are different depending on how the shapes are arranged. Ask, *Are you surprised by this? Why? Why do you think the perimeter is different for different shapes?*

170 Unit 6 Area and perimeter

6A Understanding perimeter

Discover Student Book pages 132–133 • Practice Book page 108

Specific learning focus
- Measure and calculate the perimeter of rectilinear shapes.

Global skills
- **Creative skills:** problem solving/exploring
- **Interpersonal skills:** communication/teamwork

Key vocabulary
- perimeter, centimetre (cm), estimate, rectilinear, composite shape

Resources
- mini whiteboards and markers
- cm-squared paper, rulers, scissors

Language support
Make a poster showing a range of **rectilinear** shapes with a perimeter of 24 cm to include as part of the classroom display for this unit's work. Write a definition of 'perimeter' and the formulae on the poster for students to refer to.

Introductory activity

Ask students to work in pairs and sketch as many **composite shapes** as they can with a **perimeter** of 28 cm. Explain that this means shapes made up of two or more shapes. Encourage students to explore a range of composite shapes, beyond using squares and rectangles, if appropriate. Take feedback. Ask students what they notice about the lengths of the sides of the different shapes. For example, with rectangles, the length and width together will equal half of the perimeter.

Main activity

Say, *I have a square. Each side measures 7 cm. What is its perimeter?* Ask students to write the answer on their whiteboards and then show you what they have written. Ask, *How did you work that out?* Agree that the perimeter is 28 cm because each side is 7 cm and 7 cm + 7 cm + 7 cm + 7 cm or four lots of 7 cm is 28 cm. Now say, *I have a rectangle that measures 12 cm × 8 cm. What is its perimeter?* Agree that it is 40 cm. Summarise that you can calculate the perimeter by adding the lengths of the sides and that you can also calculate the perimeter using a combination of adding and doubling, for example: double 12 + double 8 or double 12, double 8 then add. Ask, *Can you recall a formula for finding the perimeter of a square or rectangle?* Agree that for a length, l, and a width, w, a possible formula is: 2l + 2w or 2(l + w). Record all possibilities on the board. Use the formula to check the perimeters for the two shapes again, for example 2(12) + 2(8) = 24 + 16 = 40 cm.

Ask students to complete the activities on pages 132 and 133 of the Student Book in pairs. Read through each question to ensure that they are clear on what each student in a pair will be doing for each question.

Discuss the difference between an estimate and an actual perimeter. Ask, *How can you make a good **estimate**?* Next, look at the Think back formula. Ask, *For which questions could you use this formula?* Agree on question 3 because they are making rectangles. Suggest that, while they are working on question 4, they think about how they might be able to adapt their composite shapes so they can use the formula to help them find the perimeter.

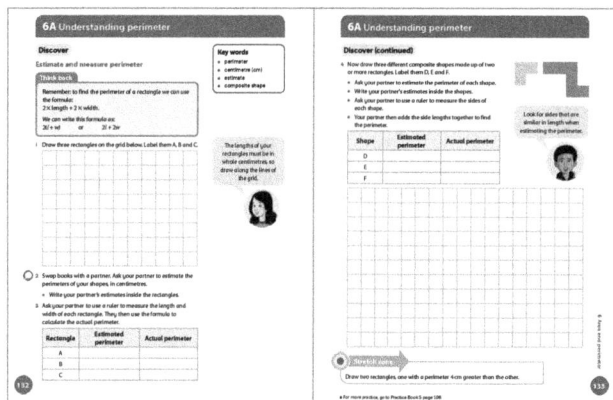

Differentiation

Supporting: Help students to check their measurements to find the perimeter.

Consolidating: Ask students to explain their method for finding the perimeter of some of their shapes.

Extending: Challenge students to compare perimeters using a formula.

Stretch zone: *Draw two rectangles, one with a perimeter 4 cm greater than the other.*

Ask students to draw a rectangle 4 cm by 2 cm, then explore how to find its perimeter using the formula. By changing the lengths of the sides, students can see how this changes the perimeter. For example, a rectangle 4 cm × 2 cm has a perimeter of 12 cm but a rectangle 4 cm × 3 cm has a perimeter of 14 cm.

Reflection time

Ask pairs, *How did you find the perimeters of your partner's rectangles?* They might explain to the rest of the class, for example, that they measured one length and one width, added them and then multiplied by 2. Some might say that they doubled each measurement and then added them.

Ask students how they estimated the perimeters of the composite shapes. What strategies did they use? Did they measure each side to work out the perimeter or did they use one side to work out the length of another?

Unit 6 Area and perimeter 171

Practice Book: Students complete Practice Book page 108. They can do this directly after the Main activity, as homework, or as the focus of a separate mathematics session to help students consolidate their learning and build fluency.

Students are asked to draw a rectangle with an area of 20 cm² as well as draw rectangles with perimeters of 16 and 24 cm. You may prefer to have students focus on perimeter only at this stage and come back to look at area once they have worked on area in more depth.

Differentiated outcomes	
All students	should draw rectangles and measure their perimeters accurately.
Most students	will draw rectangles and rectilinear shapes and estimate and measure their perimeters accurately.
Some students	may draw rectangles and rectilinear shapes and estimate and find the perimeters using the formula to find the perimeter of rectangles.

6A Understanding perimeter

Explore Student Book pages 134–135 • Practice Book page 109

Specific learning focus
- Measure and calculate the perimeter of regular and irregular polygons, including rectilinear shapes.

Global skills
- **Creative skills:** investigating
- **Interpersonal skills:** communication

Key vocabulary
- perimeter, centimetre (cm), estimate, regular, irregular, formula, formulae

Resources
- mini whiteboards and markers
- cm-squared paper, rulers, scissors

Language support
Support the language needed with sentence frames, for example:
- The perimeter of the <u>hexagon</u> is ___.
- The length of this side of the <u>triangle</u> is ___.

Answers

Student Book pages 132–133
Answers will vary. Check that students have made reasonable estimates and have calculated the perimeter of their shapes correctly.

Practice Book page 108
Answers will vary because students draw their own rectangles based on given perimeters and areas so the dimensions will vary. Check that they are correct.

Stretch zone: Students should notice that the perimeters are all even numbers.

 Introductory activity

Draw on the board a composite rectilinear figure, perhaps an L-shape, labelled with sufficient dimensions to be able to find the perimeter.

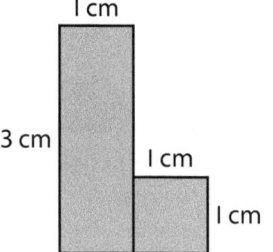

Ask students to work out the missing lengths of sides from those given and then calculate the perimeter of the shape. Discuss as a class how they worked them out. Repeat with another example.

Now give students a perimeter, say 18 cm, and ask them to draw a composite shape on squared paper with that perimeter. They should check with a partner that the shape has the correct perimeter. Ask, *Can you change your shape but keep the perimeter at 18 cm?*

 Main activity

Draw a regular triangle, pentagon, hexagon and octagon on the board. Label one side of each shape with a length. Give students a few minutes to work out the perimeters and write them on their whiteboards. Ask, *How did you work out the perimeter of the triangle?* Agree that as each side is the same length because it is a **regular** shape, they can multiply the length of the side by the number of sides (3 in the case of the triangle). Ask, *Can you make up a **formula** for finding the perimeter?* Agree that it is

172 Unit 6 Area and perimeter

3l (three times the length). Repeat for the perimeters of the pentagon (5l), hexagon (6l) and octagon (8l). Ask questions about the perimeters of these regular shapes using different lengths, such as: *My triangle has sides of 8 cm. What is its perimeter?* (3l = 24 cm). *My hexagon has sides of 12 cm. What is its perimeter?* (6l = 72 cm). Ask, *Will these **formulae** work for all shapes?* Establish that they only work for regular shapes because **irregular** shapes have sides of different lengths.

Students should now complete the activities on pages 134–135 of the Student Book. They should work in pairs and take it in turns to carry out the calculation and to check their partner's measurements and calculations. They should check to make sure that each pair of shapes they draw for page 135 have matching perimeters.

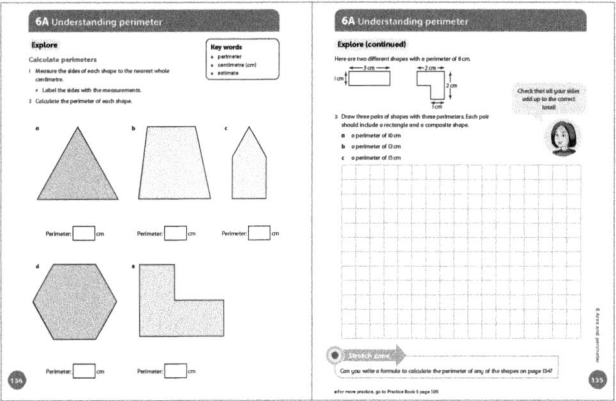

Differentiation

Supporting: Help students to check their measurements and to explain how they are finding the perimeters.

Consolidating: Ask students to explain their strategy for finding the perimeter of regular shapes and explain how they decided on the length of a side of a shape to make a shape with a set perimeter.

Extending: Challenge students to find the perimeters using a formula and make shapes that include sides of lengths to the nearest $\frac{1}{2}$ cm or $\frac{1}{4}$ cm.

Stretch zone: *Can you write a formula to calculate the perimeter of any of the shapes on pages 134?*

Students may recognise that for irregular polygons, there is no formula for perimeters, they simply add each of the side lengths to get a total. You can extend further by asking students to explore other quadrilaterals that have the same formula for finding their perimeter as for squares and rectangles, for example parallelograms, rhombi and kites.

 Reflection time

Ask, *What strategies did you use to find the perimeter of the different shapes in this lesson? Did you use the same strategy for each shape? If not, why not?* Some students will have used formulae and some will have used efficient calculation methods to add sides when a formula doesn't work.

Practice Book: Students complete Practice Book page 109. They can do this directly after the Main activity, as homework, or as the focus of a separate mathematics session to help students consolidate their learning and build fluency.

Students will need a ruler or a straight edge to complete this activity. Draw their attention to the text: 'On this page, one square represents 1 cm^2.' Can they explain what this means? Confirm that this means that if they connect four dots to make a square, it will represent an area of 1 cm^2 and each side will represent 1 cm in length.

Differentiated outcomes	
All students	should measure the perimeter of the shapes accurately with support.
Most students	will measure the perimeters of polygons accurately and draw pairs of shapes with perimeters that are equal.
Some students	may estimate accurately and find perimeters of shapes using the formula, and recognise which shapes have a perimeter formula they can use.

Answers

Student Book pages 134–135

Check that students have measured and calculated the perimeters correctly.

Practice Book page 109

Check that students have drawn four different rectilinear shapes with a perimeter of 25 cm.

Stretch zone: Each side is 6.25 cm because a square has 4 sides all the same length and $\frac{1}{4}$ of 25 cm is 6.25 cm.

Unit 6 Area and perimeter

6B Understanding area

Discover Student Book page 136 • Practice Book page 110

Specific learning focus
- Understand that shapes of different areas may have different perimeters.

Global skills
- **Creative skills:** investigating
- **Interpersonal skills:** communication
- **Self-development skills:** reflecting on learning

Key vocabulary
- perimeter, area, centimetre (cm), estimate

Resources
- cm-squared paper
- rulers, paper, sticky tape or glue

Language support
Support students' understanding of the units for perimeter and area, modelling the correct way of saying them, for example:
- Perimeter is measured in centimetres.
- Area is measured in centimetres squared.

Introductory activity

Ask students to talk in pairs and think of some practical examples of area and perimeter. Examples might include a garden surrounded by a fence (the **area** is the amount of garden and the fence is the perimeter), or a framed painting (the frame is the perimeter and the glass is the area).

Main activity

Give pairs of students sheets of cm-squared paper. Say that each square on the paper has sides that measure 1 cm so the area of each square is 1 cm². Ask them to draw a square of any size. Ask, *How can you find the perimeter and area?* Agree that for the perimeter they can find the length of the sides and multiply by 4 (4l). For the area, as they know that 1 square has an area 1 cm², they can count the number of squares to find the total area. Ask students to find the perimeters and areas of their squares individually. Then ask them to write these using the appropriate units (cm and cm²). They can then give them to a partner to check.

Ask pairs to draw as many rectangles as they can with an area of 16 cm². They should write down the perimeters of each of these rectangles. They then draw as many rectangles as they can with a perimeter of 16 cm. They should write down the areas of each of these rectangles.

They can then use this activity to help them complete the questions on page 136 of the Student Book. Before they begin, read through the instructions together and ask them to turn to their partner and predict what the answer will be to question 1, Do shapes with larger areas have larger perimeters? Ask them to explain their thinking to their partner.

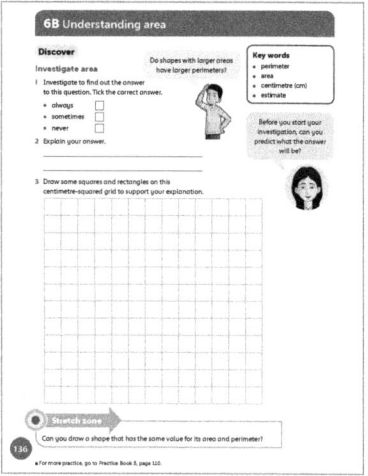

Differentiation

Supporting: Help students to record their explanation for how they are calculating the perimeters and areas.

Consolidating: Ask students to estimate before finding the perimeters and areas of their shapes and explain how they made their estimates. Were they reasonable?

Extending: Challenge students to think about what the formula for area might be.

Stretch zone: *Can you draw a shape that has the same value for its area and perimeter?*

Encourage students to work systematically to find examples, focusing on rectangles. Students may find that both a 4 × 4 and 3 × 6 rectangle have perimeters and areas with the same value.

Reflection time

Discuss the question: Do shapes with larger areas have larger perimeters? Ask students to explain their thinking to the class. Invite students to share examples of when the area gets bigger and the perimeter doesn't (and vice versa). Agree that it is sometimes true that shapes with larger areas have larger perimeters. Invite any students who completed the Stretch zone activity to share their findings.

Practice Book: Students complete Practice Book page 110. They can do this directly after the Main activity, as homework, or as the focus of a separate mathematics session to help students consolidate their learning and build fluency.

174 **Unit 6** Area and perimeter

This reflects closely what students will do in the next lesson. It may work well as a pre-learning activity or you could assign it after they have completed 6B Explore.

Differentiated outcomes	
All students	should compare perimeters and areas accurately with support.
Most students	will compare perimeters and areas accurately using appropriate units.
Some students	may find perimeters and areas and generalise their findings about the relationship between areas and perimeters.

Answers

Student Book page 136

Shapes can have areas that are the same but have different perimeters and shapes can have the same perimeters but different areas. There is no relationship between the two so sometimes a perimeter gets bigger as the area increases, and sometimes it doesn't.

Example:

A 3×10 rectangle has an area of 30 and a perimeter of 26.

A 6×6 square has an area of 36 and a perimeter of 24, which is smaller than the rectangle above that has the smaller area.

Practice Book page 110

1 $24\,cm^2$

2 $24\,cm^2$

3 $36\,cm^2$

4 $66\,cm^2$

Stretch zone: Answers will vary. Check that students have drawn rectangles with an area of $30\,cm^2$.

6B Understanding area

Explore Student Book pages 137–138 • Practice Book page 111

Specific learning focus
- Calculate rectangle areas using the formula 'length × width'.

Global skills
- **Creative skills**: investigating
- **Self-development skills**: reflecting on learning

Key vocabulary
- area, square centimetre (cm^2), formula, scale

Resources
- cm-squared paper
- rulers
- Resource sheet 6.1: tangram pieces – one per student, with the shapes cut out (and laminated, optionally)
- mini whiteboards and markers

Language support

Support the language needed to talk about area, for example:
- The area of a shape is … cm^2
- The length of the shape is … cm.
- The width of the shape is … cm.

 Introductory activity

Draw a 6 × 6 square grid on the board. Say, *This represents a square with sides measuring 6 cm. What is its area? How can you calculate its area without counting each square?* Encourage students to explain that there are six rows of six squares and six lots of 6 are 36, so the area is 36 **square centimetres (cm^2)**, which is written as $36\,cm^2$. Repeat for different squares.

 Main activity

Now say, *I have a rectangle. It is 8 cm long and 4 cm wide. What is its area?* Agree that it is four lots of 8, which is $32\,cm^2$. Write 8, 4 and 32 on the board. Ask, *What do you notice about the relationship between these three numbers?* ($8 \times 4 = 32$) *So what can you multiply to get the area of $32\,cm^2$?* Agree that you can multiply the width by the length. *Do you think this is true of all rectangles?* Check with other examples to agree it is. Ask, *So what is the formula for the area of a rectangle?* Agree that it is $l \times w$ (length × width).

Refer students to page 137 of the Student Book. Draw their attention to the box that says 1 square = $1\,cm^2$. Ask, *What does this tell us?* (1 square has an area of $1\,cm^2$) *So if we find the area of a shape by multiplying the length times the width, what is the length and width of this square?* (1 cm) Ask students to work independently to count the squares and use the formula to work out the area of each shape for question 1 and then complete question 2. They can work on the remaining questions in pairs, first explaining their answers to question 2 to their partner.

Unit 6 Area and perimeter 175

After they have completed page 137, ask students to look at question 3 on page 138 of the Student Book. Students have already used tangram pieces in Unit 4, but remind them that the set of shapes is called a tangram puzzle. Explain that it originated in China hundreds of years ago. Discuss with students the set of pieces, their names, and their areas in relation to each other and to the whole square.

Ask, *How would you find the area of each piece if the side of the square is 2 units?* Leave students to think about this and discuss it in pairs first. Move around the class, supporting them as needed.

Next, students can make their own tangram shapes using cut-out pieces from Resource sheet 6.1. Discuss what **scale** means and provide a level of support matching how much experience your students have working with scale.

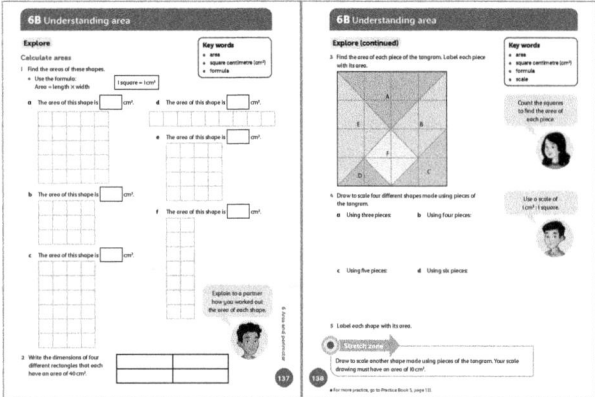

Differentiation

Supporting: Help students to find areas of the tangram pieces by counting full and part squares.

Consolidating: Ask students to explain how they calculated the area of the tangram pieces.

Extending: Challenge students to make pairs of shapes using the tangram pieces so that each shape is half the area of the whole square.

Stretch zone: *Draw to scale another shape made using pieces of the tangram. Your scale drawing must have an area of $10\,cm^2$.*

Continue to encourage students to use the tangram pieces to explore possible shapes. Ask them to explain how they know that their shape has an area of $10\,cm^2$.

 Reflection time

Ask, *What formula did you use to find the areas of the shapes in the student activity?* Write this on the board: $l \times w = 30\,cm^2$. Ask students to work in pairs to find out the possible length and width of a rectangle with an area of $30\,cm^2$. Tell them to write these on their whiteboards. Agree that these can be any combination of factor pairs for 30 (for example: 1×30, 2×15, 3×10, 5×6).

Practice Book: Students complete Practice Book page 111. They can do this directly after the Main activity, as homework, or as the focus of a separate mathematics session to help students consolidate their learning and build fluency.

Support students by prompting them to count the whole squares within each shape first, then to try to identify half-squares, quarter-squares and so on.

Differentiated outcomes	
All students	should find areas of rectangles by counting squares with support.
Most students	will explain how they calculated areas of rectangles by counting squares.
Some students	may estimate and calculate areas of shapes by counting part-squares and areas of rectangles using a formula.

Answers

Student Book pages 137–138

1. **a** $25\,cm^2$ **b** $12\,cm^2$ **c** $24\,cm^2$ **d** $9\,cm^2$
   e $16\,cm^2$ **f** $14\,cm^2$

2. Possible answers: $10\,cm \times 4\,cm$, $8\,cm \times 5\,cm$, $20\,cm \times 2\,cm$, $40\,cm \times 1\,cm$

3. $A = 4\,cm^2$, $B = 3\,cm^2$, $C = 2\,cm^2$, $D = 1\,cm^2$, $E = 4\,cm^2$, $F = 2\,cm^2$

4. Check that students have drawn the shapes correctly.

5. Check that students have labelled each shape's area correctly.

Practice Book page 111

Answers will vary because students estimate the areas and then refine their estimates by counting squares.

6C Calculating area and perimeter

Discover
Student Book pages 139–140 • Practice Book page 112

Specific learning focus
- Explore possible perimeters based on given areas of rectangles as part of a patio design.

Global skills
- **Creative skills:** problem solving/exploring
- **Real-world skills:** interpreting information
- **Interpersonal skills:** communication/teamwork

Key vocabulary
- perimeter, area, scale, rectangle

Resources
- mini whiteboards and markers
- rulers, pencils
- grid paper

Language support
Ensure that students are clear on the context of the problem in the Student Book so that they are better able to focus on the mathematics. Show them photographs of patios in a garden setting to support their understanding.

Introductory activity

Ask students to work in pairs on the following problem.

Haris wants to plant grass seed in his garden. He also wants to build a wall around his garden. His garden is a rectangle measuring 8 m by 4.5 m. He needs to know the perimeter and area so he can buy the grass seeds and bricks.

Encourage students to draw diagrams on their whiteboards to help them solve the problem. Take feedback and agree that the perimeter is 2(8 m + 4.5 m) = 25 m, and the area is 8 m × 4.5 m = 36 m². Some pairs may have counted squares on their diagram to solve the problem. Relate the counting squares to the formula.

Main activity

Explain the activity on pages 139–140 of the Student Book, developing the context and demonstrating the example on the board. Students should work in pairs to find all the possible rectangles for Sofia's patio. For each one, they need to work out the perimeter. If necessary, help them to work systematically. Ask questions to support, consolidate and challenge, for example:
- *Which multiplication facts do you know that equal 24? How can knowing these help you work out the possible side lengths of the patio?*
- *If I decide the length of one side is 4 m, what would the length of the other sides be?*
- *How can you decide what the side lengths of the patio could be?*
- *Why wouldn't 24 m be a good length for the patio?*

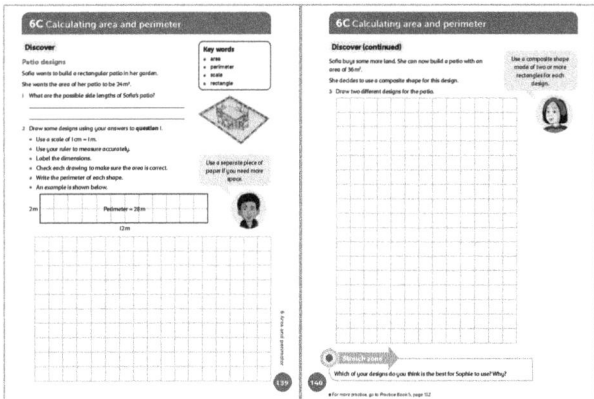

Differentiation
Supporting: Help students to find possible shapes for the patio by looking at multiplication facts.

Consolidating: Ask students to explain their choice of patio dimensions.

Extending: Challenge students to explore a variety of composite solutions for question 3 and calculate the perimeter of the shapes if 1 cm = 1 m.

Stretch zone: *Which of your designs do you think is the best for Sofia to use? Why?*

Support students by providing them with any additional vocabulary relating to the context to explain their choice.

Reflection time
Take feedback from the activity. Invite students to share some of their whole-number perimeters for Sofia's patio area. Ask, *Which is the best shape for Sofia's patio and why?* Agree as a class which solution is the 'best'. For example, Sofia could make a patio of 6 m × 4 m because this gives the nearest shape to a square and the others might be too long and thin.

Practice Book: Students complete Practice Book page 112. They can do this directly after the Main activity, as homework, or as the focus of a separate mathematics session to help students consolidate their learning and build fluency.

Check that students understand the context before they begin the activity.

Unit 6 Area and perimeter

Differentiated outcomes	
All students	should explore possible patio dimensions from multiplication facts with support.
Most students	will explain their strategies for choosing the patio dimensions with a given area.
Some students	may find a range of possible composite shapes for a given area.

Answers

Student Book pages 139–140

Possible rectangles:

1. 1 m × 24 m: perimeter 50 m, 2 m × 12 m: perimeter 28 m
   3 m × 8 m: perimeter 22 m, 4 m × 6 m: perimeter 20 m
2. Check that student's drawings match the dimensions given and are to scale.
3. Check that students' composite shapes have an area of 36 m².

Practice Book page 112

Check that students have tried different arrangements for the fence. A field of length 10 m and width 8 m has an area of 80 m².

Stretch zone: Check that students have found suitable shapes with the correct perimeter and areas.

6C Calculating area and perimeter

Explore Student Book pages 141–142 • Practice Book page 113

Specific learning focus
- Estimate and measure the perimeters and areas of rectilinear shapes.

Global skills
- **Creative skills:** exploring
- **Interpersonal skills:** communication/teamwork

Key vocabulary
- perimeter, area, scale, composite shape, rectilinear, diagonal

Resources
- rulers, pencils
- cm-squared paper

Language support

Provide stem sentences to help students explain their methods.
- I estimated the perimeter by … .
- I estimated the area by … .
- The difference between my estimate and the actual perimeter was … .
- The difference between my estimate and the actual area was … .

 **Introductory activity**

Review the formulae for the areas and perimeters of squares and rectangles. Focus on the units.

- *The perimeter is a length. You measure it in, for example, millimetres, centimetres, metres and kilometres.*
- *You measure area in squares of these lengths. You write these as, for example, mm², cm², m² and km².*

Ask pairs to think of examples of areas that are measured in the different units and record them on the board. For example, they might suggest: square millimetres (stamp), square centimetres (book), square metres (football pitch) and square kilometres (country).

 **Main activity**

Give each pair of students a piece of cm-squared paper. Ask them to draw a triangle on it. Ask them to find the perimeter and area. Take feedback. Agree that you can measure the sides with a ruler and total them for the perimeter. For the area, agree that you can count the squares, matching up part squares that are approximately a whole. Ask students to do this. Take feedback. Ensure that they use the correct units.

Explain the tasks on pages 141–142 of the Student Book. Ask students to read the text in the speech bubble on page 141. Explain that this means that in this case, each square will not actually have 1 cm sides but they should think of the squares as having 1 cm sides when they calculate each shape's perimeter and area. Students work on the first two questions independently, estimating the perimeters and areas of two shapes. Working in pairs for questions 3 and 4, they each draw their own rectilinear shapes with a perimeter of 44 cm and estimate its area. They then swap with their partner and each estimate the area, then calculate it together. They should work in pairs and check each other's solutions.

Unit 6 Area and perimeter

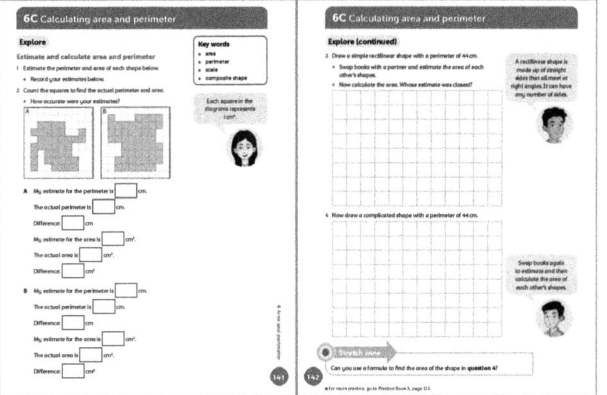

Differentiation

Supporting: Help students with their estimates and calculating the perimeters and areas.

Consolidating: Ask students to explain how they made their estimates before finding the perimeters and areas.

Extending: Ask students to describe how they calculated the perimeters and areas of all the shapes they made.

Stretch zone: *Can you use a formula to find the area of the shape in question 4?*

Students should be able to find a shape with a perimeter of 44 cm by adjusting a simpler shape. The area will be calculated by finding areas of each part of the composite shape as there is no formula available.

 Reflection time

Ask pairs for feedback on the activity. Discuss how accurate their estimates were for question 2. Ask questions such as, *How did you find the perimeters and areas of the two shapes? Did you find a quick way to count the squares?* For question 3, students should share how they designed a 'simple' shape and calculated its area. Ask students to say whose strategy they think was more efficient. Discuss their strategies for finding the area of a 'complicated' shape.

Practice Book: Students complete Practice Book page 113. They can do this directly after the Main activity, as homework, or as the focus of a separate mathematics session to help students consolidate their learning and build fluency.

Before students begin, read the instructions and check that all students understand key vocabulary: composite, irregular and **diagonal**. Provide examples to support their understanding.

Differentiated outcomes	
All students	should measure their perimeters and areas accurately with support.
Most students	will estimate and measure their perimeters accurately using appropriate units.
Some students	may estimate accurately and find the perimeters and areas using the formula, where possible.

Answers

Student Book pages 141–142

1st shape: perimeter 42 cm, area 39 cm^2

2nd shape: perimeter 38 cm, area 52 cm^2

Other answers will vary depending on students' estimates for the answers above, then the shapes students draw.

Practice Book page 113

Answers will vary because students draw their own shapes. Check that students have found the correct perimeters and areas for each shape.

Stretch zone: Check that students have drawn two composite shapes with the same area and compared their perimeters.

Unit 6 Area and perimeter   179

6 Area and perimeter

Connect Student Book page 143

Big idea
I can find areas and perimeters of rectilinear shapes by counting squares or using a formula.

Global skills
- **Creative skills:** investigating
- **Real-world skills:** interpreting information
- **Interpersonal skills:** teamwork
- **Self-development skills:** reflecting on learning

Key vocabulary
- perimeter, centimetres, metres, area, square units, cm^2, m^2, formula

Resources
- cm-squared paper
- pencils, rulers

Language support
You may want to help write the answers for those less confident with writing or in speaking English.

 **Introductory activity**

Write these two statements on the board and read them aloud.

When the perimeter of a shape gets bigger, so does the area.

When the area of a shape gets bigger, so does the perimeter.

Put students into groups of three. Ask them to discuss whether these statements are always, sometimes or never true.

Discuss as a class. Encourage students to provide examples to support their opinion.

 **Main activity**

Draw students' attention to page 143 of the Student Book. Read the problem together and discuss the context briefly as a class. Explain that the sides of the pen can have lengths that are part metres as well as whole metres.

Organise students into small groups. Give each student rulers and cm-squared paper so that they can draw their designs on it. Remind them to establish a scale first before they draw their designs (this will likely be 1 cm = 1 m). Emphasise that their measurements should be accurate if working with whole units but, if they are working in part units, they can either enlarge their scale (perhaps 4 cm = 1 m so that 1 square = 0.25 m) or do their best to estimate fractions of a whole unit.

As students complete the tasks, ask questions such as:
- *What do you think is a good design for a chicken pen?*
- *Do you think that some dimensions might work better than others? Why?*
- *Does a size of 5 m × 10 m work? Why not?*

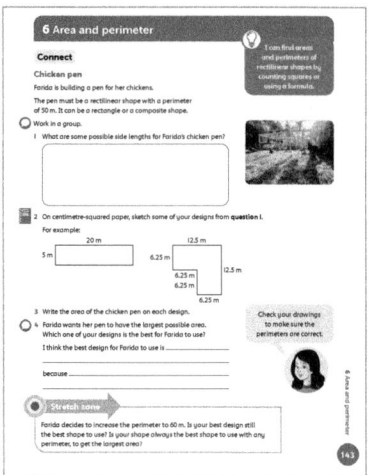

Differentiation

Supporting: Pair less-confident writers with students with strong writing skills who can support them to record their thinking for question 4.

Consolidating: Ask students to explain their choice of dimensions for the chicken pen. How do they know they are correct?

Extending: Students can research common as well as more unusual features you might include in a chicken pen. Using estimated dimensions for these items they can look at how they could adapt their design to include some of the features.

Stretch zone: *Farida decides to increase the perimeter to 60 m. Is your best design still the best shape to use? Is your shape always the best shape to use with any perimeter, to get the largest area?*

Students' answers will vary. Check their reasoning, either by looking at their written answers or asking them to share their thinking orally.

You could invite students to make up problems similar to Farida's where they need to explore different shapes with the same area but different perimeters.

 Reflection time

Invite groups to share the different areas and side lengths they found for the 50-metre perimeter chicken pen as well as designs they have chosen with the largest area. If no one has drawn a square, draw one on the board and label each side 12.5 cm. How does this area compare with their largest area?

Differentiated outcomes	
All students	should join in with group discussions.
Most students	will share ideas and explain their thinking.
Some students	will compare ideas and will support other members of their group to contribute effectively.

Answers

Student Book page 143

Possible areas:

$1 \times 24 = 24\,m^2$, $2 \times 23 = 46\,m^2$, $3 \times 22 = 66\,m^2$,
$4 \times 21 = 84\,m^2$, $5 \times 20 = 100\,m^2$, $6 \times 19 = 114\,m^2$,
$7 \times 18 = 126\,m^2$, $8 \times 17 = 136\,m^2$, $9 \times 16 = 144\,m^2$,
$10 \times 15 = 150\,m^2$, $11 \times 14 = 154\,m^2$, $12 \times 13 = 156\,m^2$

Possible chicken pen: 12 m × 13 m because it is the largest area so there is more space for the chickens to run around.

Stretch zone: Check students' reasoning used in their answers to the questions.

6 Area and perimeter

Review Student Book page 144 • Practice Book page 114

Global skills

- **Real-world skills:** presenting information/interpreting information
- **Interpersonal skills:** communication
- **Self-development skills:** reflecting on learning

Student Book

With young students, assessment activities are most effective when carried out as an everyday classroom activity. Students should be able to explore perimeter and area in different contexts and extend their previous knowledge of how to calculate perimeter and area.

Students will need rulers and pencils to complete this activity. Some students may benefit from having a separate sheet of squared paper so that they can do a rough sketch before copying it into their books. As they work on the Student Book activity on page 144, point out how perimeter and area links to other units in mathematics, such as adding sides and multiplying to find areas.

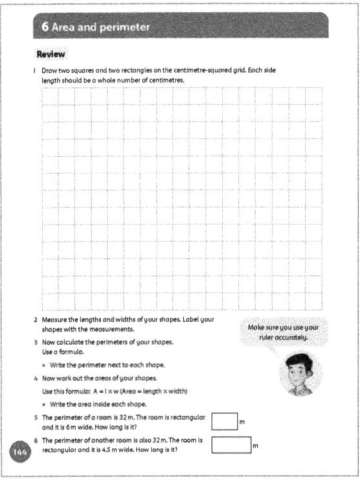

Answers

Student Book page 144

For questions 1–4, students draw their own shapes, so the answers for the perimeters and areas will vary.

1. Check that students have drawn two squares and two rectangles with whole number lengths.

2. Check that they have measured and labelled the lengths correctly.

3. Check that they have calculated the perimeter correctly.

4. Check that they have calculated the area correctly.

5. 10 m

6. 11.5 m

Practice Book

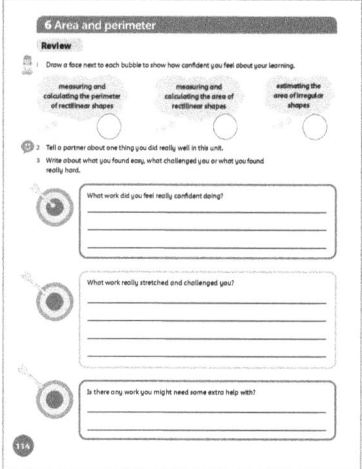

With students in the upper primary years, it is appropriate to complete this as a whole-class discussion. You may choose to keep a record of the class discussion or a copy of the Review page for your own records. Use the Student Book to briefly remind students of the areas of mathematics that they have worked on in this unit.

Unit 6 Area and perimeter 181

Ensure that students have a copy of the Student Book to support them as they discuss and answer the questions in the Practice Book.

Allow students plenty of time for discussion before asking them to share their responses with the rest of the class. If students complete this assessment at home, encourage them to discuss this with adults.

Make a note of areas that students still feel unsure about. As areas and perimeters are found in many everyday contexts, revisit these ideas regularly and build estimating and measuring into students' experiences with areas and perimeters.

Additional material

There are additional end-of-unit assessment available on the *Oxford Owl for School* website.

7 Time

Overview

Big idea

The Big idea for this unit is that time is the least perceived thing we measure, because we relate it to other things happening. With the two aspects of telling the time and recording an interval of time, measurement is involved in both. If we say 11 o'clock, we are saying that 11 hours have passed since midday or midnight.

Across this unit, students' knowledge and understanding of time will develop. They will learn about:

- time intervals: the length of time from one moment to another
- recorded time: the time at which something occurs.

Students will explore both aspects using 12-hour and 24-hour analogue (clocks with hands) and digital (numbers only) clock times. One of the best ways to support students as they learn to use measures of time is to have both analogue and digital clocks available and refer to them regularly. This is especially important in an age when students may not often see analogue clocks due to the pervasive presence of digital devices.

Look out for

- **Students who struggle to learn to tell the time and to convert between analogue and digital times.** Most people learn to tell the time eventually and usually do this because they need to in everyday situations. In this unit, students will focus on real-life situations for telling the time to support their learning.

Possible misconceptions

- **Students may think that all months have the same number of days or weeks**, which leads to difficulties with working out intervals of days or weeks. Give them plenty of experience of working with calendars.

- **Students may think that the time is same all around the world.** Use a globe to explain how it is daylight in a place when it is on the side of the earth facing the sun, and night-time when the place faces away from the sun, meaning that it is day and night at different times for different places.

- **Students may think that they can subtract the smaller time from the larger to find the difference.** Point out that the relationship between the units used means that this is not possible. Some students may find these units quite challenging to work with. Explain that time is a measurement that does not use metric units. There are 60 seconds in a minute, 60 minutes in an hour, 24 hours in a day and so on. This makes subtracting to find time differences difficult. In this unit, students will count on to find times and differences in times using number lines.

Key vocabulary

- day, week, weekend, fortnight, month, year, leap year, decade, century, millennium
- calendar, date
- a.m., p.m.
- timetable, arrive, depart
- hour, minute, second
- time interval, equivalent units, convert, equivalences, difference
- today, yesterday, tomorrow
- clock, digital time, analogue time, 12-hour time, 24-hour time, 24-hour clock

Coverage in lessons

Learning objective	E	7A	7B	7C	C	R
Solve problems involving converting between units of time.	✓	✓	✓	✓	✓	✓

7 Time

Engage Student Book page 145

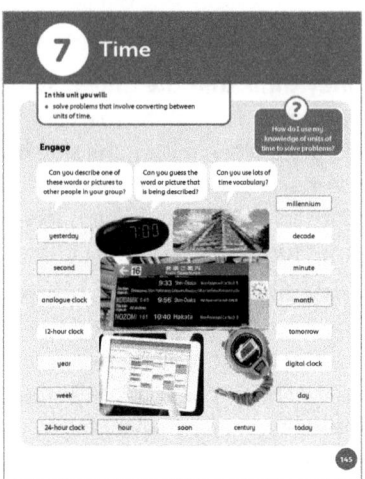

Big question
- How do I use my knowledge of units of time to solve problems?

Global skills
- **Creative skills:** problem solving
- **Real-world skills:** interpreting information
- **Interpersonal skills:** communication
- **Self-development skills:** reflecting on learning

Key vocabulary
- century, decade, year, leap year, month, week, day, hour, second, millennium, arrive, depart, 24-hour clock, time interval, today, yesterday, tomorrow

Resources
- mini whiteboards and markers
- large sheets of paper

Language support
Display the posters from the Introductory activity prominently in the classroom. Read through the sentences carefully with students, focusing on the pronunciation of the key words.

 **Introductory activity**

Ask students to work with a partner and to list on their whiteboards all the words they can think of that relate to time. Look together at page 145 of the Student Book. If you have access to an IWB, display the page. Pairs should check their list of words against the words listed on page 145. Can they use the new vocabulary?

Discuss what each word means as a class, emphasising the correct pronunciation of each. Students should share any additional time vocabulary as well.

Arrange students into mixed-attainment groups of four to create posters that include all the words on page 145 of the Student Book, written in sentences and with illustrations.

Main activity

Students stay in the same groups of four for this activity. Use the word list on page 145 of the Student Book to play a game. Students take it in turns to pick a word from the list. The rest of the group then ask questions to guess which word the student has selected. They should ask questions to which the only answers are 'yes' or 'no'. For example, they might ask:

Am you a type of clock? Do you have hands? Do you use a.m and p.m?

Are you a unit of time? Do you go by quickly? Are you longer than a day?

Support students to structure their sentences as necessary.

Differentiation
Supporting: Encourage students to read all the words from the key vocabulary list for time and to tell you their meanings.

Consolidating: Ask students to tell you another sentence containing each of the words in the key vocabulary list.

Extending: Ask students to tell you the relationships between different units of time and some conversions between different units of time.

Reflection time
Play the vocabulary game as a class. Choose a word and students ask you yes/no questions, for example: *Are you a unit of time? Are you longer than a day?* Choose a student to pick a word from the list and then you can join in, asking questions to guess the word.

7A Converting between units of time

Discover
Student Book page 146 • Practice Book page 115

Specific learning focus
- Recognise units of time and convert them to find equivalences (seconds, minutes, hours, days, months and years).

Global skills
- **Creative skills:** investigating
- **Self-development skills:** reflecting on learning

Key vocabulary
- equivalent units, equivalence, convert

Resources
- mini whiteboards and markers
- calculators

Language support
Display simple conversions as part of a working wall alongside a written description and an explanation of how to convert. For example, display:

2 minutes = 120 seconds

Two minutes equals one-hundred and twenty seconds

One minute equals sixty seconds. Two is double one so double 60 is how many seconds there are in two minutes.

Introductory activity

Use the word list on page 145 of the Student Book to describe a word by using its relationship with one of the other units of time. Students, working in pairs, write the word you are describing on their whiteboards. Allow a short amount of time for thinking and then ask students to show you their answers. If any student makes a mistake, use this as a teaching point. After a few turns, ask individual students to take over, picking the units of time to describe. Start with statements such as: 'There are 12 months in one of these.' (a year)

Main activity

Now develop the Introductory activity and make the statements more complex, for example: 'There are 48 months in 4 of these', 'There are 360 seconds in 6 of these'.

Ask students to write 'Weeks' in the middle of their whiteboards. Then ask them to write down any equivalents they know about weeks, for example 7 days make a week, 52 weeks make a year. Repeat using months, days and hours.

Give students two minutes to write facts for each of these. Invite individual students to share the facts they made. Ask the rest of the class whether they agree. Encourage them to explain their thinking.

Point out the activities on page 146 of the Student Book. Ask students to work in pairs to find **equivalences** between different units of time. Encourage them to use doubling, halving or addition to work out their answers.

Support students by asking questions such as:
- *How many minutes are there in an hour?*
- *How many months are there in two years?*
- *Which is longer, three weeks or 20 days? How do you know?*

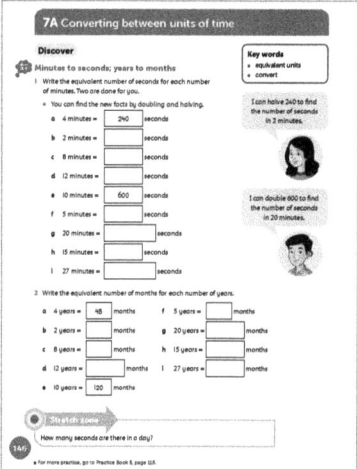

Differentiation

Supporting: Help students by building on known equivalences and modelling possible strategies for **converting** between units of time.

Consolidating: Ask students to explain their strategy for converting specific units of time.

Extending: Challenge students to create problems and solve more complex conversions involving mixed units of time.

Stretch zone: *How many seconds are there in a day?*

Students can work on this problem individually or in pairs. Ask them to explain how they converted between the two times to find that there are 86 400 seconds in a day.

Reflection time

Finish the lesson by reviewing the units of time covered. Write on the board: 1 year = 52 weeks. Ask, *What other information can you work out from this?* Write their ideas on the board, for example: 2 years = 104 weeks, 4 years = 208 weeks, 10 years = 520 weeks, 26 weeks = half a year.

Unit 7 Time

Practice Book: Students complete Practice Book page 115. They can do this directly after the Main activity, as homework, or as the focus of a separate mathematics session to help students consolidate their learning and build fluency.

Tell students that they may use a calculator to solve these problems. Students may find these questions quite challenging. For question 4, tell students to use the fact 1 year = 365 days as the start point for their calculations. You may prefer to work on these questions one at a time across the unit, possibly as a class.

Differentiated outcomes	
All students	should know the relationships between units of time.
Most students	will use the relationships between units of time to find equivalences.
Some students	may use the relationships between units of time to solve more complex equivalence problems.

Answers

Student Book page 146

1 a (Provided) 4 minutes = 240 seconds

   b 2 minutes = 120 seconds

   c 8 minutes = 480 seconds

   d 12 minutes = 720 seconds

   e (Provided) 10 minutes = 600 seconds

   f 5 minutes = 300 seconds

   g 20 minutes = 1200 seconds

   h 15 minutes = 900 seconds

   i 27 minutes = 1620 seconds

2 a (Provided) 4 years = 48 months

   b 2 years = 24 months

   c 8 years = 96 months

   d 12 years = 144 months

   e (Provided) 10 years = 120 months

   f 5 years = 60 months

   g 20 years = 240 months

   h 15 years = 180 months

   i 27 years = 324 months

Practice Book page 115

1 and **2** Answers will vary because students use their own ages and convert them into different units of time. Check that the conversions are reasonable.

3 10 080 minutes

4 31 536 000 seconds

Stretch zone: Approximately 11.5 days

7A Converting between units of time

Explore Student Book page 147 • Practice Book page 116

Specific learning focus

- Recognise units of time and convert them to find equivalences (seconds, minutes, hours, days, months and years).

Global skills

- **Creative skills:** exploring
- **Real-world skills:** interpreting information
- **Interpersonal skills:** communication

Key vocabulary

- hours, minutes, seconds

Resources

- mini whiteboards and markers
- calculators

Language support

Provide sentence frames to help students describe equivalences as well as their strategies for finding equivalences. For example:
___ is equal to __ hours and __ minutes. I know there are ___ minutes in __ hour, so ____ is equal to ____.

 **Introductory activity**

Ask students to work in pairs to estimate how many minutes they spend in school in a week and then use a calculator to check and write their answer on their whiteboards. Write the largest and smallest estimates on the board. Choose pairs to share how they made their estimate. Encourage students who used a sensible strategy to expand on it in more detail.

Now ask students to work in small groups on the question: *How many minutes do you spend in school in one day?*

They will need to use calculators for this activity. After a few minutes, ask one of the groups how they are carrying out the calculation. If needed, give the prompt that there are 60 minutes in an hour and 24 hours in a day, so they can calculate how many minutes they spend at school by converting from hours to minutes. They can do this by multiplying the number of whole hours by 60 and adding any additional minutes to find out how many minutes they spend in school in a day. They use the answer to calculate how many minutes they spend in school in a week.

 **Main activity**

Explain the activity on page 147 of the Student Book. Students work in pairs to convert numbers of minutes to hours and minutes. Look at the completed first row of question 1 together. Ask, *How could we have used what we know about minutes in an hour to work out the answer?* Agree there are 60 minutes in an hour, so they need to find out how many lots of 60 minutes there are in each amount to find the hours. Agree that what is left will be the number of minutes.

Encourage students to write the multiples of 60 on a separate sheet of paper to refer to throughout the activity. Discuss strategies for finding these. For example, they could begin by writing the multiples of 6 and then multiply each by 10. Explain as necessary, for example: *6 minutes × 2 = 12 minutes. 60 minutes is 10 times larger than 6, so if I multiply 60 × 2 my answer will also be 10 times larger, so 60 minutes × 2 = 120 minutes.*

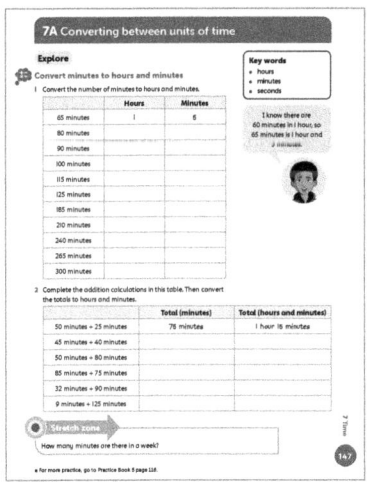

Differentiation

Supporting: Help students to create their list of multiples of 60 to support them. Revise the six times table first, as necessary, and record the products in a list, reminding them how to multiply each product by 10 to make a multiple of 60.

Consolidating: Ask students to explain their strategies for converting between hours, minutes and seconds.

Extending: Challenge students to record some of the times from the Student Book in hours, minutes and seconds.

Stretch zone: *How many minutes are there in a week?*

Students can calculate this by multiplying days (7) × hours (24) × minutes (60) × seconds (60), so 7 × 24 × 60 × 60 = 604 800.

 Reflection time

Take feedback from the activity, sharing answers and encouraging them to share their strategies. Ask questions such as:

- *How many minutes are there in 30 hours? How do you know?*
- *How many minutes are the same as 4 hours and 45 minutes? How did you work that out?*

Practice Book: Students complete Practice Book page 116. They can do this directly after the Main activity, as homework, or as the focus of a separate mathematics session to help students consolidate their learning and build fluency.

Students will continue converting numbers of seconds and hours to mixed units of time.

Differentiated outcomes	
All students	should understand the relationship between hours, minutes and seconds.
Most students	will convert numbers of minutes to mixed units of time.
Some students	may create their own problems based on converting between hours, minutes and seconds.

Answers

Student Book page 147

1. 80 minutes: 1 hour 20 minutes

    90 minutes: 1 hour 30 minutes

    100 minutes: 1 hour 40 minutes

    115 minutes: 1 hour 55 minutes

    125 minutes: 2 hours 5 minutes

    185 minutes: 3 hours 5 minutes

    210 minutes: 3 hours 30 minutes

    240 minutes: 4 hours

    265 minutes: 4 hours 25 minutes

    300 minutes: 5 hours

2. 45 minutes + 40 minutes: 85 minutes = 1 hour 25 minutes

    50 minutes + 80 minutes: 130 minutes = 2 hours 10 minutes

    85 minutes + 75 minutes: 160 minutes = 2 hours 40 minutes

    32 minutes + 90 minutes: 122 minutes = 2 hours 2 minutes

    9 minutes + 125 minutes: 134 minutes = 2 hours 14 minutes

Unit 7 Time

Practice Book page 116

1
- a 120 seconds: 2 minutes
- b 130 seconds: 2 minutes 10 seconds
- c 150 seconds: 2 minutes 30 seconds
- d 180 seconds: 3 minutes
- e 210 seconds: 3 minutes 30 seconds
- f 240 seconds: 4 minutes
- g 300 seconds: 5 minutes
- h 360 seconds: 6 minutes
- i 400 seconds: 6 minutes 40 seconds

2
- a 48 hours: 2 days
- b 50 hours: 2 days 2 hours
- c 60 hours: 2 days 12 hours
- d 72 hours: 3 days
- e 80 hours: 3 days 8 hours
- f 96 hours: 4 days
- g 100 hours: 4 days 4 hours
- h 110 hours: 4 days 14 hours
- i 124 hours: 5 days 4 hours

Stretch zone: 1000 hours = $41\frac{2}{3}$ days

7B Calculating time intervals

Discover
Student Book pages 148–149 • Practice Book page 117

Specific learning focus
- Calculate time intervals in minutes and hours using 24-hour formats.

Global skills
- **Creative skills:** problem solving
- **Real-world skills**: interpreting information

Key vocabulary
- time interval, number line, hour, minute

Resources
- sheets of paper, for example A4

Language support
Encourage students to refer to the illustrations of the word problems to support their understanding of the word problems. Provide explanations of any unfamiliar vocabulary as needed.

Introductory activity

Start to draw a double number line with 12-hour and 24-hour times on it, beginning with 12-hour times.

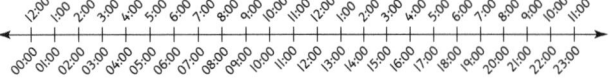

Ask students to tell you the equivalent 24-hour times to complete the number line. Say a time in one of the two ways and ask students to tell you the time in the other way. For example, say, *It's half past two in the afternoon. What time is it in 24-hour time?* Repeat with other times. Extend to asking simple word problems. For example, ask, *I left the house at thirteen thirty and returned after a half hour. What time did I arrive home in 12-hour time?*

 Main activity

Look together at the first word problem on page 148 of the Student Book. Explain that students need to work out the end time using the known **time interval** and the start time. Ask pairs to work on this together. When they have solved the problem, share methods and answers. If any pairs have made an error, use this as a teaching point. Revise again how to use a number line to work on the problem.

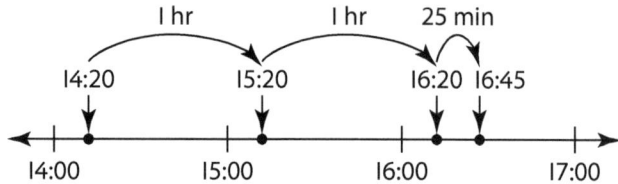

Students can then complete the remaining problems on pages 148–149 of the Student Book in pairs.

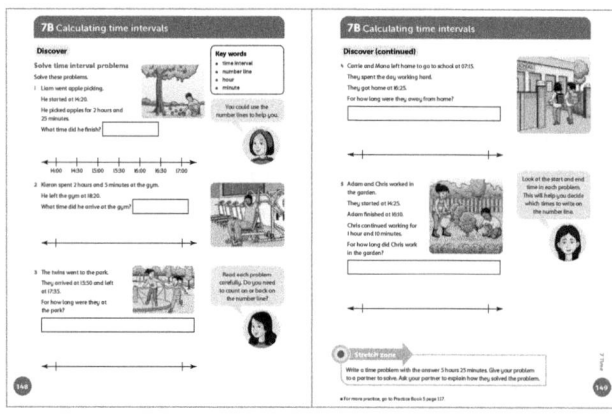

188 Unit 7 Time

Differentiation

Supporting: Work with students and model how to use a number line to carry out a calculation.

Consolidating: Ask students to explain their strategies.

Extending: Ask students to solve the problems mentally, explaining their strategies.

Stretch zone: *Write a time problem with the answer 5 hours 25 minutes. Give your problem to a partner to solve. Ask your partner to explain how they solved the problem.*

Choose any particularly successful word problems to share with the class to solve, leaving out that the answer will be 5 hours 25 minutes.

Reflection time

Choose students to model their solutions to questions 2, 3, 4 and 5. Again, take the opportunity to use any errors as teaching points, and ask students to comment on each other's methods.

Practice Book: Students complete Practice Book page 117. They can do this directly after the Main activity, as homework, or as the focus of a separate mathematics session to help students consolidate their learning and build fluency.

Encourage students to continue to draw number lines on separate sheets of paper to help them solve each problem.

Differentiated outcomes	
All students	should use number lines to solve problems with support.
Most students	will use number lines to solve problems in pairs.
Some students	may solve problems mentally and use number lines to check their answers.

Answers

Student Book pages 148–149

1 16:45

2 16:15

3 1 hour 45 minutes

4 9 hours 10 minutes

5 2 hours 55 minutes

Practice Book page 117

1 16:05

2 21:55

3 19:05

4 15:35

Stretch zone: Check that students have written suitable problems.

7B Calculating time intervals

Explore 1 Student Book pages 150–151 • Practice Book page 118

Specific learning focus
- Calculate time intervals in minutes and hours from digital time format using number lines.

Global skills
- **Creative skills:** problem solving
- **Real-world skills:** presenting information
- **Interpersonal skills:** communication/teamwork

Key vocabulary
- time interval, number line, hour, minute

Resources
- mini whiteboards and markers
- sheets of paper, for example A4

Language support

Model for students the use of language when calculating time differences on the number line, for example:
- *From 13:15 to 14:35, I jump one hour to 14:15 and then 20 minutes to get to 14:35.*
- *A film starts at 17:40 and lasts 2 hours 5 minutes. I can jump forward 2 hours to 19:40 and then jump on 5 more minutes to 19:45.*

 Introductory activity

Set the following problems.
- *Adam left home at 7:45 and arrived at school 35 minutes later. Show me the time he arrived at school.*
- *Registration was 20 minutes later and took 10 minutes. Show me when registration ended.*
- *His mathematics lesson began five minutes after registration and that lasted for one hour. Show me when the mathematics lesson finished.*

Unit 7 Time

Ask students to work on these problems in pairs, writing their answers on their whiteboards. They should show you the final times and record them in digital format. Go through the problems again. This time ask students to use a number line. Tell them to make the appropriate jumps for each length of time in the problems. Do they land on the same time as they did on their clock?

Main activity

Write on the board the 24-hour times:

08:15   11:35   13:20

Ask students to use a number line to work out how long it is from 08:15 to 11:35. Explain that this is called a **time interval**. Prompt them to count from 8:15 in whole hours and then count on any additional minutes. Ask two students to give their answers and explain how they worked them out. Agree that the time interval between 08:15 and 11:35 is 3 hours 20 minutes.

Now ask students to use a number line to work out the difference between 11:35 and 13:20 and share their answers and methods.

Students can then complete the word problems on pages 150–151 of the Student Book.

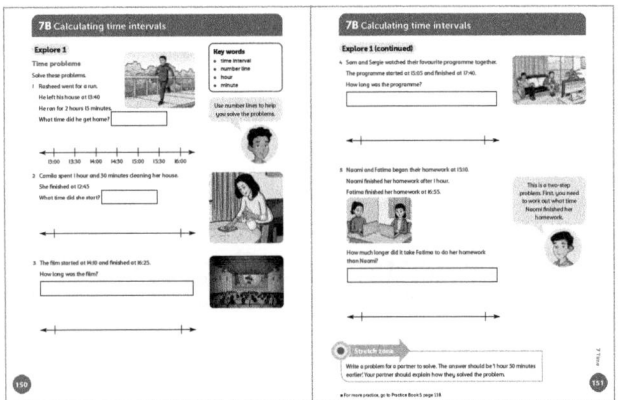

Differentiation

Supporting: Model again how to use the number line to find end times and lengths of intervals of time.

Consolidating: Ask students to explain how they used the number line to solve the problems.

Extending: Challenge students to solve the problems mentally before checking with a number line.

Stretch zone: *Write a problem for a partner to solve. The answer should be '1 hour 50 minutes earlier'. Your partner should explain how they solved the problem.*

Students can check each other's problems and compare strategies.

 Reflection time

Ask pairs of students to share their strategies and answers to the word problems. Ask how they decided what size jumps to make on their number line. How did they compare times?

Practice Book: Students complete Practice Book page 118. They can do this directly after the Main activity, as homework, or as the focus of a separate mathematics session to help students consolidate their learning and build fluency.

Encourage students to continue to draw number lines on separate sheets of paper to help them solve the problems. Students may need some additional support adding decimal numbers in the context of time. Encourage them to look for complements and near doubles and think about how they could partition to add. You may also want to remind them how to add decimal numbers using a column method.

Differentiated outcomes	
All students	should solve time word problems with support such as number lines.
Most students	will solve time word problems and explain their methods.
Some students	may create more complex time word problems.

Answers

Student Book pages 150–151
1 15:55
2 11:15
3 2 hours 15 minutes
4 2 hours 35 minutes
5 45 minutes

Practice Book page 118
1 1 hour 40 minutes
2 4 hours 17 minutes
3 08:10
4 32 minutes
5 1 minute 49.2 seconds

7B Calculating time intervals

Explore 2 Student Book page 152 • Practice Book page 119

Specific learning focus
- Calculate time intervals in seconds, minutes and hours using digital or analogue formats.

Global skills
- **Creative skills:** exploring
- **Real-world skills:** interpreting information
- **Interpersonal skills:** teamwork

Key vocabulary
- analogue, digital, difference

Resources
- mini whiteboards and markers
- analogue and digital clock faces
- 0–12 number cards (1 teacher set and sets for each pair)

Language support
Encourage those students who are less confident in English to read the sentences out in Reflection time. This allows them to practise the vocabulary.

 **Introductory activity**

Provide each student with an **analogue** clock. Call out some times, for example: *six ten a.m., ten twenty-seven in the morning, quarter past three in the afternoon.* Ask them to show the times on their analogue clocks and write the **digital** time on their whiteboards in both 12- and 24-hour time.

 **Main activity**

Display on the board a blank analogue clock and a blank digital clock. Ask a student to come forward and choose three number cards and use them to make a time, with one card as the hours and the others as the minutes. For example, if they chose 1, 2, 8, they might make 1.28 or 2.18. Discuss why they couldn't make 1.82. Ask another student to come forward and mark the hands on the analogue clock for the chosen time. They should explain how they know. Then a third student can come and fill in the time on the digital clock, again explaining how they know. Repeat this with three new number cards.

Now give each pair of students a set of number cards and ask them to complete the activities on page 152 of the Student Book. Encourage students to check with a partner that they have made possible times using the chosen number cards.

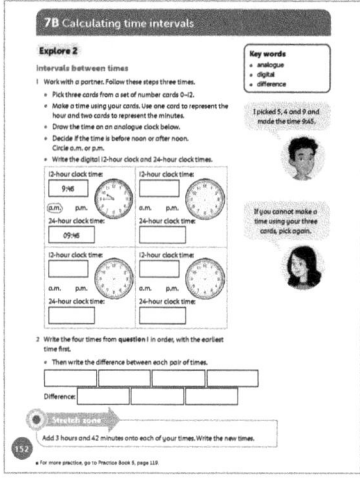

Differentiation

Supporting: Ask students to tell you aloud the times they make with the number cards. Students can continue to refer to the double time number line for additional support.

Consolidating: Ask students to tell you times that are earlier and later than the times they write.

Extending: Challenge students to make a time difference in minutes using number cards and then list pairs of times that have that difference.

Stretch zone: *Add 3 hours and 42 minutes onto each of your times. Write the new times.*

Answers will vary because students make their own times with number cards. Check that they are correct.

 Reflection time

Ask pairs of students to say a sentence using one of the times they have written. For example, a pair might say, 'I leave for school at 7.45 a.m.' Ask each pair to contribute a sentence, and the rest of the class can check each sentence for themselves.

Practice Book: Students complete Practice Book page 119. They can do this directly after the Main activity, as homework, or as the focus of a separate mathematics session to help students consolidate their learning and build fluency.

Encourage students to continue to use a number line, counting forwards and backwards to complete the table.

Differentiated outcomes	
All students	should tell and write the time in both formats.
Most students	will find time differences.
Some students	may create more complex problems involving time differences.

Unit 7 Time

Answers

Student Book page 152

Students use digit cards to make times, so answers will vary. Check that the times have been shown correctly in analogue and digital format.

Practice Book page 119

Activity	Time taken	Start time (24-hour clock)	End time (24-hour clock)
sleep	8 hours 10 minutes	22:30	06:40
have breakfast	25 minutes	07:05	07:30
walk to the bus stop	15 minutes	07:45	08:00
travel on the bus to work	15 minutes	08:05	08:20
drink coffee	10 minutes	08:25	08:35
read emails	25 minutes	08:45	09:10
go to a morning meeting	40 minutes	10:55	11:35
travel to head office	35 minutes	11:45	12:20
go to a meeting at head office	1 hour 40 minutes	12:25	14:05
have lunch	25 minutes	14:05	14:30
plan a presentation	1 hour 15 minutes	14:30	15:45
give the presentation	1 hour 10 minutes	16:00	17:10
write a report	50 minutes	17:15	18:05
travel home on the bus	15 minutes	18:10	18:25
walk home from the bus	15 minutes	18:25	18:40

Stretch zone: Check that students have completed an appropriate planner.

7C Using calendars

Discover Student Book page 153 • Practice Book page 120

Specific learning focus

- Use a calendar to calculate time intervals in days, weeks and months (using knowledge of days in calendar months).

Global skills

- **Creative skills:** problem solving
- **Real-world skills**: interpreting information

Key vocabulary

- calendar, date, difference

Resources

- calendars
- mini whiteboards and pens

Language support

Add a 12-month calendar to your working wall. Label it with statements to support students to talk about months, weeks and days, with a focus on ordinal numbers, and prepositions of position, for example: The 16th is the third Sunday of February. The 21st of May falls on a Wednesday.

 Introductory activity

Ask students to look at the **calendars** you have provided, or alternatively have them look at the calendar on page 153 of the Student Book. Some students may be used to using calendars on their tablets or laptops. Use these calendars, if appropriate.

Ask questions to monitor students' understanding of calendars. For example:

- *What is a calendar?*
- *Why are calendars useful?*
- *Which month comes first? Which is the eighth month of the year?*
- *Are there the same number of days in each month?*

Make up some questions to ask so that students can explore the calendar, for example:

- *What day of the week does the 1st of July fall on?*
- *What day of the week is the 14th of January?*
- *How many Mondays are there in March?*
- *Which is the shortest month?*
- *How many weeks are there between the 18th of May and 22nd of June?*
- *I leave on 10th of September and return the 5th of October. For how long was I gone?*
- *My holiday starts on the 22nd of June and lasts for 8 days. On what date does it end?*

Discuss their answers, including strategies for answering questions about time intervals such as counting on in whole weeks (moving down one row each time) and then counting on in days.

 **Main activity**

Ask students to work with a partner to make up five more questions about calendars and write them on their whiteboards. They should swap these questions with another pair and then check their answers.

Ask them to complete the activities on page 153 of the Student Book in pairs.

Ask students to find the length of time or difference between their birthday and another key date using the calendar. Take feedback on the strategies they used.

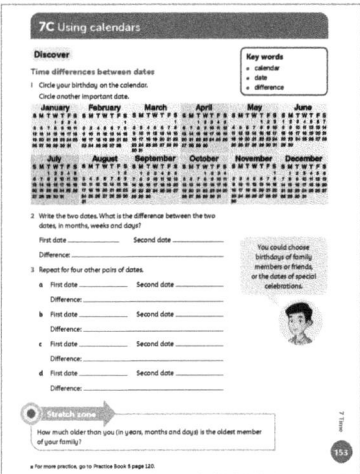

Differentiation

Supporting: Provide students with a list of key dates that they can refer to.

Consolidating: Ask students to explain how they are working out time differences.

Extending: Ask students to explain how they are working out time differences and to describe them using a range of units.

Stretch zone: *How much older than you (in years, months and days) is the oldest member of your family?*

Students will need access to calendars for multiple past years. (These are available on the internet.) If students are not sure what date to work to, provide them with a list of significant local or world figures' birthdates to choose from. Pairs of students can work together to find each other's age difference.

 Reflection time

Students should work in groups of four. They should find the time differences between all their birthdays and then put them in order, with the shortest time difference first.

Practice Book: Students complete Practice Book page 120. They can do this directly after the Main activity, as homework, or as the focus of a separate mathematics session to help students consolidate their learning and build fluency.

Differentiated outcomes	
All students	should understand how to use a calendar with support.
Most students	will use a calendar to find time differences.
Some students	may use a calendar to find time differences in different combinations of units of time.

Answers

Student Book page 153

Answers will vary because students use the dates of their own birthdays and other dates that are special to them. While students are working, ask them about these dates and check that the intervals have been calculated correctly.

Practice Book page 120

1 15 weeks 1 day
2 7 weeks 2 days
3 5 weeks 4 days
4 1 week 6 days
5 19 weeks 1 day

Stretch zone: Chidi and Farida's birthdays are closest together, 1 day apart.

Unit 7 Time

7C Using calendars

Explore
Student Book page 154 • Practice Book page 121

Specific learning focus
- Count on and back in days, weeks and months.
- Calculate time intervals in months or years.

Global skills
- **Creative skills:** problem solving
- **Real-world skills:** interpreting information
- **Interpersonal skills:** communication/teamwork

Key vocabulary
- calendar, date, year

Resources
- calendars for students to refer to
- mini whiteboards and markers

Language support
Ask students to say the dates aloud as they move around the calendar and use them in a sentence, for example: 'Three weeks and two days later, it is April the twenty-third.' Continue to model how to say dates frequently, during the activity and beyond.

 **Introductory activity**

Give each pair of students a copy of a calendar or use the one on page 154 of the Student Book. Ask them to find different dates (for example their birthdays, special festivals, 10th March) and share these with their partner. They can use the internet to research dates of special festivals or key dates and mark them on the calendar. (If they are using a calendar on a tablet, special dates may already be marked.) They should read each of the dates to their partner aloud, practising saying the date correctly. Next, ask them to choose one or two of the dates and work out, from today's date, how long it will be. Encourage them to consider what strategy would be most efficient for finding and recording the time. Share time lengths with the class as well as how they worked it out.

 Main activity

Give students a date trail to follow on the calendar on page 154, for example: Start on 14 February, move on 3 weeks, go back 2 days, go forward 24 hours, then forward another 21 days. Ask, *What strategy did you use for adding 21 days?* Agree that you can use knowledge of the seven times table to work out how many weeks to count on. Ask students to make up a date trail, write it on their whiteboards, then give it to their partner to follow.

Look together at the activity on page 154 of the Student Book and explain that students should follow the date trail on the calendar with a partner. As they work, continue to encourage them to use efficient methods when calculating how to move around the calendar, for example by asking them questions to get them thinking about equivalences:

What is another way we can describe 96 hours? What equivalence do you know that you can use to help you? (24 hours = 1 day or 72 hours = 3 days, for example)

How many days are there in a week? (7) *So, if I count on 15 days, how many rows down would that be on a calendar with how many days more left to count on?* (2 rows, 1 day)

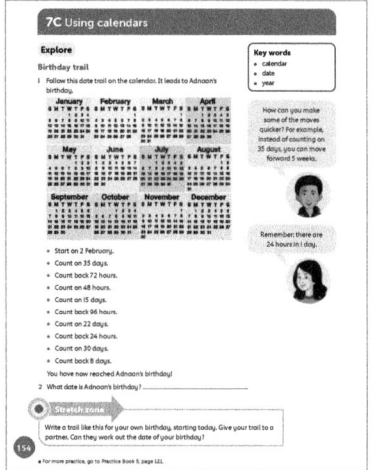

Differentiation

Supporting: Ask students to name key dates on the calendar. Practise counting on months for a set number of days, weeks and then months. Then practise counting back. Say the end date each time. Then move on to counting on to a specific date in a combination of days, weeks and months.

Consolidating: Ask students to explain their thinking as they work through the date trails.

Extending: Ask students to look at the whole date trail and estimate where they will finish on the calendar.

Stretch zone: *Write a trail like this for your own birthday, starting today. Give your trail to a partner. Can they work out the date of your birthday?*

Students can make up their own date trail for another date on the calendar.

 Reflection time

Discuss the calculations that were necessary to complete the date trail in the Student Book such as doubling, counting in multiples of 24, finding how many groups of 24 in a number and using the seven times table facts. Invite students to share the date trail to their own birthdays. The rest of the class can follow them to work out their birthdays.

194 Unit 7 Time

Practice Book: Students complete Practice Book page 121. They can do this directly after the Main activity, as homework, or as the focus of a separate mathematics session to help students consolidate their learning and build fluency.

Explain that students will need to think about how calendars work as well as everything they know about the months of the year to help them put the calendar months in order. For example, they will need to think about how many days there are in February, and on which day one month ends and another starts.

Differentiated outcomes	
All students	should understand how to use a calendar.
Most students	will follow a simple date trail.
Some students	may follow a more complex date trail and support their partners.

Answers

Student Book page 154

1 May

Practice Book page 121

Students should use the calendar for January to help them find the rest of the months. The last day of January is a Monday. This means that the first day of February must be a Tuesday. Students should look for the month that starts on a Tuesday and only has 28 days. They repeat this process for the rest of the months.

Left-hand column:

1st → April
2nd → July
3rd → January
4th → March
5th → June
6th → November

Right-hand column:

1st → February
2nd → September
3rd → May
4th → December
5th → August
6th → October

Stretch zone: September has 720 hours

7 Time

Connect Student Book page 155

Big idea

I can solve problems, involving different units of time, by converting between the units.

Global skills

- **Creative skills:** problem solving
- **Real-world skills:** interpreting information
- **Interpersonal skills:** teamwork
- **Self-development skills:** reflecting on learning

Key vocabulary

- century, decade, year, leap year, month, week, day, hour, second, millennium, arrive, depart; 24-hour clock, time intervals

Resources

- internet access or holiday brochures, magazines featuring destinations for adventure trips (optional)

Language support

Encourage students to use a bilingual dictionary to find the correct vocabulary associated with an adventure trip.

 Introductory activity

Do a quick revision of how to tell times on a digital and analogue clock, showing students times on both types and asking them to say the time. Ask further questions to get them thinking about how to work out time intervals. For example, ask, *I set off on a $1\frac{1}{2}$-hour canoe trip. If I leave at 16:15, what time will the trip end?*

 Main activity

For the activity on page 155 of the Student Book, arrange the class into mixed-attainment groups of four students. Groups will plan a school adventure trip. Encourage them to try to interpret and begin the activity without any teacher input. As students complete the tasks, ask questions, for example:

- *Have you ever planned an adventure trip? How would you start?*
- *What is a sensible length of time for each activity? Why?*
- *How can you work out the end time for each activity?*
- *How can a number line help you to work out how long an activity will take?*
- *What is the time difference between 12:36 and 17:05? How did you work that out?*
- *The hike was 1 hour 45 minutes. It finished at 18:30. What time did it begin? How did you work that out?*

When groups have finished the task, ask them to order the times on a number line and check that the lengths are correct by counting on. Tell them to work out the total length of time from the beginning of the first activity to the end of the last.

Unit 7 Time 195

Ask each group to pair up with another group and to compare their adventure trips. Ask them to compare the lengths of their activities and then to identify the longest and shortest. Invite groups to share their timetables and give information from these to the class.

Differentiation

Supporting: Help to write the answers for those less confident with writing or in speaking English.

Consolidating: Ask students to explain different aspects of their itinerary to you.

Extending: Encourage students to add additional detail to their itineraries and to support other students in their group to complete the activity.

Stretch zone: *Write two word problems about your itinerary. Give your itinerary and the problems to another group to solve.*

Remind students that they must be able to solve their own problems before giving them to another group to solve.

Reflection time

Invite each group to present their itinerary to the rest of the class. Encourage the rest of the class to ask each group in turn questions about their itinerary, focusing on timings, for example: *Which of your activities will take the longest?*

Ask students to explain their methods for working out how long the activities would take, and the start and end times.

Differentiated outcomes	
All students	should tell the time on a digital and analogue clock and say times and describe time durations with an increasing degree of accuracy.
Most students	will work out time durations correctly.
Some students	may describe time durations in more than one way.

Answers

Student Book page 155

Answers will vary according to the activities chosen. Check that the length of each activity fits the stated start and end times.

7 Time

Review   Student Book page 156 • Practice Book page 122

Global skills

- **Real-world skills:** presenting information/ interpreting information
- **Interpersonal skills:** communication
- **Self-development skills:** reflecting on learning

Student Book

With young students, assessment activities are most effective when carried out as an everyday classroom activity. Students will need calendars to complete these questions as well as an analogue or digital clock. Watch as students calculate the number of months, days and minutes for the questions, and observe their strategies for working them out.

Support students if they get stuck on working out start and end times, explaining that, for example, if they are unsure of when they woke up this morning they can estimate.

Ask students additional questions to assess their understanding, for example:

- *What time does the clock show in your Student Book? What time will it in a quarter of an hour?*
- *Look at the calendar page in your Student Book. Tell me two months it could show. Tell me two months it could not. How do you know?*

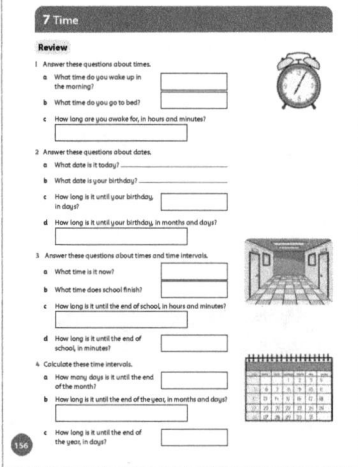

196   Unit 7 Time

Answers

Student Book page 156

1 Answers will vary for individual students.

2 Answers will vary depending on when the activity is completed and when each student's birthday falls.

3 Answers will vary depending on when the activity is completed and the school end time.

4 Answers will vary depending on which month the activity is completed.

Practice Book

With students in the upper primary years, it is appropriate to complete this as a whole-class discussion. You may choose to keep a record of the class discussion or a copy of the Review page for your own records. Use the Student Book to briefly remind students of the areas of mathematics that they have worked on in this unit.

Ensure that students have a copy of the Student Book to support them as they discuss and answer the questions in the Practice Book.

Allow students plenty of time for discussion before asking them to share their responses with the rest of the class. If students complete this assessment at home, encourage them to discuss this with adults.

Make a note of areas that students still feel unsure about, for example counting on days on a calendar, or working out the time difference between the start and end of events. Continue to give students plenty of opportunities to calculate durations of events and time differences between dates in everyday contexts.

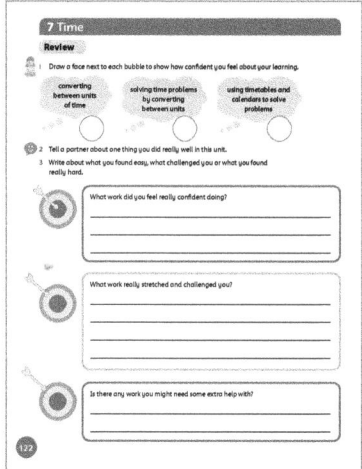

Additional material

There are additional end-of-unit assessment available on the *Oxford Owl for School* website.

Unit 7 Time

8 Geometry – properties of shapes

Overview

Big idea

The Big idea for this unit is that we use the language of shape to classify shapes into categories, and classification is a key intellectual process to help us make sense of 2D and 3D shapes in our environment.

Throughout the unit, students develop their awareness of the attributes that shapes have.

For 2D shapes, they explore how many sides or angles the shapes have, what size they are and how to measure them. 2D shapes made of straight sides are called **polygons**, which means 'many knees', referring to their many angles.

Students need to use ideas of symmetry and transformation when exploring 2D shapes. The coordinate grid is an invaluable aid for this learning.

For 3D shapes, we can sort and compare shapes by their faces, edges and vertices, and recognise the 2D nets that can be folded to make the 3D shapes. In 3D, flat-faced shapes are called **polyhedra** for 'many faces'.

Look out for

- **Students who find it difficult to visualise shapes.** Provide plenty of opportunities for them to do this. For example, ask students to visualise an irregular pentagon and then to draw it in the air with a finger before drawing it on paper. You can provide practice for this skill by asking them to carry out other visualisation activities, such as visualising a square, and then cutting off its corners and drawing what they can now 'see'.
- **Students who may not appreciate multiple lines of symmetry or that lines of symmetry can be horizontal or diagonal as well as vertical.** Give examples in all orientations. Encourage students to turn shapes around.

Possible misconceptions

- **Students may see regular examples of shapes as the only example of a shape.** For example, they do not recognise an irregular octagon as an octagon. Expose students to a wide variety of examples of shape. Encourage students to describe shapes as 'regular' or 'irregular'.
- **Students may not understand the definitions of shape properties.** For example, a rectangle is any four-sided shape with four right angles, and this includes a square.
- **Students may think that all polygons have the same number of lines of symmetry as they do number of sides.** Use counter-examples such as scalene triangles or irregular hexagons to show that this is not always true.
- **Students may think identical angles are unequal because the lengths of the lines are different in each.** Encourage students to measure the angle using a protractor to focus them on what part is the angle.

Key vocabulary

- 3D, three-dimensional
- cube, cuboid, pyramid, sphere, hemisphere, cone, cylinder, prism, tetrahedron, triangular prism, faces, edges, vertices, curved surfaces
- 2D, two-dimensional; circle, semicircle, ellipse, triangle, equilateral triangle, isosceles triangle, scalene triangle, right-angled triangle, square, rectangle, oblong, pentagon, hexagon, heptagon, octagon, polygon, polyhedron, quadrilateral
- angle, degrees, acute, obtuse, reflex, protractor
- mirror line, reflection, apex, nets, angles at a point, angles in a straight line, reflective symmetry, rotational symmetry, lines of symmetry
- parallel, perpendicular, regular, irregular, diagonal, horizontal, vertical, sides

Coverage in lessons

Learning objective	E	8A	8B	8C	8D	8E	C	R
Identify 3D shapes, including cubes and other cuboids, from 2D representations.	✓			✓				✓
Identify all lines of symmetry in 2D shapes and in patterns.			✓					
Use knowledge of reflective symmetry to complete symmetrical patterns.			✓					
Know angles are measured in degrees: estimate and compare acute, obtuse and reflex angles.		✓			✓	✓	✓	✓
Draw given angles, and measure them in degrees (°).					✓	✓	✓	
Identify: • angles at a point and one whole turn (total 360°) • angles at a point on a straight line and $\frac{1}{2}$ a turn (total 180°) • other multiples of 90°.					✓	✓	✓	✓
Use the properties of rectangles to deduce related facts and find missing lengths and angles.				✓				
Distinguish between regular and irregular polygons based on reasoning about equal sides and angles.	✓	✓		✓			✓	✓

Unit 8 Geometry – properties of shapes

8 Geometry – properties of shapes

Engage Student Book page 157

Big question
- How can I use properties of shapes to describe 2D and 3D shapes?

Global skills
- **Creative skills:** investigating
- **Interpersonal skills:** communication
- **Self-development skills:** reflecting on learning

Key vocabulary
- 3D shape, three-dimensional shape, cube, cuboid, pyramid, sphere, hemisphere, cone, cylinder, prism, tetrahedron, triangular prism, faces, nets, edges, vertices, curved surfaces
- 2D shape, two-dimensional shape, circle, semicircle, ellipse, triangle, equilateral, isosceles, scalene, right-angled, square, rectangle, oblong, pentagon, hexagon, heptagon, octagon, polygon, polyhedron, quadrilateral, mirror line, reflection, reflective symmetry, line of symmetry

Resources
- mini whiteboards and markers
- 3D shapes (a cube, cuboid, square-based pyramid, sphere, hemisphere, cone, cylinder, triangular prism, tetrahedron) – 1 set per pair
- large sheets of paper

Language support
Students may need support in forming questions correctly. Provide them with question frames as well modelling how to answer yes/no questions, such as:

Does it have *five* sides? Yes, it does. No, it doesn't.

Is it an *equilateral triangle*? Yes, it is. No, it isn't.

Create posters of all the shapes named in the key vocabulary to display on your working wall. Include images of shapes in the environment, with their names and properties.

Introductory activity

Give each pair of students a set of 3D shapes. If you do not have a set of 3D shapes, refer to the illustrations of 3D shapes on page 157 of the Student Book. If you have access to an IWB, display the page.

Students should choose a shape and describe its properties to their partner without saying which shape it is. Their partner has to name the shape. From previous stages, students should already know: cube, cuboid, pyramid, sphere, hemisphere, cone, cylinder, triangular prism and tetrahedron.

Main activity

Move on to 2D shapes. Each student should draw a 2D shape on their whiteboard and not show it to their partner. Their partner has to guess the shape by asking questions to which the only answer is 'yes' or 'no'.

Expect students to name and describe the following shapes: circle, semicircle, triangle, equilateral triangle, isosceles triangle, scalene triangle, square, rectangle, oblong, pentagon, hexagon, heptagon octagon, quadrilateral. Encourage them to describe the shapes' properties in terms of the number of sides, lines of symmetry and the number and types of angle.

Discuss the words **polyhedron** and **polygon**. Establish that a polyhedron is any 3D shape that has flat faces and that a polygon is a 2D closed shape with three or more straight sides. Ask students to give you examples and non-examples, from the shapes they have looked at in this lesson.

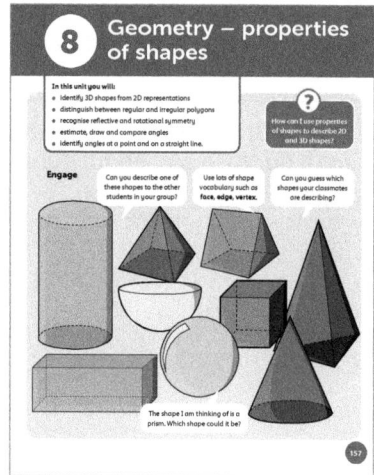

Differentiation

Supporting: Support students to help them to recognise the names and properties of common 2D and 3D shapes, and to formulate questions about shapes correctly.

Consolidating: Ask students to describe the properties of 2D and 3D shapes, including symmetry and regularity.

Extending: Ask students to find examples of 2D and 3D shapes in the environment, including all of those on the key vocabulary list, and describe their properties.

Reflection time

Take feedback from the activity. Invite students to name and describe 3D shapes they can see around the classroom, including sharing the 2D names of the faces.

Ask questions to assess students' understanding, for example:

- *Which shapes have quadrilateral-shaped faces?*
- *What are the names of the triangular faces on the pyramid?* (isosceles triangles)

8A Regular and irregular polygons

Discover Student Book page 158 • Practice Book page 123

Specific learning focus
- Identify and describe properties of triangles and classify them as isosceles, equilateral or scalene and draw examples of each.

Global skills
- **Creative skills:** exploring
- **Real-world skills:** presenting information/interpreting information
- **Interpersonal skills:** communication/teamwork

Key vocabulary
- equilateral, isosceles, scalene, right-angled, regular, irregular, perpendicular and parallel

Resources
- mini whiteboards and markers
- rulers, set squares

Language support
Model the pronunciation of 'equilateral triangle', 'isosceles triangle', 'right-angled triangle' and 'scalene triangle'.

Make sure that the posters from the Engage lesson are used to support language development.

 **Introductory activity**

Ask pairs to sketch a range of polygons on their whiteboards. Give them a few minutes to describe, in groups of three or four, shapes they drew. Let other students in the class give each shape's name from the description. Discuss the fact that the shape name refers to its properties. For example, any three-sided shape is a triangle. Discuss **regular** and **irregular** shapes and **perpendicular** and **parallel** sides, by asking questions, for example:

- Who drew some irregular shapes?
- How do you know that these shapes are irregular?
- What shapes did you draw that have perpendicular sides?
- Which shapes that you drew have parallel sides?
- What is the same and what is different about parallel and perpendicular sides?

 **Main activity**

Focus on triangles. Invite students to sketch on the board any triangles they drew during the Introductory activity. Ensure that there is a variety of **isosceles**, **equilateral**, **scalene** and **right-angled** triangles. Ask:

- What is the same about these triangles?
- What is different about the triangles?
- Is it possible to have a triangle that has a right angle?
- Is it possible to have triangles with two right angles
- What can you tell me about the other two angles?

Give students time after each question to discuss it quickly in pairs before sharing answers. Students should provide examples to support their answers, either using triangles already drawn on the board or by drawing new ones.

Explain the activity on page 158 of the Student Book. Students work in pairs. Encourage them to draw their triangles as accurately as they can, using rulers to measure lengths and also to draw any right angles using set squares.

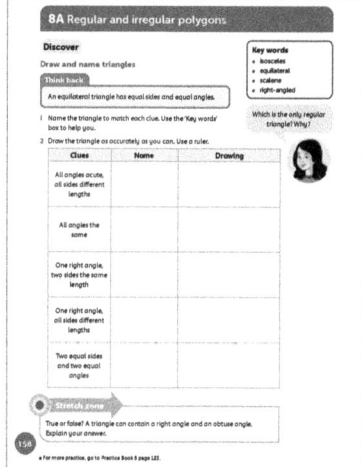

Differentiation

Supporting: Support students to remember the names and properties of each type of triangle, showing them a variety of examples of each and emphasising their properties.

Consolidating: Ask students to find examples of different triangles in the environment.

Extending: Challenge students to draw another example of each type of triangle.

Stretch zone: *True or false? A triangle can contain a right angle and an obtuse angle. Explain your answer.*

Students should say that this is false, perhaps referring to the sum of the angles for a triangle; that is, a right angle measures 90° and an obtuse angle measures greater than 90°, but less than 180°. A triangle has three angles with a sum of 180°, so with one angle of 90° it is impossible for the shape to contain an obtuse angle.

Unit 8 Geometry – properties of shapes

Reflection time

Ask small groups of students to create posters that define:

- an equilateral triangle
- an isosceles triangle
- a right-angled triangle
- a scalene triangle.

These posters can be displayed for the whole unit as part of your working wall.

Practice Book: Students complete Practice Book page 123. They can do this directly after the Main activity, as homework, or as the focus of a separate mathematics session to help students consolidate their learning and build fluency.

Students will need a ruler for this activity.

Differentiated outcomes	
All students	should draw a range of triangles accurately.
Most students	will draw and name all the triangles.
Some students	may describe a wide range of properties of the triangles.

Answers

Student Book page 158

Check that students' drawings of the triangles match the descriptions.

Scalene triangle

Equilateral triangle

Right-angled isosceles triangle

Right-angled scalene triangle

Isosceles triangle

Practice Book page 123

Check that students have written two properties for each named triangle. Answers will include:

1 scalene triangle: all angles acute, all angles different, all sides different lengths

2 equilateral triangle: all angles acute, all angles the same, all sides the same length

3 right-angled isosceles triangle: one right angle, two angles the same, two sides the same length

4 isosceles triangle: two angles the same, two sides the same length

5 Check that students have drawn an equilateral triangle, an isosceles triangle and a scalene triangle.

Stretch zone: False. A triangle can never contain two right angles. The sum of the three angles of a triangle are 180°. Two right angles measure 180° so it is not possible.

8A Regular and irregular polygons

Explore Student Book pages 159–160 • Practice Book page 124

Specific learning focus

- Draw and describe triangles and quadrilaterals.

Global skills

- **Creative skills:** investigating
- **Real-world skills:** interpreting information

Key vocabulary

- equilateral, isosceles, scalene, acute, obtuse, reflex

Resources

- mini whiteboards and markers
- rulers, set squares

Language support

Add 'the properties of symmetry' and 'perpendicular lines' to the posters students made in 8A Discover.

 **Introductory activity**

As a class, take suggestions for all the different types of triangles students can think of. They should give the name and describe a property of each.

Draw this Carroll diagram on the board:

	Contains a right angle	Does not contain a right angle
Contains an **obtuse angle**		
Does not contain an obtuse angle		

Ask students to work in pairs. They should copy the Carroll diagram onto a whiteboard and take it in turns to write the name of a triangle in a space. Tell them to try to add at least two examples to the spaces.

Unit 8 Geometry – properties of shapes

Ask individual students to come to the front to draw the triangles and write the triangles' names onto the large Carroll diagram on the board. As they add each triangle, they describe its properties.

After some time, ask students whether there are any other types of triangles they want to add to the diagram. They should notice that there are no triangles in one of the cells. Can they say which one and reason why? Discuss as a class.

Main activity

Discuss all the properties students can think of for quadrilaterals. Prompt them to think about number of sides, side lengths, angles, lines of symmetry and so on.

Ask students to draw a regular quadrilateral and an irregular quadrilateral on their whiteboards and describe their properties as accurately as possible to their partner. Invite pairs to the front of the class to share the drawing and descriptions of their quadrilaterals. Ask the class to see whether they can describe them in any additional ways.

Students should then complete, in pairs, the activities on pages 159 and 160 of the Student Book.

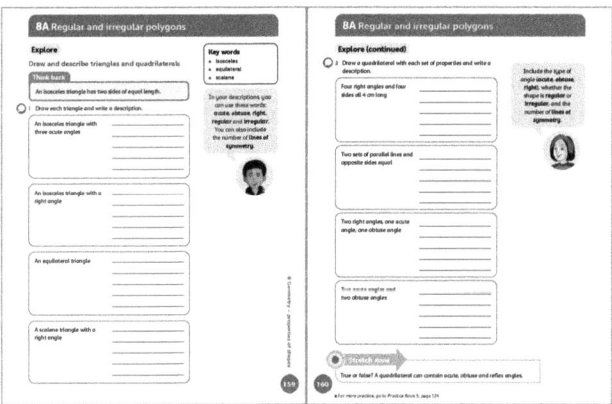

Differentiation

Supporting: Support students to remember the names and properties of triangles and quadrilaterals. Encourage them to refer to their Student Book glossary and the working wall in the first instance.

Consolidating: Ask students to describe the properties of triangles and quadrilaterals, including referring to symmetry and regularity.

Extending: Ask students to find examples of different triangles and quadrilaterals in the environment.

Stretch zone: *True or false? A quadrilateral can contain acute, obtuse and reflex angles.*

Ask students to first define each type of angle. Ask students to draw examples to justify their answer of true or false. They may answer 'true' if they recognise that a dart can have a reflex angle, and it could have both a reflex and an obtuse angle, for example as in the following diagram.

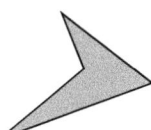 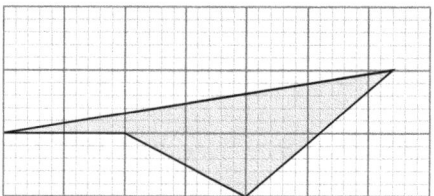

Reflection time

Take feedback from the activity. Invite students to draw some of their shapes on the board as accurately as they can. Ask other students to describe the shapes. Ensure that they talk about the types of angles and the number of lines of symmetry.

Practice Book: Students complete Practice Book page 124. They can do this directly after the Main activity, as homework, or as the focus of a separate mathematics session to help students consolidate their learning and build fluency.

Students will need a ruler for this activity and may also prefer to continue to work with a set square.

Differentiated outcomes	
All students	should draw a range of triangles and quadrilaterals accurately with support.
Most students	will draw and name all the triangles and quadrilaterals.
Some students	may describe a wide range of properties of the triangles and quadrilaterals.

Answers

Student Book pages 159–160

Students' drawings and descriptions should match the names of the triangles and quadrilaterals.

Practice Book page 124

Check that students have written two properties for each named polygon. Answers will include:

1 trapezium: one pair of parallel sides, 2 acute angles, 2 obtuse angles.

2 irregular hexagon: 6 sides, some of different lengths; 6 angles, some different sizes.

3 regular octagon: eight angles the same, eight sides the same length.

4 square: four right angles, four sides the same length.

5 regular decagon: ten angles the same, ten sides the same length.

6 right-angled trapezium: two right angles, one pair of parallel sides, four sides different lengths.

Stretch zone: True. Check that each student's sketch of an octagon has three right angles.

Unit 8 Geometry – properties of shapes

8B Symmetry in polygons

Discover Student Book page 161 • Practice Book page 125

Specific learning focus
- Recognise reflective symmetry in regular polygons.
- Create patterns with two lines of symmetry.

Global skills
- **Creative skills:** investigating
- **Real-world skills:** presenting information
- **Interpersonal skills:** communication

Key vocabulary
- mirror line, horizontal, vertical, reflective symmetry, reflection

Resources
- string
- different-coloured counters
- plain paper, lined paper, small mirrors, coloured pencils

Language support

Provide students with sentence frames to help them describe their shapes, for example:

My shape is a _____ regular/irregular _____.

It has ____ sides.

It has ____ obtuse angles.

It has ____ acute angles

It has ____ reflex angles.

It has ____ line(s) of symmetry.

 Introductory activity

Give pairs a piece of string and some coloured counters. Tell them to place the string **horizontally** on their table and explain that it will act as their **mirror line**. Tell one student to place a selection of counters on one side of the mirror line. The other student then places identical counters on the other side to make a symmetrical pattern. Give them a few minutes to record their pattern on plain paper. Ask them to check to make sure that the pattern on the two sides of their mirror line are the same. Invite pairs to share their patterns. Each time, ask the rest of the class, *Were they successful? Can you see the pattern's reflection on the other side of the mirror line? Is this a symmetrical pattern?*

Now ask students to position the string **vertically** and repeat the activity. This is more difficult. Tell the pairs that they can use a mirror to help, or they can check that the counters are the same distance – measured perpendicularly – as the counterpart the other side of the mirror line. Demonstrate how to do this.

As before, ask students to check to make sure that the two sides of their mirror line are the same. Invite pairs to share their patterns. Each time, ask the rest of the class, *Were they successful? Is this a symmetrical pattern?*

 **Main activity**

Refer students to the activity on page 161 of the Student Book and choose a student to model the activity, drawing a shape on the board and then reflecting it along the mirror line to make a new shape. Discuss as a class the name of the new shape and its properties.

- *How many sides did our original shape have? How many does it have now?*
- *Can you describe this angle? Are there any more examples of this type of angle?*
- *Our shape has a least one line of symmetry. How do I know? Can you see any more?*

All students will need coloured pencils and a notebook or lined paper to complete the activity in the Students Book in pairs.

Note whether students are finding it difficult to visualise that a shape is 'flipped' in the mirror line. While students are working, ask them to explain how they know how to draw the shape as it will look on the other side of the mirror line. Use small mirrors to help students visualise the reflection.

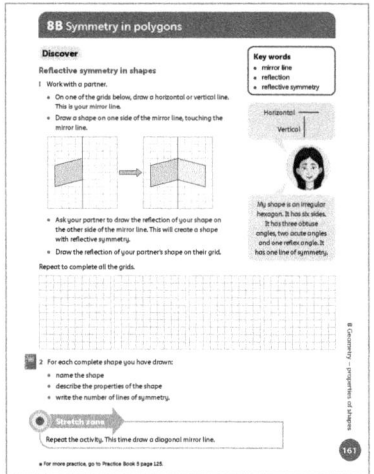

Differentiation

Supporting: Help students to describe the symmetry of the patterns and new shapes they are creating.

Consolidating: Ask students how they know a pattern or shape is symmetrical.

Extending: Challenge students to describe how to identify a mirror line on a symmetrical pattern.

Stretch zone: *Repeat the activity. This time draw a diagonal mirror line.*

Check that students correctly draw a diagonal mirror line and reflect their shape correctly along it. They should also describe the new shape's properties.

204 Unit 8 Geometry – properties of shapes

Reflection time

Invite pairs of students to share the new shapes they made, drawing them on the board. For each shape, ask the class: *Where is the mirror line? Can you describe your new shape's properties? Does it have* **reflective symmetry***? How do you know?*

Students could also find examples of Islamic or Rangoli patterns. (These are usually symmetrical.) They can make copies of these patterns or design their own patterns to display in the classroom.

Practice Book: Students complete Practice Book page 125. They can do this directly after the Main activity, as homework, or as the focus of a separate mathematics session to help students consolidate their learning and build fluency.

Discuss what impact having two lines of symmetry will have when they are making their pattern. For the second question, tell students that they can use part squares and diagonal lines of symmetry but they must ensure that their pattern is symmetrical.

Differentiated outcomes	
All students	should create simple symmetrical patterns with support.
Most students	will recognise and create more complex symmetrical patterns.
Some students	may be able to describe the properties of symmetry in detail.

Answers

Student Book page 161

Answers will vary because students create their own symmetrical shapes in pairs. Check that that the shapes are symmetrical and that they have added the correct details of their shape in question 2.

Practice Book page 125

1 Check that students have completed the pattern correctly.

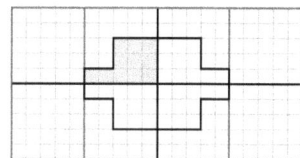

2 Students create their own patterns with two lines of symmetry, so answers will vary. Check that the shapes do have two lines of symmetry.

Stretch zone: Check whether students have made a 10-square shape with two lines of symmetry; they can do this using half squares.

8B Symmetry in polygons

Explore Student Book page 162 • Practice Book page 126

Specific learning focus

- Recognise reflective and rotational symmetry in regular polygons.

Global skills

- **Creative skills:** investigating
- **Real-world skills:** presenting information
- **Interpersonal skills:** communication

Key vocabulary

- mirror line, reflective symmetry, line of symmetry, (the order of) rotational symmetry

Resources

- set of equilateral triangles cut out of card
- card, A4 paper, large sheets of paper
- rulers
- scissors

Language support

Focus on the key vocabulary 'reflective symmetry' and 'rotational symmetry'. Repeat these phrases when working with pairs. Ask students who are confident in the use of English to work with other students who are less confident, so that good linguistic models are available.

 Introductory activity

Give each pair a card equilateral triangle. Ask one of the pair to draw around the triangle and put a dot on one of its corners (to help them keep track of whether they rotate their triangle 360 degrees). Tell them to rotate the triangle about its centre one whole turn and to note how many times it fits into the outline drawn at the beginning. (It will be three times.)

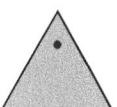 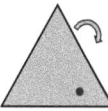 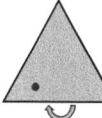 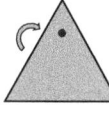

Start position   Order 1   Order 2   Order 3

The other student repeats this to check. Explain that the number of times that the triangle can fit onto the original shape before returning to the beginning is called the order of **rotational symmetry**. The order of rotational symmetry of an equilateral triangle is 3.

Unit 8 Geometry – properties of shapes

Repeat the activity, this time using a square. Students will find that the square has 4 lines of symmetry and order of rotational symmetry 4.

 Main activity

Draw students' attention to page 162 of the Student Book, where they will be exploring the number of **lines of symmetry** of regular polygons. Students could use mirrors or cut out and fold the shapes to support them in their investigation.

As pairs work, ask them to explain their strategy for finding lines of symmetry. Are some shapes easier than others? Why?

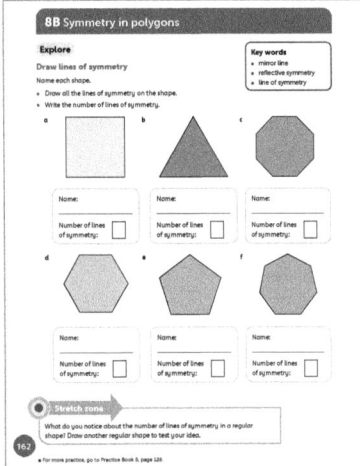

Differentiation

Supporting: Ask students what they understand by reflective symmetry.

Consolidating: Ask students to explain how they find the lines of symmetry in these regular polygons. What do they find easy? What is hard?

Extending: Challenge students to find the order of rotational symmetry for the remaining four polygons in the Student Book.

Stretch zone: *What do you notice about the number of lines of symmetry in a regular shape? Draw another regular shape to test your idea.*

Students should notice that the number of lines in a polygon is the same as its number of sides.

 Reflection time

Ask students to share their answers for each shape and agree on the number of lines of symmetry. Ask one student to draw their answers on the board. Others could use visualisation to decide the order of rotational symmetry. If necessary, make a paper copy of the shape and show the class what the order of symmetry is.

Students can find examples of shapes with rotational symmetry in the environment and photograph them to make a display. Alternatively, they can find images on the internet and make a poster.

Practice Book: Students complete Practice Book page 126. They can do this directly after the Main activity, as homework, or as the focus of a separate mathematics session to help students consolidate their learning and build fluency.

Students will need a ruler to draw the shapes. They may also want to make each shape out of card so that they can physically rotate it as they did in the Introductory activity.

Differentiated outcomes	
All students	should understand what line symmetry means and find some of the lines of symmetry of regular polygons.
Most students	will be able to find all the lines of symmetry of regular polygons.
Some students	may also be able to name the order of rotational symmetry of other regular polygons.

Answers

Student Book page 162

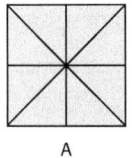 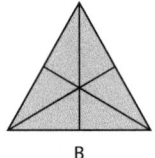 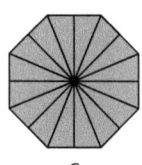
A B C

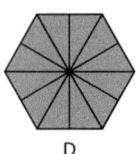 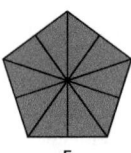 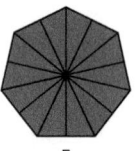
D E F

a square: 4 lines of symmetry

b equilateral triangle: 3 lines of symmetry

c regular octagon: 8 lines of symmetry

d regular hexagon: 6 lines of symmetry

e regular pentagon: 5 lines of symmetry

f regular heptagon: 7 lines of symmetry

Practice Book page 126

Check that students' sketches of the shapes are correct.

Polygon	Rotational symmetry	Lines of symmetry
equilateral triangle	3	3
square	4	4
rectangle	2	2
regular pentagon	5	5
regular heptagon	7	7
isosceles triangle	1	1
scalene triangle	1	0

Stretch zone: True. Check that students have drawn at least one scalene triangle to support their answer.

8C Identifying 3D shapes

Discover Student Book page 163 • Practice Book page 127

Specific learning focus
- Visualise 3D shapes from 2D drawings and nets.

Global skills
- **Creative skills:** exploring
- **Interpersonal skills:** communication

Key vocabulary
- face, prism, pyramid, apex

Resources
- 3D shapes: sphere, cylinder, cone, cube, cuboid, tetrahedron, square-based pyramid, triangular prism, pentagonal prism, hexagonal prism
- examples of different boxes that can be opened up to give a range of nets (include boxes that are different prisms and pyramids)
- A4 paper, sticky tape, rulers
- scissors
- modelling clay

Language support
Include drawings of common 3D shapes as part of your working wall. Add speech bubbles next to them to model questions relating to their properties, including key vocabulary, for example: *How many triangular faces does this prism have? How many vertices does this pentagonal prism have?*

Introductory activity

Remind students that a prism is a shape where the cross-section is the same shape as the ends, for example a cuboid. Give students a piece of modelling clay. Ask them to make a cylinder. Discuss its properties (curved surface, edges, vertices, whether it is a **prism** or not) and where you can see cylinders in real life. Ask them to hold it up in front of them and ask, *Which 2D shape can you see?* Students might see a circle if they hold the end facing them, or they might see a rectangle if it is side on.

Repeat for a cuboid. Agree that the number of **faces**, edges and vertices are the same as for a cube, the only difference is the shape of the faces. These are either six oblongs or four oblongs and two squares. Next, ask students to turn their cuboid into a square-based **pyramid**. Agree that the shape has five faces (four triangular, one square), eight edges and five vertices.

 **Main activity**

Ask students, in pairs, to look around the room and list all the 3D shapes they can see. If you have time, it is much better if you can take the students on a shape walk around the school or local area. List the shapes on the board. Ensure that the following are included: sphere, cylinder, cone, cube, cuboid, tetrahedron, square-based pyramid, triangular prism, pentagonal prism, hexagonal prism. Show your collection of shapes. Ask students to match them with the shape names on the board. Ask, *What is the face of a 3D shape?* Agree that a face is a flat surface. A sphere doesn't have a face, it has a curved surface. A cone has one face and one curved surface. A cylinder has two faces and one curved surface.

Ask, *What is the edge of a 3D shape?* Establish that an edge is where two faces or surfaces meet. A sphere doesn't have an edge. Then ask, *What is a vertex?* Establish that a vertex is where three or more edges meet. Agree that spheres, cylinders and cones don't have vertices. The pointed part of a cone is called the **apex**.

Ask students to complete page 163 of the Student Book. Explain the task, which is to look at a set of 2D shapes and to identify the possible 3D shapes they can be. Have a variety of 3D shapes available for students to look at to support them in identifying possible shapes.

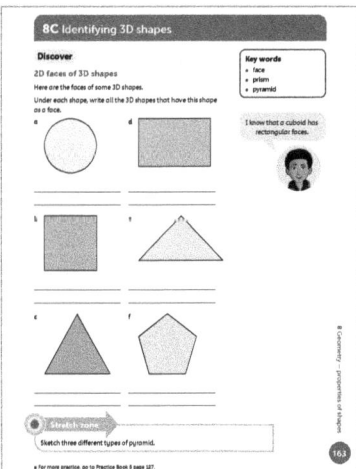

Differentiation

Supporting: Support students in remembering the names of 2D and 3D shapes.

Consolidating: Ask students to say what is the same and what is different about two 3D shapes.

Extending: Challenge students to think of 3D shapes that have combinations of 2D faces, for example at least one triangular and one hexagonal face.

Stretch zone: *Sketch three different types of pyramid.*

Students could try to do a 3D sketch or draw a net. Students should be aware of different base shapes such as pentagons and hexagons.

Unit 8 Geometry – properties of shapes

Reflection time

Look at each 2D shape from the Student Book in turn. Ask, *What 3D shape could this be?* Agree that: the circle could be a sphere, cylinder or cone; the oblong could be a cuboid or triangular prism; the square could be a cube, cuboid or pyramid; the triangle could be a tetrahedron, pyramid or triangular prism.

Draw some other 2D shapes on the board, for example a pentagon and hexagon. Ask, *Can you tell me what 3D shapes these can be?* (pentagonal and hexagonal pyramids and prisms)

Practice Book: Students complete Practice Book page 127. They can do this directly after the Main activity, as homework, or as the focus of a separate mathematics session to help students consolidate their learning and build fluency.

Encourage students to use a dictionary or their glossary so that they are recording the names of each of the shapes correctly.

Differentiated outcomes	
All students	should name a range of 3D shapes and describe their faces.
Most students	will name and recognise the faces of a wide range of 3D shapes.
Some students	may visualise and draw accurate faces for a range of 3D shapes.

Answers

Student Book page 163

Examples of possible answers:

circle: cylinder, cone

rectangle: cuboid, cylinder, triangular prism

square: cube, cuboid, pyramid, triangular prism

triangle: tetrahedron, pyramid, triangular prism

pentagon: dodecahedron, pentagonal pyramid, pentagonal prism

Practice Book page 127

Check that students' names and properties for the shapes are correct.

1 triangular prism: two identical triangles and three identical rectangles.

2 tetrahedron: four identical triangles.

3 cylinder: two identical circles and one curved surface.

4 cone: one circle and a curved surface.

5 cuboid: two identical rectangles and four identical rectangles

Stretch zone: False. Students could draw pyramids with any polygon as its base.

8C Identifying 3D shapes

Explore Student Book page 164 • Practice Book page 128

Specific learning focus
- Visualise 3D shapes from 2D drawings and nets.

Global skills
- **Creative skills:** investigating
- **Real-world skills:** interpreting information

Key vocabulary
- 3D shape, faces, nets

Resources
- paper, squared paper or card, sticky tape
- rulers
- scissors
- modelling clay, toothpicks

Language support

Students can also include nets to the shape posters they made at the start of the unit as part of the working wall display. Encourage them to label and describe their nets, for example: 'This net folds up to make a square-based pyramid. It is made up of 4 triangular faces and a square face.'

 **Introductory activity**

Ask students to make a triangular prism using modelling clay and toothpicks. Show them a real one or draw one on the board to help them.

Discuss its properties, including the shapes of the faces. Ask, *What do you think its **net** looks like?* Remind students of what a net is as necessary. Ask pairs to sketch it, using squared paper if necessary. They should cut the net out, fold and stick it together to make the shape. *Is it possible to have more than one type of net?* Compare the nets they made, presenting several options. For example:

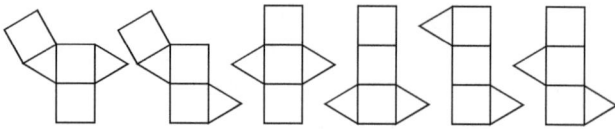

208 Unit 8 Geometry – properties of shapes

 **Main activity**

Look together at page 164 of the Student Book. Explain that the activity involves identifying what 3D shape is made by folding a series of nets. Tell students that they can make the nets and fold them to check their answers. Alternatively, they could build them with modelling clay and toothpicks.

As pairs work, ask them to describe the faces of each of the nets to their partner.

Point out the lower speech bubble. Challenge them to see how many different nets for a cube they can find, once they have completed the Main activity.

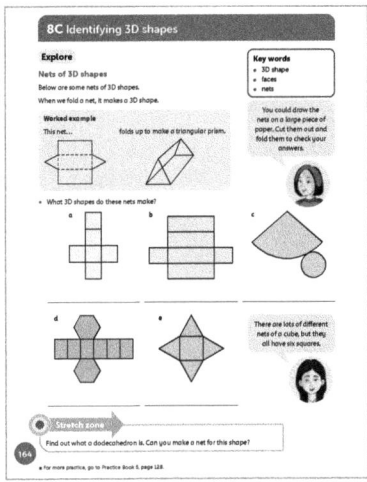

Differentiation

Supporting: Give students nets that match those in the Student Book for them to fold and then name the shape. Ask, for example, *What faces can you see on the net? What type? How many? Can you match nets to 3D shape models?*

Consolidating: Ask students to visualise and describe the nets of a range of 3D shapes.

Extending: Ask students to visualise, describe and draw the nets of a range of 3D shapes. Can they make more than one for the same shape in some cases?

Stretch zone: *Find out what a dodecahedron is. Can you make a net for this shape?*

Find an image of or show students a model of a dodecahedron, if necessary. Ask them to describe and count its faces before trying to draw its net. It may be helpful to have 2D shapes to trace around to make their net-making easier.

 Reflection time

Hold up a series of 3D shapes and ask students whether they can identify which of the nets on page 164 the shapes came from. Prompt students to say what they can identify easily from the nets by asking, for example, *Are faces or edges easier to count on a net? Why?*

Practice Book: Students complete Practice Book page 128. They can do this directly after the Main activity, as homework, or as the focus of a separate mathematics session to help students consolidate their learning and build fluency.

Students will need a ruler or a straight edge to complete this activity.

Differentiated outcomes	
All students	should be able to fold a net to make a 3D shape and name it.
Most students	will name and recognise the nets of a range of 3D shapes.
Some students	may be able to create more than one possible net for a range of 3D shapes.

Answers

Student Book page 164

Shapes made by the nets are:

a cube

b cuboid

c cone

d hexagonal prism

e square-based pyramid

Practice Book page 128

Check that students' nets for the shapes are correct. Nets may vary.

1 triangular prism: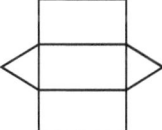

2 tetrahedron:

3 square-based pyramid: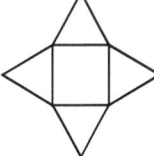

4 cuboid: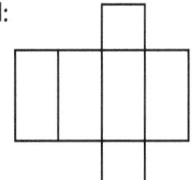

Stretch zone: True. The parts of a net each make one face or curved surface in the shape.

8D Angles

Discover Student Book pages 165–166 • Practice Book page 129

Specific learning focus
- Draw and classify angles as acute, right, obtuse or reflex.

Global skills
- **Creative skills:** investigating
- **Real-world skills:** presenting information/interpreting information
- **Interpersonal skills:** communication

Key vocabulary
- acute angle, obtuse angle, right angle, reflex angle, degrees, protractor

Resources
- large sheets of paper
- mini whiteboards and markers
- photographs of shapes in the environment (extension activity only)

Language support
Remind students how to read, record and say the degree symbol (°), modelling and emphasising the correct pronunciation.

Introductory activity

Organise students into small mixed-attainment groups for this activity. They should create a poster that tells the reader everything they know about angles. Let groups start working and then list the key vocabulary on the board and ask them to include these words on their poster if they can remember what they mean.

Each group should feed back to the whole class. Create a whole-class poster to display as part of the working wall for this unit. Include examples of **acute, obtuse, right,** and **reflex angles**, along with definitions of each that you decide upon as a class. For example, students may be unfamiliar with reflex angles as angles that measure more than 180 **degrees** but less than 360 degrees.

Main activity

Ask students to work in pairs. Each pair should draw a right angle, acute angle, reflex and obtuse angle. They can draw their angles and then check each other's and explain to their partner how they know they are correct. They should then draw a shape on their whiteboards that contains at least one right angle, one acute angle, one reflex angle and one obtuse angle. Share several of these shapes, copying them onto the board, and asking students to identify which angle is an example of each type.

This leads into students completing the activities on pages 165–166 of the Student Book. They can do these individually, then discuss and check their answers with a partner.

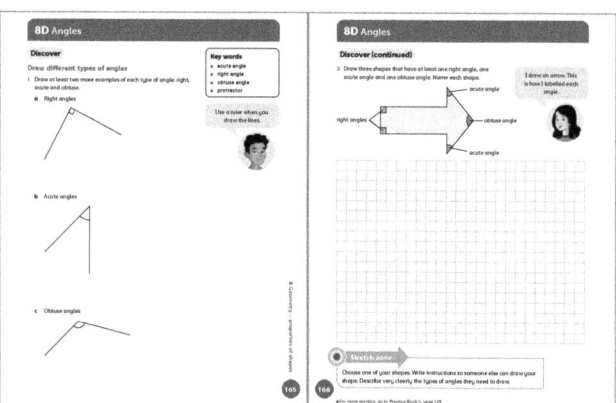

Differentiation

Supporting: Give students flashcards with written definitions of each type of angle as well as visual examples to refer to throughout this lesson and 8D Explore 1.

Consolidating: Ask students to show you examples of acute angles, right angles, reflex angles and obtuse angles and to describe them.

Extending: Ask students to estimate the size of acute angles and obtuse angles in their drawings. They can then measure to check their accuracy.

Stretch zone: *Choose one of your shapes. Write instructions so someone else can draw your shape. Describe very clearly the types of angles they need to draw.*

Encourage students to record their instructions in a numbered or bulleted list. They may want to include illustrations to support their instructions.

Reflection time

Invite students to share the angles and shapes they drew in their Student Book. Ask the class to identify the different types of angle. Invite students to draw some angles on the board for the class to name.

As an extension activity, ask students to create an angles poster using photographs of shapes in the environment. They can label these, using the key angle vocabulary from this lesson.

Practice Book: Students complete Practice Book page 129. They can do this directly after the Main activity, as homework, or as the focus of a separate mathematics session to help students consolidate their learning and build fluency.

Students will need to draw angles using a **protractor** in this activity. You may prefer to delay completion of this activity until after 8D Explore 1, when students will have had more practice working with a protractor.

Differentiated outcomes	
All students	should identify and draw acute, right, reflex and obtuse angles.
Most students	will identify, draw and define acute, right, reflex and obtuse angles.
Some students	may identify and define and estimate the size of acute angles, right angles and obtuse angles.

Answers

Student Book pages 165–166

Check that students have correctly drawn at least two more examples of right, acute and obtuse angles, and three shapes that each contain at least one right angle, one acute angle and one obtuse angle inside them.

Practice Book page 129

Check that students have drawn the angles to within 2° of accuracy.

1 65°   acute
2 90°   right angle
3 170°   obtuse
4 35°   acute
5 85°   acute
6 105°   obtuse

Stretch zone: Students answers will vary, depending on the accuracy of their drawing.

8D Angles

Explore 1 Student Book page 167 • Practice Book page 130

Specific learning focus

- Understand and use angle measure in degrees.
- Measure angles to the nearest 5° using a protractor.
- Identify, describe and estimate the size of angles and classify them as acute, right, reflex or obtuse.

Global skills

- **Creative skills:** investigating
- **Real-world skills:** interpreting information

Key vocabulary

- angle, estimate, degrees

Resources

- mini whiteboards and markers
- A4 paper, rulers, pencils, protractors

Language support

Ask students questions to encourage them to use the key vocabulary as as they measure angles, for example:
- *What unit do we measure angles with?*
- *What equipment can we use?*
- *What type of angle is this?*

 **Introductory activity**

Remind students of the work from 8D Discover by asking, *What can you tell me about angles?* (An angle can be a right angle, acute, reflex or obtuse.) Ask students to draw some of each of these on their whiteboards.

What does a right angle measure? (90°) *Can you use this information to work out what a straight line measures?* Give students some time to discuss in pairs and then discuss as a class. Agree that a straight line measures the same as two right angles (180°). Ask them to use this information to work out the inside of a circle (four right angles). You can ask them to cut four right angles from the corners of a piece of paper and put them together and sketch a circle.

 **Main activity**

Give each student a protractor. Ask them to look at it talk to their partner about what they notice. *How many scales can you see? How can you describe the scales? How are they the same and different from a number line?* Share, as a class, their observations.

Count along the scale from 0 in multiples of 10, saying that the angles are getting bigger. Repeat, counting back from 180 degrees. Call out various multiples of 10 degrees for them to find, for example: 60°, 120°. Repeat for multiples of 5°. Give each student a piece of A4 paper, ruler and pencil. Ask them to draw a line. Then tell them to put the straight side of their protractor on the line they drew. Ask them to put a mark beside the 55° mark on their protractor. Repeat this for other numbers.

Unit 8 Geometry – properties of shapes

Point out to students that a protractor contains two scales and therefore two measures of 55. Remind them that the angle is a measure of turn and that they place the protractor so that the centre mark is on the vertex. They then follow the scale that measures the turn from 0 degrees.

Finally, demonstrate how to measure an angle to 5° with the protractor. Give students different angles to measure, such as 45°, 120°. Students should work in pairs so that they can check one another's accuracy using the protractor.

Students should then complete the activities on page 167 of the Student Book.

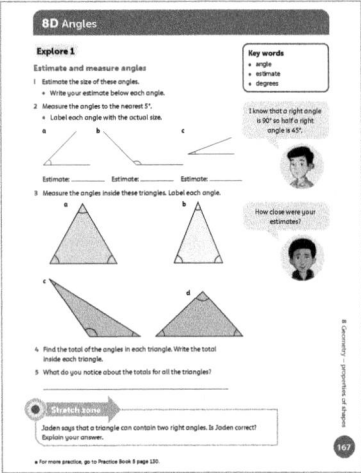

Differentiation

Supporting: Model the use of a protractor for students, taking them through each step carefully.

Consolidating: Ask students to describe how to use a protractor. How do they know which scale to use?

Extending: Ask students to estimate the angles before they measure them and try to measure to the nearest degree.

Stretch zone: *Jaden says that a triangle can contain two right angles. Is Jaden correct? Explain your answer.*

Jaden is not correct. Students should refer to triangles only having 180° in total in their explanation.

 Reflection time

Look together at questions 4 and 5 on page 167 of the Student Book. Ask students to share what they found about the total of the angles in each triangle. (The total of the angles for each is 180°.) Ask, *Do you think this will always happen?* Give students some paper. Ask them to draw a triangle and cut it out. They then cut off the corners and lay them in a straight line. Agree that a straight line has 180° and that this is the same as the inside angles of a triangle.

Practice Book: Students complete Practice Book page 130. They can do this directly after the Main activity, as homework, or as the focus of a separate mathematics session to help students consolidate their learning and build fluency.

Students will need to a ruler and a protractor to complete this activity. You may choose to demonstrate how to draw angles greater than 180 degrees with a protractor before they begin the activity.

Differentiated outcomes	
All students	should measure angles with support to a reasonable degree of accuracy.
Most students	will measure angles accurately.
Some students	may estimate and measure angles accurately.

Answers

Student Book page 167

1–3 Check that students have correctly measured the angles and that their estimates are reasonable.

4 The totals for every triangle will be 180°.

5 All triangles will have an angle sum of 180°.

Practice Book page 130

Check that students have drawn each reflex angle accurately.

Stretch zone: Check that students have drawn a 280° angle reasonably accurately.

8D Angles

Explore 2
Student Book page 168 • Practice Book page 131

Specific learning focus
- Understand and use angle measure in degrees.
- Measure angles to the nearest 5° and use this to draw triangles.

Global skills
- **Creative skills:** exploring

Key vocabulary
- angle, acute, obtuse, right, reflex

Resources
- paper, rulers, pencils, protractors

Language support

Provide sentence frames to support students to compare and describe their shapes.

My shape is the same because it _____.

My shape is different because it _____.

 **Introductory activity**

Review students' knowledge about different types of triangles. Ask them to describe the properties of scalene, right-angled, isosceles and equilateral triangles. Choose students to come forward to draw each one on the board and describe its properties, for example they should say that an equilateral triangle has 3 equal length sides and 3 equal acute angles.

 **Main activity**

Ask students to draw a triangle as follows: it has one side of 6 cm and one angle of 45°. Students should be prompted to start with a line of 6 cm and then use their protractors to measure the angle from one end of the line. They can complete the triangle with any other side lengths and resulting angles.

Ask students to share the triangles they drew. Ask, *Are your triangles all the same? How are they different? How are they the same?*

Now ask students to complete the activities on page 168 of the Student Book. Students should work in pairs and then compare their triangles. *Which ones did they draw the same? Which did they draw differently? Why are these different?*

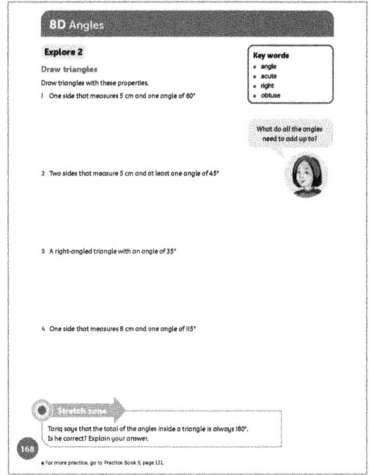

Differentiation

Supporting: Help students to measure lengths of sides and use a protractor to measure angles accurately when drawing the triangles.

Consolidating: Ask students to explain how they completed their triangles. *What was easy about drawing the triangles? What was challenging? Why?*

Extending: Challenge students to draw a triangle with a base of 10 cm and a side of 4 cm, with an angle of 25° between the two sides. Can they do it in more than one way?

Stretch zone: *Tariq says that the total of the angles inside a triangle is always 180°. Is he correct? Explain your answer.*

Students may explain this by drawing different examples of triangles and measuring the angles to find the total.

 Reflection time

Discuss with the class how they were able to complete the triangles from the information given. Can they give you any examples where there was more than one option for completing the triangle? Ask students to suggest other similar problems and see whether they can all be drawn.

Practice Book: Students complete Practice Book page 131. They can do this directly after the Main activity, as homework, or as the focus of a separate mathematics session to help students consolidate their learning and build fluency.

Students will need a protractor and a ruler to complete this activity. Ask students to explain how you find the perimeter and area of a rectangle before they begin the activity.

Differentiated outcomes	
All students	should measure and draw triangles with support.
Most students	will measure and draw triangles independently, measuring angles with a reasonable amount of accuracy.
Some students	may explore more than one option for each triangle from the information given.

Unit 8 Geometry – properties of shapes

Answers

Student Book page 168
Check that students have drawn triangles that match the information given.

Practice Book page 131
Check that students have drawn each shape accurately.

Stretch zone: False. A rectangle with an area of 24 cm² would have a perimeter of 20 cm, 22 cm, 28 cm or 50 cm. It is possible some students may say 'true' if they consider side lengths that are not whole centimetres. Mark as correct if their explanation is correct.)

8E Angle sums

Discover Student Book page 169 • Practice Book page 132

Specific learning focus
- Calculating the angle sums, at a point or on a straight line or in a triangle.

Global skills
- **Creative skills:** investigating
- **Real-world skills:** presenting information

Key vocabulary
- angles at a point, angles on a straight line, degrees

Resources
- large, class protractor or protractor for IWB
- mini whiteboards and markers
- paper, rulers, pencils, protractors

Language support
Continue to model and encourage correct pronunciation of key vocabulary.

 Introductory activity

Draw a right angle on the board and ask students to tell you how many degrees it measures. Now add a line to divide the angle into two equal angles and ask what the sum of the two angles must be. (90°) Then change the line to a different angle, dividing the right angle into, say, approximately 60° and 30°. Use a board protractor to measure one of the angles accurately and then ask the students to say how much the other angle must be and how they know. For example, if the larger angle is 60°, the smaller will be 90° − 60° = 30°. Repeat for two different partitions of 90° and ask different students to explain how they know one part if you have measured the other.

 Main activity

Ask students to draw on their whiteboards a right angle, and then to add two more lines from the angle to divide it into three smaller angles. Ask them to estimate and then measure any two of the angles and write them down. For example, they may have 54° and 24°. Can they calculate what the remaining angle must be? (54° + 24° = 78°, 90° − 78° = 12°). They should then measure it to check.

Now ask students to draw a straight line and add another line from the centre to make an angle with the straight line. Ask students what the two angles should total and agree that **angles on a straight line** will total 180°. Students can then estimate and measure one of the angles and calculate the other before measuring to check.

Students now complete the activities on page 169 of the Student Book. For each question, ask them to estimate the angles before measuring.

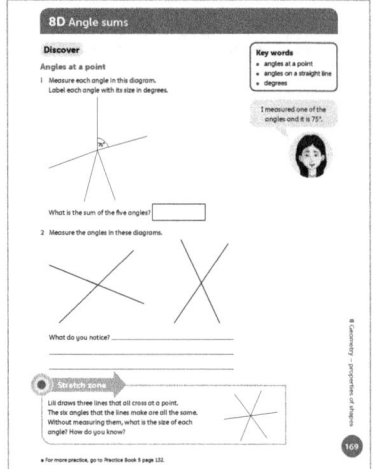

Differentiation

Supporting: Help students measure angles accurately and calculate missing angles.

Consolidating: Ask students to explain their estimates of angles.

Extending: Ask students to partition a straight line angle into four or more angles to estimate their size and then calculate.

Stretch zone: *Lili draws three lines that all cross at a point. The six angles that the lines make are all the same. Without measuring them, what is the size of each angle? How do you know?*

Unit 8 Geometry – properties of shapes

Students should use their knowledge of there being 360° in a full rotation to find out that each of the six angles will measure 60°.

 ## Reflection time

Look at the three diagrams from the Student Book. What did students notice about the angles for each of the diagrams? Agree that the opposite angles were equal and that all angles total 360°. Say, *The sum of* **angles at a point** *is 360 degrees.* Allow time for students to confirm that this is true.

Draw a star made from four lines that cross through a single point. Explain that each angle is equal. Say, *What have we learned today that could help us work out what each angle measures?*

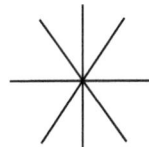

Leave pairs some time to discuss and work out the answer. They might divide 180 by 4 or 360 by 8 to find that each angle measures 45 degrees. Check by measuring using the class protractor, if appropriate.

Practice Book: Students complete Practice Book page 132. They can do this directly after the Main activity, as homework, or as the focus of a separate mathematics session to help students consolidate their learning and build fluency.

Students will need a protractor and a ruler to complete this activity.

Differentiated outcomes	
All students	should measure or calculate missing angles with support.
Most students	will estimate and measure or calculate missing angles.
Some students	may estimate and measure or calculate missing angles and share their reasoning.

Answers

Student Book page 169

1 360°

2 The opposite angles are equal and all four angles total 360°.

Practice Book page 132

Check that students have drawn each set of angles accurately.

Stretch zone: 72°

8E Angle sums

Explore Student Book page 170 • Practice Book page 133

Specific learning focus
- Calculating the angle sums at a point, in a triangle or a rectangle.

Global skills
- **Creative skills:** exploring
- **Real-world skills:** presenting information

Key vocabulary
- angles at a point, angles on a straight line, internal angle

Resources
- mini whiteboards and markers
- paper, rulers, pencils, protractors

Language support

Add angle rules to your working wall.
- Angles in a triangle add up to 180°.
- Angles on a straight line add up to 180°.
- Angles at a point add up to 360°.

Include visual examples to support each and write as a calculation next to each example.

 ### Introductory activity

Draw a right-angled triangle on the board. Ask students how many degrees there are in the triangle, referring them to the Think back statements on page 170 of Student Book, if needed. Ask a student to estimate the size of one of the acute angles, then measure it to check. Then ask students to calculate the size of the other acute angle. They may subtract the known acute angle from 90°, or they may add the right angle and the measured acute angle and subtract the total from 180°. Discuss each strategy in turn.

Unit 8 Geometry – properties of shapes 215

Main activity

Draw an isosceles triangle.

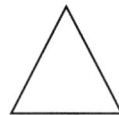

Ask students to estimate the angles. Now measure one angle on the isosceles angle. Ask, *Can you calculate the remaining angles now you know this one?'* Ask students to offer an answer and to describe their strategy and reasoning. If the known angle is unique, the other two must be equal and can be calculated, but if the known angle is one of the equal angles, the others can be easily calculated.

Repeat for a scalene triangle.

This time, ask for estimates for all three angles first. Then ask, *How many angles do I need to measure until can calculate the rest?* Students may explain that you need to know two of the angles to be able to calculate the third one. Now measure any two of the angles, and let students calculate the missing one.

Students can now complete the activities on page 170 of the Student Book, using their knowledge of angles in triangle, and angles on a straight line or at a point.

Use questions to support students' angle knowledge, for example:

- *How many degrees are there in a triangle?*
- *How many degrees are there in a quadrilateral?*
- *What do angles on a straight line add up to?*
- *What do you know about opposite angles at a point? How do you know?*
- *What information do you have that can help you work out the size of the missing angles?*

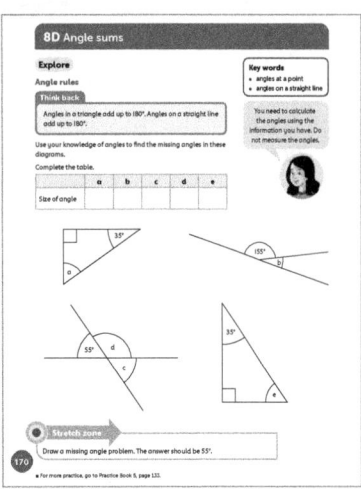

Differentiation

Supporting: Help students to measure angles accurately and calculate missing angles.

Consolidating: Ask students to explain their estimates of angles and how they worked out the size of the missing angles.

Extending: Challenge students to draw a triangle with an obtuse angle to estimate and then calculate the remaining angles.

Stretch zone: *Draw a missing angle problem. The answer should be 55°.*

Students might choose to make the 55° angle one in a triangle, or part of a straight line or around a point.

Reflection time

Ask students to think back over their calculations of missing angles. Ask them to share what they have learned about angles in different contexts, for example using the fact that there are 180° in a triangle to calculate missing angles.

Practice Book: Students complete Practice Book page 133. They can do this directly after the Main activity, as homework, or as the focus of a separate mathematics session to help students consolidate their learning and build fluency.

Students will need a protractor and a ruler to complete this activity.

Differentiated outcomes	
All students	should measure or calculate missing angles with support.
Most students	will estimate and measure or calculate missing angles.
Some students	may estimate and measure or calculate missing angles and explain their reasoning.

Answers

Student Book page 170

	a	b	c	d	e
Size of angle	55°	25°	55°	125°	55°

Practice Book page 133

Check that students have drawn the quadrilaterals correctly and measured their angles accurately.

Stretch zone: They all add up to 360°.

8 Geometry – properties of shapes

Connect Student Book page 171

Big idea
I can use a wide range of properties of shapes, such as their faces, their angles and their symmetry, to describe 2D and 3D shapes.

Global skills
- **Creative skills:** problem solving
- **Real-world skills:** presenting information/interpreting information
- **Interpersonal skills:** communication
- **Self-development skills:** reflecting on learning

Key vocabulary
- 2D shape, sides, angles, obtuse angle, acute angle, line of symmetry

Resources
- paper, rulers, pencils, protractors
- assorted boxes or containers

Language support
Encourage students to refer to their glossaries or the working wall if they are unsure of any vocabulary and to support them in writing their instructions.

Introductory activity
Explain that many boxes come 'flat packed' so they are easier to transport. They need to be assembled later. Show students various boxes. Ask them to think about what the boxes would look like flat or as a net. They sketch the net of one box then swap sketches with a partner. Can partners identify the box from the sketch? How?

Main activity
Students work in pairs for the activities on page 171 of the Student Book. Encourage them to interpret and begin the activity without your input. As they work, ask questions, for example:
- *Have you included all of the properties in your shape?*
- *Can you have more than one obtuse angle? Why?*
- *How many lines of symmetry does your shape have?*
- *Are your instructions for your partner clear?*

When students have finished the tasks, ask them to check that their partner has drawn a shape that fits the criteria. Tell them to check that the acute and obtuse angles are less than and greater than 90° respectively.

Ask them to join another pair and compare their shapes. Ask them to compare the angles and then to identify the smallest and largest angle, using a protractor to check.

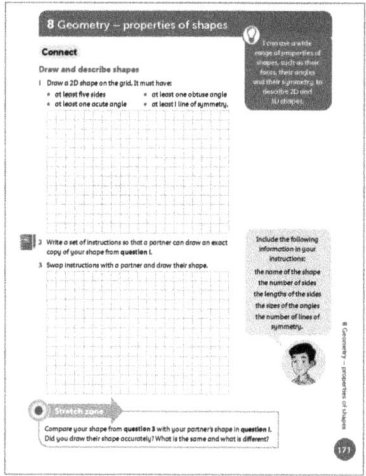

Differentiation
Supporting: Help students who are less confident with writing or speaking English to complete the sets of instructions.

Consolidating: Ask students, in pairs, to discuss and compare shapes.

Extending: Ask students to list their own set of criteria to use to draw a shape. Then they give it to their partner to use. Pairs compare shapes and discuss differences.

Stretch zone: *Compare your shape from question 3 with your partner's shape in question 1. Did you draw their shape accurately? What is the same and what is different?*

Students can verify whether their instructions were completed by their partner.

Reflection time
Choose some pairs to share their shapes and describe the properties in detail to the class. Next, ask selected students to read out their instructions and let the rest of the class try to draw the shapes on their whiteboards.

Use questions to review their knowledge of shapes, for example: *Can a triangle have angles that add to 100°? Can a rectangle have four lines of symmetry?*

Differentiated outcomes	
All students	should draw 2D shapes from a set of instructions including types of angles with support.
Most students	will draw 2D shapes from a set of instructions including types of angles.
Some students	may draw 2D shapes and write clear a set of instructions including types of angles, side properties and symmetry.

Answers

Student Book page 171

Check that students have drawn suitable shapes according to the properties and written a set of clear instructions.

8 Geometry – properties of shapes

Review Student Book page 172 • Practice Book page 134

Global skills

- **Creative skills:** problem solving
- **Real-world skills:** interpreting information
- **Interpersonal skills:** communication/teamwork
- **Self-development skills:** reflecting on learning

Student Book

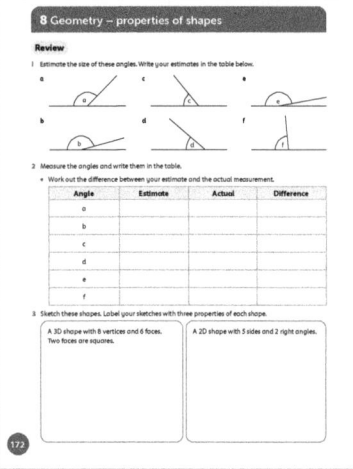

Answers

Student Book page 172

1. Check that students have estimated appropriately.
2. Check that students have estimated and measured appropriately and calculated the differences.
3. Check that students have drawn suitable shapes to include the stated properties.

Practice Book

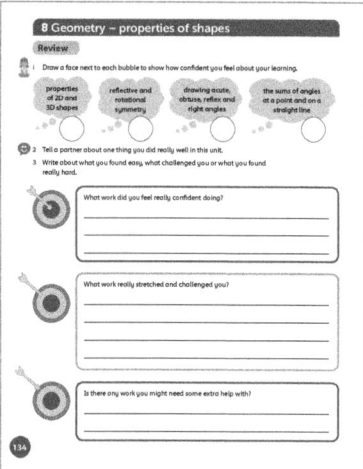

With young students, assessment activities are most effective when carried out as an everyday classroom activity. Students should have pencils, grid paper and protractors available to support them. They can also refer to the glossary at the back of the Student Book.

Students should complete the activity on page 172 of the Student Book individually as it is a summative assessment.

Watch as students estimate the size of an angle using their knowledge that the angles on a straight line add to 180 degrees to support their estimation. They measure the actual size of angles using a protractor and calculate the difference between its actual and estimated size. Observe their strategies for working the angles out. For question 3, some students may find that it easier to draw their 3D shape on grid paper and then cut it out and stick it in their Student Book.

Ask students further questions as they work through the activity to further assess their understanding, for example:

- *If you know the size of angle A, what does this tell us about the size of the other unmarked angle?*
- *How many acute angles does your 2D shape have?*
- *How large is the largest angle? How did you work that out?*
- *How much is the sum of the angles in your 2D shape?*
- *What do we call a shape that has the same face on both ends?*

With students in the upper primary years, it is appropriate to complete this Practice Book review as a whole-class discussion. You may choose to keep a record of the class discussion or a copy of the Review page for your own records. The Review provides an opportunity for students to reflect on their learning from the unit, to discuss any areas of mathematics that they feel went particularly well, and any areas that they feel less confident about. Ensure that all students have a copy of the Student Book as a reminder of the areas of mathematics that they have worked on in this unit.

Allow students plenty of time for discussion before asking them to complete the Practice Book page individually, and then, if appropriate, to share their responses with the rest of the class. If students complete this self-assessment at home, encourage them to discuss this with adults.

Continue to encourage students to discuss and notice shapes and angles across all subjects and in everyday contexts, describing them using mathematical terms across the year.

Additional material

There are additional end-of-unit assessment available on the *Oxford Owl for School* website.

9 Geometry – position and direction

Overview

Big idea

The Big idea for this unit is the transformation of shapes by movement. A shape plotted on a coordinate grid can be moved by translation, or slide, across the grid in any given direction, or it can be reflected in a line of symmetry.

Translations leave a shape looking identical but in a new position, as it slides a number of units in the horizontal direction and also a number of units in the vertical direction. The translation is defined by those units in the x and y directions. The size of a translated shape does not change, nor does its orientation.

When a shape is reflected, it 'flips' over to become a mirror image of itself but remains the same size. A reflection is defined by the location of the mirror line, which can be vertical, horizontal or any other straight line on the coordinate grid.

Look out for

- **Students who have difficulty translating a shape.** Reinforce the idea of translating each vertex of the shape in turn.
- **Students who forget how to read coordinates.** It may help to encourage them to think of a way of remembering, for example by thinking 'along the corridor and up the stairs'.
- **Students who find it difficult to visualise the activities, particularly when reflecting over a diagonal mirror line and rotating.** Give students lots of opportunities to do this. For example, you could ask them to draw and cut out a shape and reflect or rotate it in different ways. They can then close their eyes and visualise what they did.

Possible misconceptions

- **Students may think a left turn always takes them to facing west, for example.** Reinforce the idea that a left turn will take you towards a different compass direction, depending which way you are facing before turning.

Key vocabulary

- position, movement
- above, under, beside
- direction, turn, right, left
- clockwise, anti-clockwise
- coordinates, coordinate pairs, coordinate grid, horizontal axis, vertical axis, x-axis, y-axis, brackets, pattern, repeated
- mirror line, line of symmetry, reflective symmetry, reflection, rotation, rotational symmetry
- compass point, north, south, east, west, translate, translation

Coverage in lessons

Learning objective	E	9A	9B	9C	C	R
Identify, describe and represent the position of a shape following a reflection or translation, using the appropriate language, and know that the shape has not changed.	✓	✓	✓	✓	✓	✓

9 Geometry – position and direction

Engage Student Book page 173

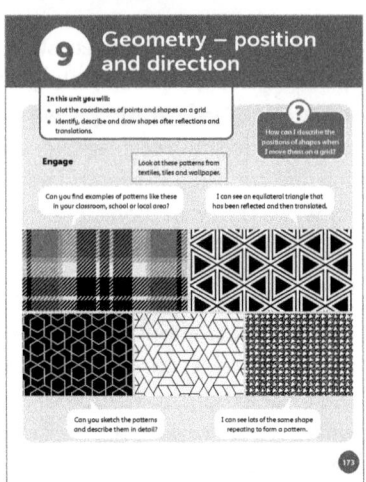

Big question
- How can I describe the positions of shapes when I move them on a grid?

Global skills
- **Creative skills:** exploring
- **Interpersonal skills:** communication
- **Self-development skills:** reflecting on learning

Key vocabulary
- coordinates, grid, reflections, translation, pattern, repeated

Resources
- paper and coloured pencils

Language support
Help students to describe the position of shapes and how they move under reflection and translation. Encourage them to use both prepositional language and mathematical language.

Introductory activity
Ask students, in pairs, to look together at the images on page 173 of the Student Book. They look carefully at the images and identify shapes they recognise. Ask them to describe the shapes as fully as possible. For example, instead of 'triangle', they should say 'equilateral triangle'. Ask them to choose a shape and then look for where it appears again in the pattern. Can they describe how the shape could have moved from one location to the other – has it been **reflected** or **translated**? Remind students if necessary that a translation means sliding a shape to a new position on a grid; the shape stays the same size and does not turn.

Main activity
Focus together now on the image of the equilateral triangles (top-right image) on page 173 of the Student Book. If you have access to an IWB, display the page. Point to one of the black equilateral triangles. Ask students to put their finger on the same triangle in their books, then see whether they can point to another triangle that is a reflection of the first one. Can they also see where the reflection line would be? Now ask them to look for other triangles that could be obtained by reflecting the first, and each time identify the reflection line. How many triangles in the image are reflections of the first one?

Now repeat, but look for triangles that are translations of the first. Can the triangle be translated in different directions to make new ones? How many triangles in the image are translations of the first one?

Put students into small groups to have a similar discussion about one of the other patterns. They can then try to sketch the pattern on a sheet of paper.

Differentiation
Supporting: Help students understand how one shape translates or reflects onto another using a mirror, for example.

Consolidating: Ask students to describe translations and reflections they can seen in the images in the Student Book.

Extending: Challenge students to describe any shapes that can be reached by doing a translation and then a reflection, or two translations.

Reflection time
Bring the lesson to a close by having a summary discussion about what students have noticed, learned and understood about the images and their repeated patterns. If students have noticed any similar patterns in the classroom, they can share these, along with any patterns they have sketched.

9A Coordinates

Discover Student Book page 174 • Practice Book page 135

Specific learning focus
- Read, write and plot coordinates in the first quadrant.

Global skills
- **Creative skills:** exploring
- **Real-world skills:** presenting information

Key vocabulary
- coordinate pairs, *x*-axis, *y*-axis, horizontal axis, vertical axis

Resources
- plain paper, squared paper, pencils
- a map or plan with coordinates, ideally of your local area

Language support
Draw a large coordinate grid on a poster and annotate it so that the key vocabulary can be part of a classroom display during the unit.

 Introductory activity

Ask students to work in small groups. Give each group a map. Ask them to find the coordinates. Establish that they are on the **horizontal axis (*x*-axis)** and the **vertical axis (*y*-axis)**. Demonstrate how you write **coordinate pairs**. To do this, find a feature on the map. Write the coordinates on the board, such as (3,6). Explain that we always write coordinates inside brackets with a comma separating them. Explain that you write the horizontal coordinate first and then the vertical coordinate. Ask students to find the coordinate that you wrote on the board on the map.

 **Main activity**

Ask students to locate town names, features or landmarks on their maps. Ask them to do this individually and to write the coordinates in the correct way on plain paper. Then ask them to give their paper to a partner. Their partner can then identify what place or landmark is at these coordinates.

Students should work in pairs on the activity on page 174 of the Student Book. Demonstrate the activity using the example. Remind students to read the number on the horizontal *x*-axis first, then the number on the vertical *y*-axis, and then to draw brackets around the coordinates.

Students should plot the second set of coordinates individually and then check one another's coordinates before joining them up to find out what shape they make.

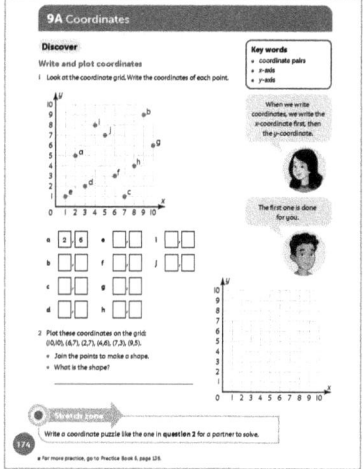

Differentiation

Supporting: Point to intersections on the grid and ask students to tell you the coordinates. Show students how to slide their finger along the grid lines to help them read the coordinates.

Consolidating: Ask students to tell you how they remember the way to write coordinates.

Extending: Ask students to create their own shapes using coordinates on the grid in the Student Book.

Stretch zone: *Write a coordinate puzzle like the one in question 2 for a partner to solve.*

Check that students have written their coordinates correctly and that they can describe the shape that the points make when joined up.

 Reflection time

Give each student a piece of squared paper. Ask them to draw a grid similar to the one in the Student Book. Model this on the board. Call out various coordinates. Ask students to find them on their grids and to mark them with a circle. Tell them to list the coordinates in the correct format, for example (8,9). Next, ask them to work with a partner to take it in turns to give coordinates. Ask them both to plot the coordinates onto their grids. Then ask them to compare their results, checking to see whether they agree. Invite some pairs to share their coordinates.

Practice Book: Students complete Practice Book page 135. They can do this directly after the Main activity, as homework, or as the focus of a separate mathematics session to help students consolidate their learning and build fluency.

Ask students to read the names of the fruits and vegetables aloud to model and practise their pronunciation. Tell students to keep the drawing of the fruits very simple and to focus on getting the coordinates correct.

Unit 9 Geometry – position and direction

Differentiated outcomes	
All students	should understand how to write coordinates.
Most students	will write and locate coordinate points accurately.
Some students	may draw a new shape using their own coordinates and list the coordinates correctly.

Answers

Student Book page 174

1. **a** (2,5) **b** (9,9) **c** (7,1) **d** (3,2)
   e (1,1) **f** (6,3) **g** (10,6) **h** (8,4)
   i (4,8) **j** (5,7)

2. Crosses should be on the correct coordinates, ensure that students plot the horizontal first.

The shape is an irregular hexagon.

Practice Book page 135

tomato	(5,0)
watermelon	(3,7)
squash	(2,3)
lemon	(0,7)
strawberry	(10,7)
pineapple	(7,6)
apple	(4,9)
garlic	(8,1)
pepper	(0,0)

Stretch zone: Students' answers will vary because they each choose a position for the first fruit. They should all get three fruits in a straight line.

9A Coordinates

Explore
Student Book page 175 • Practice Book page 136

Specific learning focus
- Read and plot coordinates in the first quadrant.

Global skills
- **Creative skills:** problem solving
- **Real-world skills:** presenting information

Key vocabulary
- coordinate pairs, x-axis, y-axis, vertex/vertices

Resources
- plain paper, pencils, squared paper, rulers

Language support
Model the key vocabulary carefully and frequently: 'coordinate', 'x-axis', 'y-axis', 'axes', 'vertex/vertices'. The pronunciation and spelling of these words do not follow patterns students will be more familiar with.

Introductory activity

Students work in pairs. Give each pair a piece of squared paper. Ask them to draw a vertical axis on the left of their paper covering ten squares and a horizontal axis of the same length from the base of the vertical axis. Tell them to label the vertical axis from 0 to 10 on the lines and repeat for the horizontal axis.

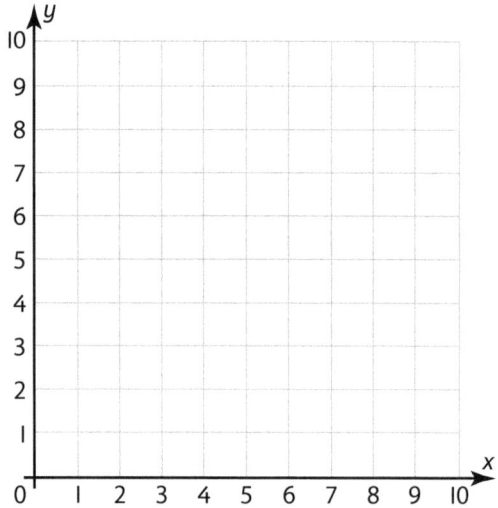

Explain to students that the axes are given names to help us refer to the coordinates that can be plotted. The horizontal axis is called the x-axis and the vertical axis is called the y-axis. Give an example of a pair of coordinates, for example the point (4,7). Tell students that the first coordinate, 4, is the x-coordinate and is counted along the horizontal x-axis. The second coordinate, 7, is the y-coordinate and is counted vertically in the direction of the y-axis. Tell them to label the axes y and x.

Ask students to plot eight different coordinates onto their grid, for example (6,3), (2,9) and six more. Tell them to join these together to make a shape. Ask, *What shape is this?* There may be several responses. The most likely one will be an irregular octagon.

Unit 9 Geometry – position and direction

 **Main activity**

Ask pairs to draw another grid. This time, give the following coordinate pairs: (2,5), (4,7), (4,3). Ask students to work out the fourth coordinate to make a square and to plot it onto their grid. Then ask them to join the coordinates to see whether they were correct. Repeat on the same grid for an isosceles triangle (5,8), (9,9), (?) and an irregular pentagon (6,1), (5,2), (7,4), (9,2), (?).

Look together at page 175 of the Student Book and explain what students need to do.

Ask students questions to support their learning as they work in their pairs, for example:

- *How do you read coordinates? What can you think of to help you to remember this?*
- *How do you record a coordinate? Why do you think the brackets are important?*
- *What size is an obtuse angle?*
- *Can you recall what a scalene triangle is?*

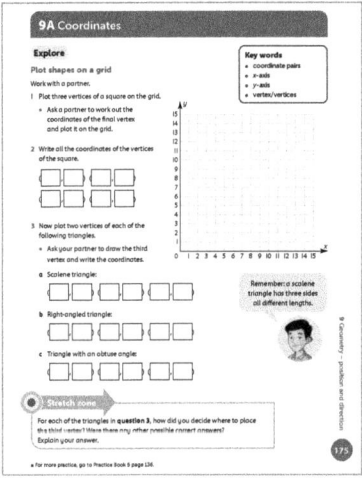

Differentiation

Supporting: Model how to find the first coordinate on the *x*-axis and place your finger on it. Then find the second coordinate on the *y*-axis, slide your fingers up and along to find the point where they intersect and then plot it.

Consolidating: Point to intersections on the grid and ask students to tell you the coordinates.

Extending: Ask students to create their own shapes using coordinates on the grid. Encourage them to explore irregular quadrilaterals, pentagons, hexagons and octagons.

Stretch zone: *For each of the triangles in question 3, how did you decide where to place the third vertex? Were there any other possible correct answers? Explain your answer.*

Students should give a clear explanation of how they chose the third **vertex** for each triangle. They should include their coordinates for each triangle as well as an example of at least one more possible coordinate for their third vertex. For the scalene triangle, there are numerous possibilities. For example, if the first two **vertices** are at (1,1) and (4,3), the third could be at (5,9) or (5,10).

For a right-angled triangle with vertices at (5,2) and (8,2), the third vertex could be at (8,4) or (5,4), for example.

For a triangle with an obtuse angle, with vertices at (4,1) and (9,2), the third vertex could be at (14,4) or (14,7), for example.

 Reflection time

Invite students to share the shapes they drew. Draw a grid on the board. Plot the coordinates for two corners of an equilateral triangle onto it. Ask students to visualise: *Where is the third coordinate?* Try out their suggestions. Ask, for example: *Why did you choose that coordinate? What helped you to complete the shapes? Is there anything else?*

Practice Book: Students complete Practice Book page 136. They can do this directly after the Main activity, as homework, or as the focus of a separate mathematics session to help students consolidate their learning and build fluency.

Students will need a ruler or a straight edge to complete this activity. Encourage students to refer to their glossaries to remind them of any properties of the shapes they will be plotting, if necessary.

Differentiated outcomes	
All students	should understand how to write coordinates.
Most students	will write and locate coordinate points on a grid accurately to make a 2D shape.
Some students	may draw a variety shapes accurately using their own coordinates.

Answers

Student Book page 175

1. Answers will vary but check that the missing vertex is identified correctly.
2. Check that the coordinates for the square have been written correctly.
3. Check that the triangles have been drawn and the coordinates of the vertices written correctly.

Practice Book page 136

Answers will vary because students use their own coordinates to draw the shapes. Check that all the required shapes have been plotted correctly:

1. square: (2,18)
2. isosceles triangle: the missing point could be anywhere that has 15 as the horizontal coordinate, for example (15,5), (15,12)
3. rectangle: (8,6)

Unit 9 Geometry – position and direction

4 hexagon: it is most likely that the students give the missing coordinate as (0,2) but it could in fact be anywhere that isn't in line with two of the given coordinates.

Check that the shapes have been drawn correctly on the coordinate axes.

Stretch zone: Check that students have written the correct coordinates for their own shapes.

9B Reflection

Discover
Student Book page 176 • Practice Book page 137

Specific learning focus
- Predict where a polygon will be after reflection where the mirror line is parallel to one of the sides, including where the line is diagonal.

Global skills
- **Creative skills:** investigating

Key vocabulary
- line of symmetry, reflect, *x*-axis, *y*-axis

Resources
- squared paper or blank coordinate grids, pencils
- rulers
- mirrors

Language support
Read the text in the green box on page 176 of the Student Book together. Ask questions to check that students have understood. Ask, for example, *Can you show me a horizontal mirror line? Which mirror line is y = 5?*

Introductory activity

Ask the class, *Can you explain what reflection is?* Take students' suggestions, for example: 'It is like flipping a shape over a mirror line.'

Ask pairs of students to discuss where they can find examples of reflections in everyday life. They may suggest, for example, patterns on rugs or carpets, clothing, wallpaper or wrapping paper. Look around the classroom to see whether they can spot any reflections.

Main activity

Give each pair of students a piece of squared paper and a mirror. Ask them to draw a coordinate grid or give them a blank one to work with. Explain that they should label it with numbers as they did in 9A Explore. Ask students to draw a triangle plotting the vertices at coordinates (2,9), (1,6) and (3,6). Next, ask them to draw a mirror line from (3,0) to (3,10) and to reflect their triangle across it

by placing the mirror along the mirror line to see where the reflection will be. Now, ask them to draw a diagonal mirror line from (0,0) to (10,10) and to reflect the second triangle they drew across it.

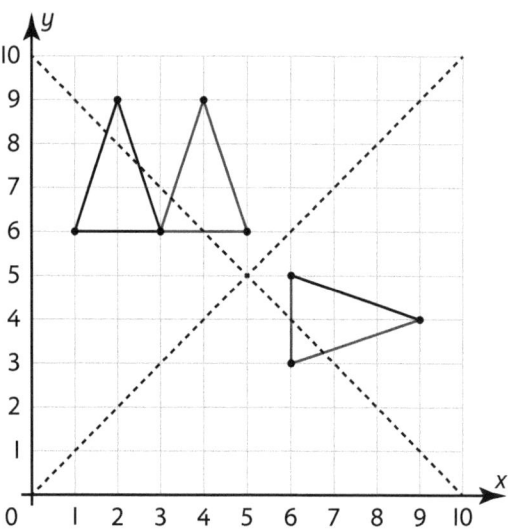

Now ask students to draw another grid. Repeat this activity for a rectangle at (6,9), (7,9), (6,5), (7,5) and ask them to draw a horizontal mirror line from (0,5) to (10,5). Now ask them to reflect the rectangle in the mirror line. Then, they draw a diagonal mirror line from (0,0) to (10,10) and reflect the original rectangle along this line. Students can use their mirrors to check that their partner has reflected their shapes correctly.

Refer students to page 176 of the Student Book. They should continue to use a mirror placed on the mirror line to help them reflect each shape. After they have reflected the first shape, ask pairs to discuss how they could find a reflection without using a mirror. Point out the text in the speech bubble to prompt their discussion.

Pairs should then reflect the shapes, using a mirror or using a mirror line to check whether they are correct, as appropriate.

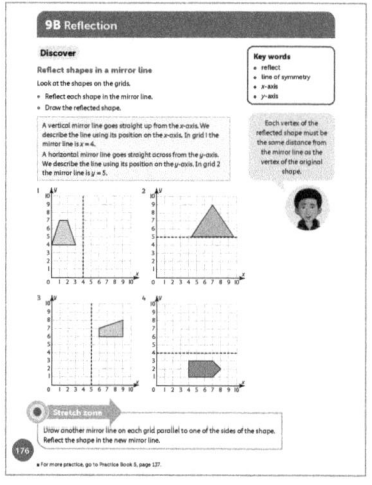

224 Unit 9 Geometry – position and direction

Differentiation

Supporting: Support students to use a mirror to find reflections.

Consolidating: Ask students how they know where to place a reflection with and without a mirror.

Extending: Encourage students to use the mirror to check their reflections and ask them to explain in detail how they reflected one of the shapes without a mirror.

Stretch zone: *Draw another mirror line on each grid parallel to one of the sides of the shape. Reflect the shape in the new mirror line.*

Remind students what parallel means, if necessary. Check that they have reflected the shape correctly along the new mirror line.

Reflection time

Ask students to draw their reflections on a large grid on the board. Ask them to describe how they could find the reflection without using a mirror. They may suggest using tracing paper or cutting out a copy of the shape and then folding it over the mirror line. They may also suggest working out how far each coordinate point is from the mirror line by counting whole and part squares then 'flipping' the points the same distance on the other side of the mirror line.

Practice Book: Students complete Practice Book page 137. They can do this directly after the Main activity, as homework, or as the focus of a separate mathematics session to help students consolidate their learning and build fluency.

Encourage students to work in pencil and use a ruler or straight edge to join up their points. You may want to ask students to record the coordinates of the original shapes as well as their reflections.

Differentiated outcomes	
All students	should reflect shapes in a mirror line using mirrors.
Most students	will begin to understand how to reflect in a diagonal mirror line.
Some students	may understand that you measure the distance of equivalent points perpendicular to the mirror line.

Answers

Student Book page 176

Check that the reflections have been drawn in these positions:

1 reflection of trapezium has vertices at: (8,4), (5,4), (7,7), (6,7)

2 reflection of triangle has vertices at: $(4\frac{1}{2},5)$, (7,1), $(9\frac{1}{2},5)$

3 reflection of trapezium has vertices at: (4,7), (4,6), (1,6), (1,8)

4 reflection of pentagon has vertices at: (4,5), (4,7), (7,5), (7,7), (8,6)

Practice Book page 137

Check that the reflections have been drawn in these positions:

1 reflection of quadrilateral has vertices at: (1,8), (3,9), (4,8), (2,6)

2 reflection of triangle has vertices at: (6,1), (10,0), (9,4)

3 reflection of trapezium has vertices at: (6,8), (8,6), (9,8), (8,9)

4 reflection of pentagon has vertices at: (2,4), (1,3), (1,1), (3,1), (4,2)

Stretch zone: The two reflections are reflections of each other and the second reflection is the same as the starting shape, but translated.

Unit 9 Geometry – position and direction

9B Reflection

Explore 1
Student Book page 177 • Practice Book page 138

Specific learning focus
- Predict where a polygon will be after reflection where the mirror line is parallel to one of the sides, including where the line is diagonal.

Global skills
- **Creative skills:** investigating
- **Real-world skills:** presenting information/interpreting information

Key vocabulary
- line of symmetry, reflection

Resources
- squared paper, coloured pencils
- polygon cut out of card
- card copy of the arrow from question 3 on page 177 of the Student Book
- mirrors

Language support
In Reflection time, set up 'talk partners' so that one partner can provide a good model of spoken English. When you work with individuals, make sure that you use the key phrase 'reflection in the … axis'.

 Introductory activity

Review reflecting in a mirror line from 9B Discover. Explain that it is possible to reflect a shape in different mirror lines. Draw a set of axes on the board and draw a horizontal and vertical line to split the grid into four sections as on page 177 of the Student Book.

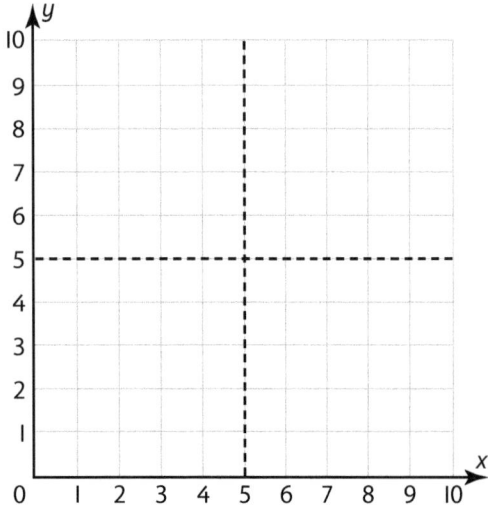

Take your card polygon and stick it on the board in the first quadrant so that one of its vertices or sides touches the horizontal line that you drew. Ask a student to come and draw where they think the reflection in the horizontal line will be. How do they know? Check the answer by 'flipping' the shape in the horizontal line.

Now reflect the original shape in the vertical line. *Where will the reflection be? How do you know?* Look at the coordinates for each shape's vertices. *Can you notice anything about the x-coordinates when reflecting in a horizontal mirror line? What about the y-coordinates when reflecting in a vertical mirror line?*

Main activity

Ask students to draw a set of axes like those on page 177 of the Student Book on a piece of squared paper. They can use these axes to support them in the activity, which they can work on in pairs. As you move around the class, ask students what they notice about the reflections. Support them in noticing that the vertices of the shapes are always the same distance from the axis on which they are reflecting.

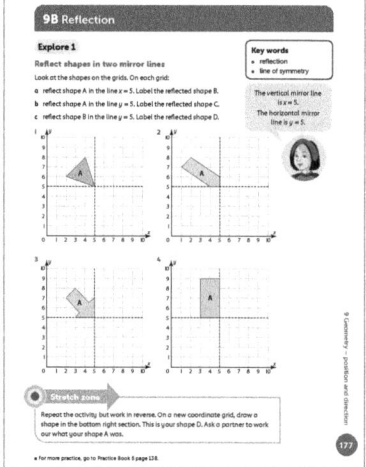

Differentiation

Supporting: Support students to use a mirror to reflect shapes. They can also cut out the shapes if they need help with visualising the reflections and use mirrors to check.

Consolidating: Ask students to tell you the coordinates for their reflected shapes.

Extending: Encourage students to measure to reflect shapes and use mirrors to check their reflections.

Stretch zone: *Repeat the activity but work in reverse. On a new coordinate grid, draw a shape in the bottom right section. This is your shape D. Ask a partner to work out what your shape A was.*

Students should explain how they found shape A by reflections.

Reflection time

Use your card copy of an arrow to illustrate the answer to question 3 on page 177 of the Student Book. Ask

226 Unit 9 Geometry – position and direction

students to talk to a partner about anything they noticed about the reflections they were making. Take feedback, drawing on as many different pairs as possible. Students may, for example, comment on how all the arrowheads meet in a point at (5,5), or they may notice that they could reflect both shapes above $y = 5$ together.

Practice Book: Students complete Practice Book page 138. They can do this directly after the Main activity, as homework, or as the focus of a separate mathematics session to help students consolidate their learning and build fluency.

Students will need a ruler and coloured pencils to complete this activity.

Differentiated outcomes	
All students	should find reflections using mirrors.
Most students	will begin to predict the coordinates of a reflected shape and check with a mirror.
Some students	may understand that you measure the distance of equivalent points perpendicular to the mirror line and use this knowledge to reflect shapes in the mirror line.

9B Reflection

Explore 2 Student Book page 178 • Practice Book page 139

Specific learning focus
- Recognise reflective symmetry in regular polygons.

Global skills
- **Creative skills:** investigating
- **Real-world skills:** presenting information/interpreting information
- **Interpersonal skills:** communication/teamwork

Key vocabulary
- reflection, line of symmetry, mirror line

Resources
- mini whiteboards and markers
- mirrors, scissors, pencils, rulers, squared paper

Language support
Support the language needed to describe the shapes, for example:

A polygon is a closed shape with more than three sides.

Answers

Student Book page 177

1. B is (5,5), (6,8), (8,6)
   C is (5,5), (2,4), (4,2)
   D is (5,5), (6,2), (8,4)
2. B is (5,5), (5,6), (6,5), (9,7), (8,8)
   C is (5,5), (4,5), (5,4), (2,2), (1,3)
   D is (5,5), (5,4), (6,5), (8,2), (9,3)
3. B is (5,5), (5,7), (7,5), (8,7), (7,8)
   C is (5,5), (3,5), (5,3), (2,3), (3,2)
   D is (5,5), (5,3), (7,5), (8,3), (7,2)
4. B is (5,5), (5,9), (7,5), (7,9)
   C is (5,5), (3,5), (5,1), (3,1)
   D is (5,5), (7,5), (5,1), (7,1)

Practice Book page 138

Check that students have drawn the shape in the correct positions after the translations.

Stretch zone: Students should be able to comment about coordinates that change or remain unchanged after reflections.

 **Introductory activity**

Leaving time for students to discuss first in pairs, ask, *What is a polygon?* (a closed shape with three or more sides, all straight sides) They then complete the following sentences: 'The sides and angles of a regular polygon are …'; 'An irregular polygon has …'. Discuss their answers as a class.

Next, ask pairs of students to sketch a variety of regular and irregular polygons on their whiteboards. Invite students to draw their sketches on the board and to name the shapes. Discuss the properties of each of the shapes. Include:

- number of sides and vertices (equal or not)
- types of angle (acute, obtuse, right)
- **lines of symmetry**.

Invite students to draw the lines of symmetry on the shapes on the board, where appropriate. Ensure that the following shapes are drawn: equilateral, isosceles and scalene triangle, square, rectangle, other quadrilaterals, pentagon, hexagon, heptagon and octagon.

 **Main activity**

Allow plenty of time for students to complete the investigations on page 178 of the Student Book. Go through the instructions together. Discuss what is meant by predicting where a reflection will be. Can they explain

Unit 9 Geometry – position and direction

how they could do this? Agree that they can work out how far the vertices of the original shape are from the **mirror line**. They then plot the coordinates of each vertex of the reflected shape the same distance from the mirror line but the other side of it.

Remind students how to use a mirror to check reflections. Each pair will need scissors, squared paper, pencil, ruler and a mirror. The pairs should collaborate, checking one another's solutions.

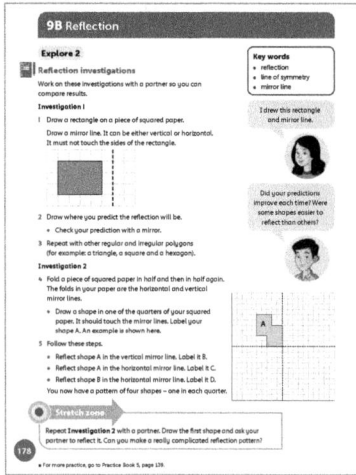

Differentiation

Supporting: Help students to use mirror lines parallel to a side of a polygon.

Consolidating: Ask students to use mirror lines that are not parallel to a side of a polygon.

Extending: Challenge students to use a range of mirror lines not parallel to a side of a polygon.

Stretch zone: *Repeat Investigation 2 with a partner. Draw the first shape and ask your partner to reflect it. Can you make a really complicated reflection pattern?*

Check that students have reflected the pattern correctly. Can they explain how they know that they have done it correctly?

 Reflection time

Take feedback from the investigations in the Student Book. Invite students to show the shapes they drew and what they looked like after reflecting them in the horizontal and vertical mirror lines. Invite other students to show and describe how they made their symmetrical patterns.

Practice Book: Students complete Practice Book page 139. They can do this directly after the Main activity, as homework, or as the focus of a separate mathematics session to help students consolidate their learning and build fluency.

If appropriate, ask students to describe to a partner or an adult how they know whether the reflection is correct or incorrect.

Differentiated outcomes	
All students	should recognise symmetry when the mirror line is parallel to a side of a polygon.
Most students	will recognise symmetry when the mirror line is not parallel to a side of a polygon.
Some students	may recognise symmetry wherever mirror lines are placed.

Answers

Student Book page 178

Students experiment with other regular and irregular polygons, so answers will vary.

Practice Book page 139

1 incorrect

2 incorrect

3 correct

4 correct

Check that for the incorrect reflections, students have drawn the correct version.

Stretch zone: True. Reflection only changes the position and orientation of a shape, it doesn't alter the shape's size.

Unit 9 Geometry – position and direction

9C Translations

Discover
Student Book page 179 • Practice Book page 140

Specific learning focus
- Understand translation as movement along a straight line.
- Identify where polygons will be after a translation and give instructions for translating shapes.

Global skills
- **Creative skills:** exploring
- **Real-world skills:** presenting information/interpreting information
- **Interpersonal skills:** communication/teamwork

Key vocabulary
- translation, translate, position, coordinate grid

Resources
- counters
- squared paper

Language support
Model how to give translation instructions and provide students with a list of key vocabulary with visual clues to support their understanding. For example:

Translate/Slide your counter (1, 2, 3, 4 and so on) _____ (up, down, left, right).

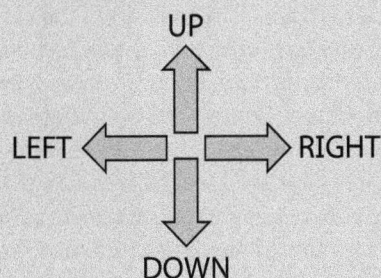

Introductory activity
Give pairs of students a counter and a piece of squared paper. Ask them to place the counter in the top left square. Call out instructions, for example: *Slide the counter 4 squares to the right, 3 squares down, 2 left, 1 up.* After all the instructions, students can compare the position of their counter with other students. Ask, *Are all your counters in the same **position**?*

Repeat, with another example.

If possible, on a grid background on the board, hold a 'counter' on a position on the grid and mark your start position. Explain that you are going to translate it, but this time you would like students to describe each translation. Move the counter slowly and then mark your new position. Choose a student to describe the translation.

Main activity
Ask students to do this activity with a partner. One gives the instructions, keeping a note of them by marking the movement on their paper. The other moves their counter according to the instructions. When finished, they can compare finishing positions to ensure that the 'mover' followed the instructions correctly. They then swap roles. Invite pairs to demonstrate what they did by giving their instructions for the class to follow.

Tell students that this movement is called a **translation**. Say, *When we slide an object from one position to another, we **translate** it*. Ask students to discuss where they can see translations in everyday life or in patterns. For example, wallpaper designs are often made from translations and so are designs on clothing and curtain materials.

Students should complete the activity on page 179 of the Student Book individually. Once they have completed question 2, they should give their instructions to a partner, who follows these and draws the shape in the end position on their grid.

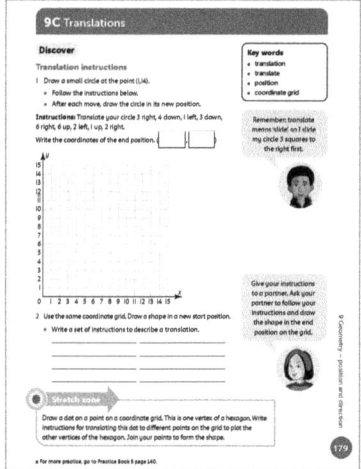

Differentiation

Supporting: Give students a sentence frame to support them with writing instructions. Some students may benefit from physically moving an object to complete the activities in the Student Book.

Consolidating: Give students instructions to follow to translate a shape. Can they tell you the coordinates of its end position?

Extending: Ask students to give you instructions to translate a shape on a grid. Make deliberate mistakes to see whether they correct you.

Stretch zone: *Draw a dot on a point on a coordinate grid. This is one vertex of a hexagon. Write instructions for translating this dot to different points on the grid to plot the other vertices of the hexagon. Join your points to form the shape.*

Unit 9 Geometry – position and direction

Students could swap their instructions and then join the points to see whether the instructions are correct.

Reflection time

Invite students to describe their translations for other students in the class to follow. Ask students to give the coordinates of the starting and end position.

Practice Book: Students complete Practice Book page 140. They can do this directly after the Main activity, as homework, or as the focus of a separate mathematics session to help students consolidate their learning and build fluency.

You may want to extend this activity by asking students to record the coordinates of the translated shapes.

Differentiated outcomes	
All students	should follow instructions to translate a shape accurately with support.
Most students	will give and follow instructions accurately.
Some students	may describe translations using coordinates.

Answers

Student Book page 179

1 (9,14)

2 For students' own instructions, check that the instructions describe their translation accurately.

Practice Book page 140

Check that the translations have been drawn in these positions:

blue shape has vertices at:
(12,1), (17,1), (19,3), (17,5), (12,5)

green shape has vertices at:
(4,7), (9,7), (11,9), (9,11), (4,11)

Stretch zone: True. Check that students' sketches support their answer.

9C Translations

Explore Student Book pages 180–181 • Practice Book page 141

Specific learning focus
- Understand translation as movement along a straight line.
- Identify where polygons will be after a translation and give instructions for translating shapes.

Global skills
- **Creative skills:** problem solving/exploring
- **Real-world skills:** presenting information/ interpreting information
- **Interpersonal skills:** communication/teamwork

Key vocabulary
- translation, instruction

Resources
- A4 paper, squared paper, coloured pencils

Language support

Check that students know which direction is left and right so that they can give clear instructions. Model again how to give translation instructions.

Introductory activity

Ask a student to move from one part of the classroom to another. Give **instructions** for them, such as *Move straight, left, forwards, right, backwards*. Next, ask another student to make up their own route around the classroom. Tell the rest of the class to describe the students' routes. Give students half a piece of A4 paper. Ask them to write down a translation instruction, such as *6 steps to the right*. Invite three students to stand in different parts of the classroom. Ask students, one at a time, to read the instruction they wrote. Tell the three students to move accordingly. When a student is stopped by a piece of classroom furniture, they are out. Continue until there is only one student standing. Repeat this with a different three students.

 Main activity

Explain the tasks on pages 180–181 of the Student Book. Work through an example together to ensure that students understand what to do. Students can work independently on the first part of the activity and in pairs on the second part.

230 Unit 9 Geometry – position and direction

While students are working, listen to the instructions they give and check whether they can follow instructions carefully.

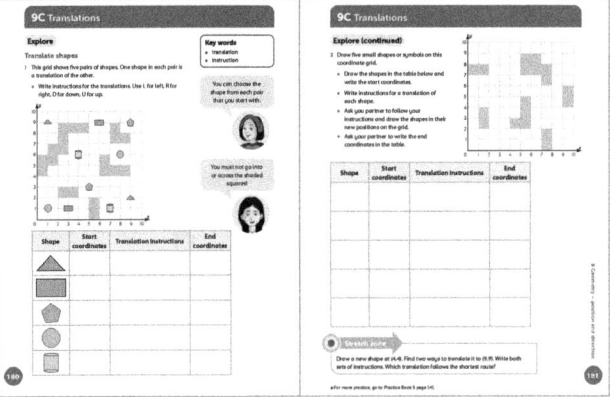

Differentiation

Supporting: Provide cut-outs of the shapes that can be physically manipulated on coordinate grids like the ones in the Student Book.

Consolidating: Ask students to explain how they followed the instructions to translate the shapes and record the end coordinates of the vertices.

Extending: Choose two points on the grid on page 181 of the Student Book. Ask students to write instructions with the fewest steps possible to get a shape from one point to the other.

Stretch zone: *Draw a new shape at (4,4). Find two ways to translate it to (9,9). Write both sets of instructions. Which translation follows the shortest route?*

Check that students' instructions end at (9,9) and that they have identified the shorter route correctly.

Reflection time

Take feedback from the activities in the Student Book. Invite students to share the translations they used to move the shapes. Were they the same? Did some students find different ways? Together, work out all the possible ways to translate some of the shapes.

Practice Book: Students complete Practice Book page 141. They can do this directly after the Main activity, as homework, or as the focus of a separate mathematics session to help students consolidate their learning and build fluency.

Students will need coloured pencils and a ruler to complete this activity. They will also need to work with a partner, or an adult, to complete the Stretch zone activity.

Differentiated outcomes	
All students	should write translation instructions accurately with support.
Most students	will give and follow translation instructions accurately.
Some students	may understand translations using coordinates.

Answers

Student Book pages 180–181

1 Note that students can start with either of the two shapes so the opposite translation to those listed will also be correct.

Shape	Start	Translation	End
triangle	(1,9)	8 right, 7 down	(9,2)
rectangle	(6,9)	3 left, 8 down	(3,1)
pentagon	(5,3)	4 right, 6 up	(9,9)
circle	(1,1)	7 right, 5 up	(8,6)
cylinder	(4,6)	3 right, 5 down	(7,1)

2 Students choose their own routes around the grid, so answers will vary. Check that the written instructions work for each shape.

Practice Book page 141

Check that students have drawn the shape in the correct positions after the translations. The positions will depend on students' choices of mirror lines.

Stretch zone: Check that students' instructions are accurate.

Unit 9 Geometry – position and direction

9 Geometry – position and direction

Connect   Student Book page 182

Big idea

I can use coordinates to describe the positions of shapes and objects after I have reflected them or translated them.

Global skills

- **Creative skills:** problem solving
- **Real-world skills:** presenting information
- **Self-development skills:** reflecting on learning

Key vocabulary

- coordinates, *x*-axis, *y*-axis, translation, reflection

Resources

- plain paper, squared paper

Language support

Check that students understand the context of each word problem and any related vocabulary so that it does not interfere with their ability to solve the problem.

Introductory activity

Ask students to draw a numbered grid, from 0 to 10 on both the *x*-axis and *y*-axis. Ask them to plot the following points and join them to make a shape: (1,1), (3,1), (1,5), (3,3).

Now ask them to carry out the following moves: a translation 5 squares to the right, followed by a reflection in $y = 5$, then a translation 5 squares to the left.

Ask, *What move will bring the final shape back to where it started?* Students complete this in pairs, discussing how to complete each move and checking with each other. After a short time, draw students together to discuss their results and allow them to explain their methods.

Main activity

Discuss with students situations in real life where translations and reflections are used and why they may be necessary. Examples could include designing buildings to be symmetrical, border patterns or painting the word 'AMBULANCE' back to front on a vehicle so motorists can see it the right way round in their mirror when driving.

Now ask students to look at the activities on page 182. They plot a shape and then move it in the context of two landing areas for parachute jumps. They can work in pairs as they plot the landing areas and discuss how they do this.

Unit 9 Geometry – position and direction

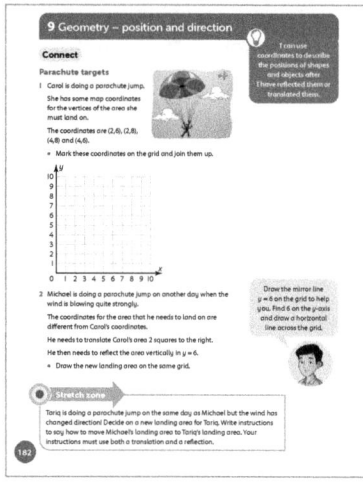

Differentiation

Supporting: Take students through the problem-solving steps, helping them to identify key information and work out what they will need to do to solve the problem.

Consolidating: Ask students to explain their reasoning for how they completed the translation and reflection in the Introductory activity.

Extending: Challenge students to write their own similar word problems to those in the Student Book. Encourage them to include both a translation and a reflection.

Stretch zone: *Tariq is doing a parachute jump on the same day as Michael but the wind has changed direction! Decide on a new landing area for Tariq. Write instructions to say how to move Michael's landing area to Tariq's landing area. Your instructions must use both a translation and a reflection.*

Encourage students also to describe their strategy for deciding on a new landing area and to add the new landing area to their grid.

 Reflection time

Ask selected students to explain the steps they used to plot each coordinate and move the landing area. Use questions to review their knowledge of translations and reflections, for example: *Can any coordinates stay the same when a shape is translated? Can any coordinates stay the same when a shape is reflected? Can you explain your answer?*

Differentiated outcomes	
All students	should translate or reflect a shape with support.
Most students	will translate or reflect a shape.
Some students	may construct their own problems involving translating or reflecting a shape, and explain their reasoning.

Answers

Student Book page 182

Carol's coordinates form a square matching the coordinates stated (horizontal axis first).

Michael's coordinates: (6,6), (6,4), (4,4), (6,4)

9 Geometry – position and direction

Review Student Book page 183 • Practice Book page 142

Global skills

- **Creative skills:** problem solving
- **Real-world skills** interpreting information
- **Interpersonal skills:** communication/teamwork
- **Self-development skills:** reflecting on learning

Student Book

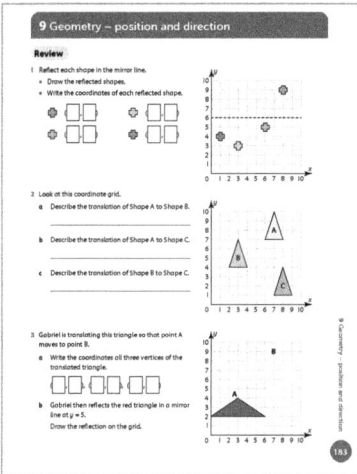

With young students, assessment activities are most effective when carried out as an everyday classroom activity. Students should have pencils, rulers and squared paper available. They can also refer to the glossary at the back of the Student Book if they are unsure of the meaning of any key words.

Watch as students draw and reflect shapes using coordinates, identify lines of symmetry and describe translation of shapes, and observe their strategies for working them out. Extend by asking a range of questions about plotting coordinates and translating shapes.

Students should complete the Review activity on page 183 of the Student Book individually as it is a summative assessment.

Answers

Student Book page 183

1 Check that students have drawn the shapes correctly. The new coordinates are: (1,8), (3,9), (6,7), (8,3).

2 **a** A to B: 4 squares left, 3 squares down

   b A to C: 1 square right, 6 squares down

   c B to C: 5 squares right, 3 squares down.

3 **a** (7,9), (4,7), (10,7)

   b Check that the shape is drawn correctly, with reflected coordinates (4,6), (0,8), (6,8)

Practice Book

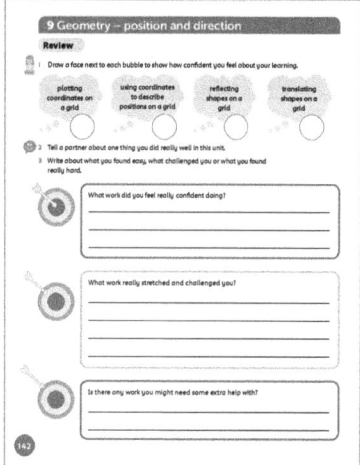

With students in upper primary years, it is appropriate to complete this Practice Book review as a whole-class discussion. You may choose to keep a record of the class discussion or a copy of the Review page for your own records. The Review provides an opportunity for students to reflect on their learning from the unit, to discuss any areas of mathematics that they feel went particularly well, and any areas that they feel less confident about. Ensure that all students have a copy of the Student Book as a reminder of the areas of mathematics that they have worked on in this unit.

Allow students plenty of time for discussion before asking them to complete the Practice Book page individually, and then, if appropriate, to share their responses with the rest of the class. If students complete this self-assessment at home, encourage them to discuss this with adults.

Make a note of areas that students still feel unsure about, for example plotting coordinates correctly, not confusing the x-axis and y-axis and being able to translate and reflect shapes accurately. Give students plenty of further opportunities to plot coordinates and translate or reflect coordinates and shapes. You could look for opportunities across subject areas. For example, opportunities might include playing a translation game in PE, creating repeating patterns in Art, or map reading in Geography.

Additional material

There are additional end-of-unit assessment available on the *Oxford Owl for School* website.

10 Statistics

Overview

Big idea

The Big idea for this unit is that data can be presented in different ways. This is part of the 'data handling cycle' along with collecting, organising and interpreting data.

Data can be either discrete, which means separate, or continuous, which means it can take any value, whether whole or not. Both types of data can be recorded in a frequency table, which records how many. For discrete data such as number of vehicles in a traffic survey, this could be 4 cars, 7 lorries, 1 bus and so on. For discrete data, such as students' heights, the data can be recorded in groups, for example from 1 m to 1.1 m, 1.1 m to 1.2 m and so on. The representation of the data in this unit revisits pictograms for discrete data and line graphs for continuous data. When reading a line graph, interpretation is needed to make sense of what the line means in between the plotted points.

Interpreting data can also mean reading information from tables and, in this unit, timetables give students opportunities to interrogate data and make sense of it in real-life contexts such as planning journeys.

Probability is about the presentation of data on the 'likelihood' of events occurring. Probability is a measurement of likelihood based on belief. How strongly do we believe that an event will happen? This can range from 'impossible' through to 'certain'. We compare events by saying which is more likely, and attributing a number to this, in the range 0 to 1. The probability of some events can be calculated by considering all possible outcomes, such as choosing a card randomly from a set of digit cards.

Look out for

- **Students who may choose an inappropriate scale or number the axes incorrectly.** Encourage them to check carefully. Students should always use squared paper for drawing graphs, not paper that is blank or lined.

- **Students who do not fully understand the meaning of each section of a Carroll or Venn diagram.** By working through examples and discussing which criteria fit each data item, students will gain deeper understanding of how the diagrams represent data.

- **Students who find it difficult to read axes on graphs that go up in intervals of more than one.** It may help to link axes, particularly vertical ones, to number lines that go up in steps of different sizes. In this way, students can see that the axes on graphs are basically number lines.

- **Students who make only very simple statements such as 'three children like cakes'.** Encourage them to think more deeply and draw more complex conclusions, for example: 'Five more children like biscuits than cakes, so it's a good idea for Jim to buy biscuits for the party.'

Possible misconceptions

- **Students may think that a line graph is accurate at all points along the line.** Give opportunities to examine line graphs and discuss the interpretation of what the line can represent between the marked points.

Key vocabulary

- tally, list, frequency table, frequency, line segment, plot
- Carroll diagram, Venn diagram, pictogram, bar chart, bar line graph, line graph, time graph
- timetable, time interval, duration, time difference
- label, title; vertical axis, horizontal axis, axes, x-axis, y-axis
- most popular, least popular, mode
- survey, data, discrete data, continuous data, frequency, likelihood, probability, chance, even chance, event

Coverage in lessons

Learning objective	E	10A	10B	10C	10D	C	R
Solve comparison, sum and difference problems using information presented in a line graph.		✓	✓			✓	✓
Complete, read and interpret information in tables, including timetables.	✓	✓	✓	✓	✓	✓	✓
Use the language of probability to describe the likelihood of an event occurring.					✓		

10 Statistics

Engage Student Book page 184

Big question
- How can I present data using tables, timetables and line graphs? Can I present the same data in different ways?

Global skills
- **Creative skills:** investigating
- **Real-world skills:** presenting/interpreting information
- **Interpersonal skills:** communication/teamwork
- **Self-development skills:** reflecting on learning

Key vocabulary
- data, tally, frequency table, Carroll diagram, Venn diagram, pictogram, bar chart, bar line graph, line graph, vertical axis, horizontal axis, axes, mode, discrete data, continuous data

Resources
- none needed

Language support
Use the examples of charts and graphs to model key vocabulary. Remind students of the features of each type of data representation and model the types of questions you can ask to interrogate the data, for example:
- *Which was the most popular fruit?*
- *How many people had red or blue bikes altogether?*

 Introductory activity

Ask students, in groups, to look at the different examples of representing data on page 184 of the Student Book. They should try to think of questions they can ask and answer using the information on the charts and graphs. Each group should come up with three questions they can put to the whole class.

 **Main activity**

Look together at page 184 of the Student Book. If you have access to an IWB, display the page. Students continue to work in small groups to discuss the questions in the first four speech bubbles. You can allocate each group to a different speech bubble to start with, then move them on to a new one when they are ready.

Once each group has completed their discussion around the four speech bubbles, everyone should look at the last speech bubble and work on it in pairs before sharing some of their information with the class.

As students are discussing in their groups, ask questions to assess their understanding of statistics, for example:
- *Which shapes can you see in the Carroll diagram that are not triangles and do not have a right angle?*
- *What is the difference between the number of red bikes and orange bikes. How did you work that out?*
- *What information is missing from the pictogram? What could it tell us?*

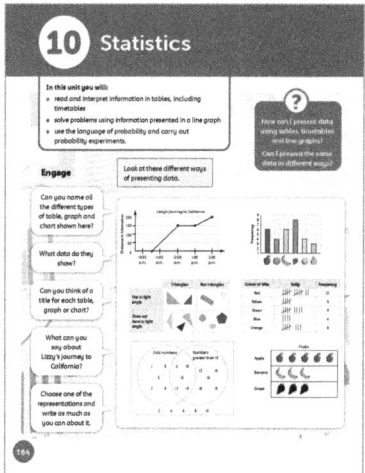

Differentiation
Supporting: Help students to interpret the graphs and charts and answer the questions the group asks about them.

Consolidating: Ask students to name the graphs and charts they are selecting and describe how they are used.

Extending: Ask students to write one additional question that can be answered by each graph or chart and describe how it can be answered.

 Reflection time

Together, discuss the graphs and charts in the Student Book. Ask students to tell you the names of each and to describe them. Invite individual students to identify one graph or chart, describe what it is about and give the class some information that it shows.

10A Frequency tables and bar charts

Discover Student Book page 185 • Practice Book page 143

Specific learning focus
- Collect and organise relevant data and display it a frequency table.

Global skills
- **Creative skills:** investigating
- **Real-world skills:** presenting information/ interpreting information
- **Interpersonal skills:** teamwork

Key vocabulary
- data, discrete, frequency table, bar chart

Resources
- none needed

Language support
Add an example of a completed tally frequency table to your class display. Next to the table, include the question that was posed as well as a fictional context for why the question was asked. Include questions for students to discuss relating to the data collected.

Introductory activity

Tell students that in this lesson they will be looking at **frequency tables**. Ask, *Do you know what a frequency table is?* Establish that it is a table that shows information about how often an item or event happens. Draw this table on the board to demonstrate:

Favourite Sport	Number of students
Baseball	22
Soft ball	18
Cricket	31
Soccer	14
Athletics	15

First ask, *What question do you think was asked? Who might have asked this question? Who did they ask? Why might they ask?*

Ask students to talk to a partner and discuss the information in the table. Take feedback, asking, *What information can you tell me from the frequency table?*

 Main activity

Put students into mixed-attainment groups of three or four. Each group should agree on a question they want to find the answer to. They should design a suitable questionnaire and carry out a survey. They record the results of the survey on a copy of a frequency table on page 185 of the Student Book.

As students work in their groups, ask them questions such as:

- *What type of answer are you expecting from your question? Is it a yes no question, for example?*
- *How do you record a number less than 5 using tally marks? How about more than 5?*
- *What does frequency mean?*

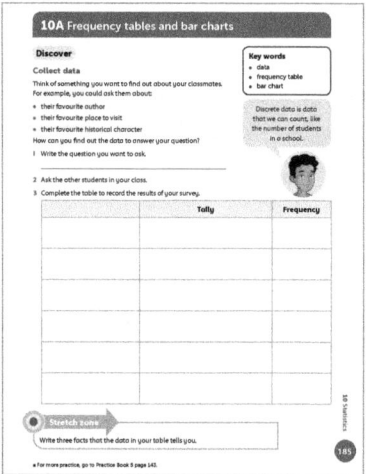

Differentiation

Supporting: Remind students of how to tally their findings. Model how to count up in fives and then ones to find the frequency.

Consolidating: Ask students to explain and show other students how to construct a frequency table from their data.

Extending: Ask students to interpret the data in the frequency table in at least three different ways. These might be: finding the most popular, the difference between two sets of data, the total number of people surveyed.

Stretch zone: *Write three facts that the **data** in your table tells you.*

Check that students' data supports their facts. Ask students to reflect on whether their data makes them curious about asking any further questions. How would they collect his data? Who would they ask?

 Reflection time

Invite students to share their results from their survey. Discuss their results and ask questions to interrogate the data. The range of questions will vary according to the topic of the survey, but more general questions could include: *Which was the most/least popular…?, How many people chose….?, How many more people chose … than … ?*

Practice Book: Students complete Practice Book page 143. They can do this directly after the Main activity, as homework, or as the focus of a separate mathematics session to help students consolidate their learning and build fluency.

You may prefer to complete this activity after students have looked more closely at **bar charts** in 10A Explore. If not, review the features of bar charts, discussing how the amount of data collected will help you decide on an appropriate scale.

Differentiated outcomes	
All students	should ask an appropriate question, collect data and use it to create and complete a frequency table with support.
Most students	will create and complete frequency tables, as well as supporting other students to do so.
Some students	may interpret the data in the frequency table in a variety of ways.

Answers

Student Book page 185

Answers will vary because groups carry out their own surveys and record the results in frequency tables. Check that the facts that students have said can be found from the information in the table are shown there.

Practice Book page 143

	Day	Frequency
1	Monday	52
2	Tuesday	38
3	Wednesday	30
4	Thursday	36
5	Friday	54

6 Check that the bar chart shows the data in the frequency table correctly, especially that students have chosen an appropriate scale for the y-axis and used it correctly.

Stretch zone: A table or bar chart would be most useful for the data. Check that students can explain their choice.

10A Frequency tables and bar charts

Explore Student Book page 186 • Practice Book page 144

Specific learning focus
- Answer a set of related questions by collecting, selecting and organising relevant data; draw conclusions from their own and others' data and identify further questions to ask.

Global skills
- **Creative skills:** exploring
- **Real-world skills:** presenting/interpreting information
- **Interpersonal skills:** communication/teamwork

Key vocabulary
- frequency table, bar chart

Resources
- mini whiteboards and markers

Language support
Prior to the lesson, consider what the most popular snacks may be among students and find visual examples for students to refer to, as well as their English translation, if necessary.

 **Introductory activity**

Give this statement, or similar: *I think potato chips are the most popular snack in this class.* Discuss this statement as a class. Ask, *How could we find out whether I am correct?*

Take suggestions. Agree that they could survey everyone in the class, by asking, *What is your favourite snack?*

Ask students to take it in turns to write their favourite snack on the board and make a tally to show any students who also state that it is their favourite. Look at the votes collected and discuss. *How many favourites do we have? Does this clearly show which is the most popular? What could we say about our most popular snack? Was I correct? Is it potato chips?*

Say, *What if I wanted to serve only two types of snack at a class party? Could I ask a similar, but different question to help me decide on two types?* Agree that limiting the number students can choose from makes it easier to make this decision. Agree on a question to ask, for example: *Which of these five snacks would you most like to eat at a class party?* (Choose five based on the data already collected.)

As before, students record their choice on the board as a tally. Display this information on a frequency table.

 **Main activity**

Ask students, in pairs, to create a bar chart on their whiteboards to show the data they collected about snacks. Quickly discuss what features they need to include, drawing on previous learning. Fill in any gaps in knowledge, as necessary.

Unit 10 Statistics

Ask pairs to share their bar charts. It is likely that all pairs will have used a scale of 1 unit to represent 1 person. Ask them to describe what it shows. Other students can ask questions about their data or their charts.

Use the frequency table to model how to create a bar chart with one unit representing two people.

Next, point out the Think back example on page 186 of the Student Book. Compare the frequency table and bar chart and identify where the information appears on both.

Ask students to work in pairs to create a bar chart using the data they collected in the previous lesson. Once they have completed the charts, come together to share their work and discuss similarities and differences in how they chose to display the data in a bar chart.

As students work, move around the class asking questions such as:

- *What can you tell me about a bar chart?*
- *What do the bars on your bar chart represent?*
- *How are frequency tables and bar charts the same? How are they different?*

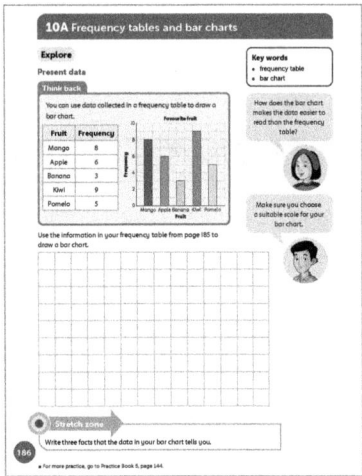

Differentiation

Supporting: Encourage students to refer to the bar chart and frequency chart on page 186 of the Student Book from the outset to help them create their first bar chart.

Consolidating: Ask students to model how to create bar charts using a scale of 1 to 2.

Extending: Ask students to explore different scales for bar charts. Ask them to think about why one scale might be better than another.

Stretch zone: *Write three facts that the data in your bar chart tells you.*

Encourage them to write facts that use a variety of skills, for example finding difference, comparing, and totalling. Check that students' data supports their facts.

 Reflection time

Ask pairs to discuss what they have discovered about the class in the last two lessons. Is there anything that surprised them? Did they expect the responses that are represented in the data? Share the results of the discussions.

Practice Book: Students complete Practice Book page 144. They can do this directly after the Main activity, as homework, or as the focus of a separate mathematics session to help students consolidate their learning and build fluency.

If time permits, discuss the different types of food, showing students visual examples of each. Do they like them? Which would be their favourite? Why?

Differentiated outcomes	
All students	should create bar charts using a scale of 1 to 1 with support.
Most students	will create bar charts using a scale of 1 to 2.
Some students	may use a range of appropriate scales for their bar charts, based on the amount of data collected.

Answers

Student Book page 186

Answers will vary because students use the information they and other groups recorded in 10A Discover to draw bar charts. Check that the bar charts accurately represent the data collected.

Practice Book page 144

Food	Number of children choosing it
lentil soup	5
pizza	4
kebab	3
hummus	4
roast chicken	5

Check that their bar charts accurately represent the data provided, and that students have chosen an appropriate scale.

Stretch zone: The bar chart will have much less data and it would be hard to make a general statement about favourite foods.

10B Line graphs

Discover Student Book page 187 • Practice Book page 145

Specific learning focus
- Answer a set of related questions by collecting, selecting and organising relevant data; draw conclusions from their own and others' data and identify further questions to ask.

Global skills
- **Creative skills:** exploring
- **Real-world skills:** presenting/interpreting information

Key vocabulary
- line graph, discrete data, continuous data, line segment, plot

Resources
- squared paper, pencils and rulers

Language support
Remind students how to read and say temperatures in degrees Celsius as well as 24-hour times.

Introductory activity

Ask students, *What can you tell me about line graphs?* If necessary, remind students that a **line graph** is a graph that shows change over a period of time. They should mention things such as: it will have a title, it has an *x*- and a *y*-axis, the unit of time is usually on the *x*-axis and the quantity is on the *y*-axis, it tells you what is being measured and how much, the points on the graph are joined up to make a line.

Ask students to work with a partner and think of things that happen over time, for example a plant growing, or temperature rising and falling.

Draw the following line graph on the board.

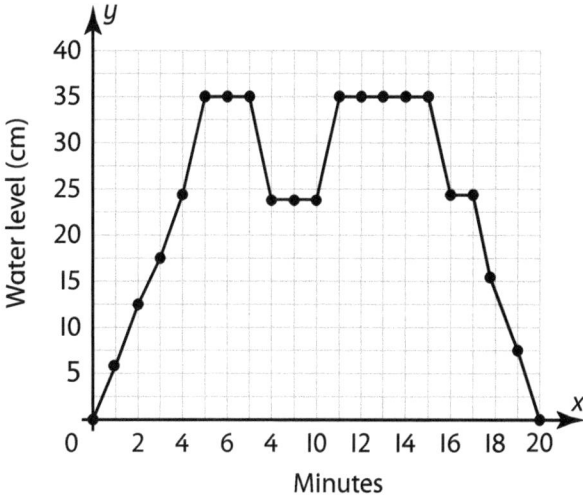

Changes in water levels in a sink

Ask students to tell you what the graph shows. Point to unmarked positions along the axes. Ask students to estimate what goes there. Ask them to talk to a partner about what is happening in the graph. Listen to their ideas. Look at the times where there is no movement of water. Why do they think there are changes in the water and then none? Students should recognise that the water level starts at 0 cm because the sink is empty. The line goes up as the sink fills up; it goes down as it empties; the line is straight when the water level does not change. They may suggest that the water level went down and then up again when the water got cold and more hot water was added.

Ask pairs to write down three statements they can make from looking at the line graph, for example: 'The highest water level reached is 35 cm.' 'There was a bit less than 25 cm in the sink after 16 minutes.'

 Main activity

Explain to students that line graphs are drawn from a series of data collected over a period of time. *Can you say what the time interval was for the example in the Introductory activity?* (minutes) *How often was the depth of the water **plotted**?* (each minute) Point out that the **line segments** between plotted points are only an indication of the data during that time and are not exact.

Introduce the concept of discrete and continuous data. Explain that data that can be counted (number of students who prefer apples, number of times they can jump in a minute) is **discrete data**, but data that is measured is described as **continuous data** because there are no gaps on the measuring scale. However, point out that we round continuous data to a suitable value before graphing.

Carefully explain the activity on page 187 of the Student Book. Remind students that they need to label the axes and give their graph a title. Students should work in pairs to support each other and to check for accuracy.

Unit 10 Statistics 239

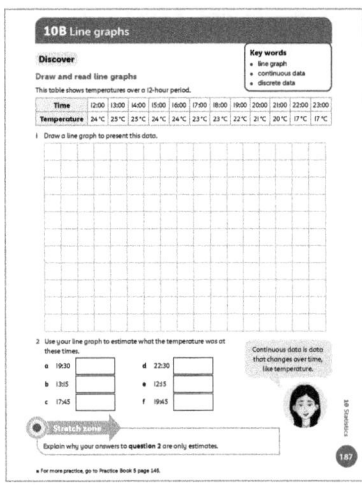

Differentiation

Supporting: Support students to draw and label the *x*- and *y*-axis and plot the first couple of temperatures before they continue independently to complete the line graph.

Consolidating: Ask students to explain how they plan to estimate the temperatures that fall between plotted points. Why do they think they will be correct?

Extending: Ask students to interpret the graph for you in a range of ways using all the key vocabulary.

Stretch zone: *Explain why your answers to question 2 are only estimates.*

Students may say that the temperature can only be an estimate because it was measured every hour and the times shown are between hours.

 Reflection time

Invite students to show the line graphs they drew in their Student Books. Invite pairs to make up questions to ask. The rest of the class can use the line graphs they made to answer these questions.

Practice Book: Students complete Practice Book page 145. They can do this directly after the Main activity, as homework, or as the focus of a separate mathematics session to help students consolidate their learning and build fluency.

Students will need a ruler to draw their line graphs. Remind them to join up their points as accurately as possible. Encourage them to refer back to the examples in their Student Books to support them with drawing a line graph.

Differentiated outcomes	
All students	should be able to plot and interpret line graphs with support.
Most students	will be able to plot and interpret line graphs.
Some students	may be able to plot and interpret line graphs in a range of ways.

Answers

Student Book page 187

1 Answers will vary because students draw their own line graphs. Check that their line graphs accurately represent the data.

2 **a** 19:30: 21.5 °C   **b** 13:15: 25 °C
   c 17:45: 23 °C   **d** 22:30: 17 °C
   e 12:15: 24.25 °C **f** 19:45: 21.25 °C

Practice Book page 145

Check that students have drawn and labelled the line graph correctly.

Temperature estimates:

1 15:00   42 °C
2 20:00   39 °C
3 00:00   33 °C
4 05:00   30 °C

Stretch zone: The temperature changed from 41 °C to 35 °C, but it went up before falling. It took 15 hours to fall from the highest to the lowest.

10B Line graphs

Explore Student Book pages 188–189 • Practice Book page 146

Specific learning focus
- Interpret lines graphs and use the data represented to answer questions.

Global skills
- **Creative skills:** investigating
- **Real-world skills:** interpreting information
- **Self-development skills:** reflecting on learning

Key vocabulary
- line graph, time graph

Resources
- squared paper, pencils
- globe or map

Language support
Add an example of a labelled line graph to your classroom display. Include a definition of continuous data. Students should also add any vocabulary to their glossaries that has been introduced throughout the unit.

Introductory activity

Draw this example of a line graph on the board.

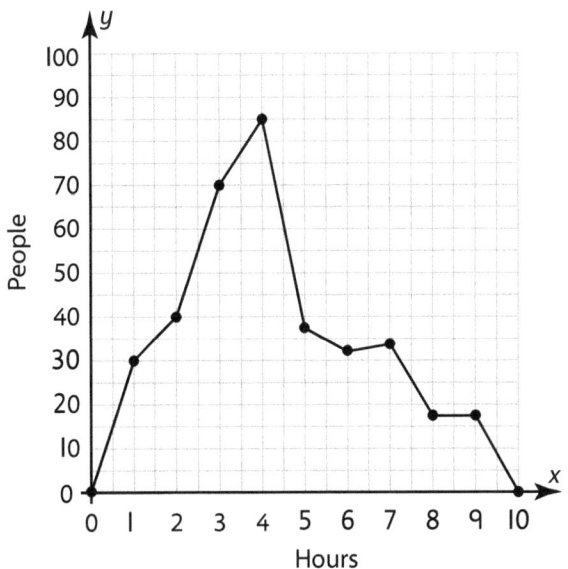

Ask students in pairs to discuss what is the same and what is different from the other charts they have been using. Take feedback. Ask students to explain what it shows (movement of people over a period of hours). Give students a few minutes to make up a story about the line graph and come up with a title. Invite them to share their stories. Discuss the idea of discrete data and continuous data.

- Discrete data is a set of data values that can be counted, for example shoe size, test scores.
- Discrete data is represented on, for example, bar line graphs, pictograms.
- Continuous data is something that is happening over a period of time.
- Continuous data is represented on a line graph.

Discuss the intervals of the vertical axis. Ask students to estimate how many people were doing something at different hourly points. Discuss the points between the hours. Establish that there are still people doing something. Agree that this is continuous data because something is happening all the time during the 10-hour period. Ask students to work in pairs on the activity on page 188 of the Student Book so that they can discuss what the line graph shows.

Main activity

Ask students to look at the activity on page 189 of the Student Book. Ask, *What does this line graph show us?* Agree that it shows the temperatures over a 14-hour period in Nairobi. Ask, *Do you know where Nairobi is?* Establish that it is in Kenya, a country in Africa. Show students its location on a world map or globe. Ask them to tell you what the temperature is at different hours. Ask, *How can you tell what the temperature is when the point is not on a labelled interval?* Agree that it is half-way between two labelled points. Ask them to work out the number between the two labelled numbers.

Students work on the activities on page 189 of the Student Book individually. Once they have written three more questions about the graph, they can swap with a partner to answer. Alternatively, they can work on all questions in pairs and then swap questions with another pair.

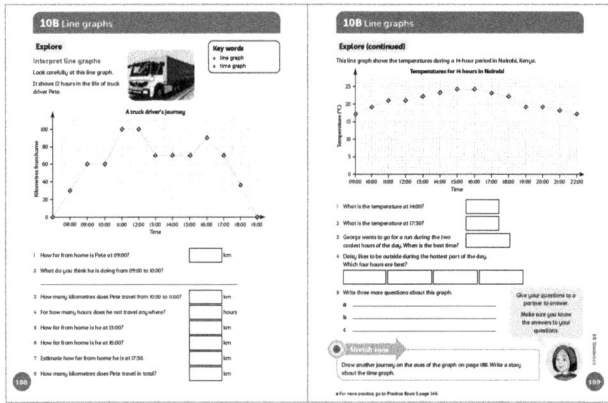

Differentiation

Supporting: Ask students to interpret the line graph for you. Provide support as needed.

Consolidating: Ask students to justify their interpretation of the data.

Extending: Ask students to interpret the graph for you in a range of ways.

Stretch zone: *Draw another journey on the axes of the graph on page 188. Write a story about the **time graph**.*

Students make up their own story for the line graph in the Student Book.

 Reflection time

Discuss the questions in the Student Book. Invite students to share their answers. Next, invite other students to ask the questions they made up for the rest of the class to answer. Finish the lesson by reviewing the different ways of representing data they have learned in this unit.

Practice Book: Students complete Practice Book page 146. They can do this directly after the Main activity, as homework, or as the focus of a separate mathematics session to help students consolidate their learning and build fluency.

Discuss what the frequency table tells them. Ask questions to assess students' understanding of the data as well as line graphs. Ask, for example, *What is the difference in the puppy's mass between 4 and 12 weeks old? Do you need to draw the line graph to answer questions 2 to 5? How could a line graph help you to interpret the data that is in the frequency table?* Encourage students to refer back to the line graphs in their Student Books to help them to draw their graphs.

Differentiated outcomes	
All students	should be able to interpret line graphs with support.
Most students	will be able to interpret line graphs.
Some students	may be able to interpret line graphs in a range of ways.

Answers

Student Book page 188

1. 60 km
2. Example: resting, eating
3. 40 km
4. 4 hours
5. 70 km
6. 90 km
7. 50 km
8. 240 km

Student Book page 189

1. 23 °C
2. 22.5 °C
3. Between 20:00 and 22:00
4. 14:00 and 15:00, 15:00 and 16:00, 16:00 and 17:00, 17:00 and 18:00
5. Check that students' questions reflect data represented in the graph.

Practice Book page 146

1. Check that students have drawn and labelled the line graph correctly.
2. 16.5 kg
3. 20.5 kg
4. 16–17 kg
5. 19 weeks
6. 55 kg

Stretch zone: Check that students have written a suitable question and found the answer.

10C Timetables

Discover Student Book page 190 • Practice Book page 147

Specific learning focus
- Read timetables using the 24-hour clock.

Global skills
- **Creative skills:** exploring
- **Real-world skills:** interpreting information

Key vocabulary
- timetable, time interval

Resources
- mini whiteboards and markers
- local bus or train timetables

Language support

Review how to read 24-hour times correctly. Continue to reinforce this by saying times slowly, clearly and frequently throughout the lesson.

 **Introductory activity**

Draw a 24-hour clock number line on the board.

Unit 10 Statistics

Organise students in pairs and give each pair a bus **timetable** (or use the bus timetable on page 190 of the Student Book if you cannot find any local timetables). Ask students to look at the bus timetable and write on their whiteboards three statements about journey times, using the time line for support if necessary. For example, if they are using the timetable on page 190, they could write: It takes 20 minutes for Bus 1 to drive from the bus station to the shopping centre.

Ask pairs to share their statements. As the class to check that they are correct. Ask them some additional questions about the timetable, for example: *I catch the bus at 10:20. What stop did I get on at?* Include some questions that involve converting to 12-hour times, for example: *I need to be at the airport by 1 p.m. Which bus will get me there closest to that time?*

Main activity

Ask students to look at the activity on page 190 of the Student Book. They need to work out the total time it takes for each bus to travel from the bus station to the train station. Tell them that if they would like to they can use number lines, as you did in the Introductory activity.

As students work, ask them further questions relating to the timetable. For example:

- What is the time difference between 10:47 and 13:52? How did you work that out? Is there another way?
- Bus 3 is running 25 minutes late. What time does it get to the school?

Once pairs have worked out how long it takes for each bus to travel from the bus station to the train station, ask them to write their own word problem using the bus timetable. They should work out the answer and then swap questions with another pair. They can then share answers and check to see whether they are correct.

Differentiation

Supporting: Help students to write down statements about each bus.

Consolidating: Ask students to use the timetables to calculate the **time intervals**.

Extending: Ask students to use the timetables to calculate the differences between the time intervals.

Stretch zone: *Kimi needs to catch a bus to the airport. Her flight departs at 12:00. Which bus should she catch from the bus station? Explain your answer.*

Students can use the information in the Student Book to work out the time it takes each bus to go from bus stop to bus stop. Does the time length between stops vary or is it always the same?

Reflection time

Refer to the bus timetable in the Student Book again. Ask, *Which bus had the shortest total journey time? Which had the longest? What was the difference between the shortest and longest time?* Select pairs to share their word problems for the rest of the class to answer.

Practice Book: Students complete Practice Book page 147. They can do this directly after the Main activity, as homework, or as the focus of a separate mathematics session to help students consolidate their learning and build fluency.

Encourage students to draw a 24-hour time number line marking intervals to at least the half hour. Encourage them to count on to find the times each bus arrives at each stop. Explain that the answers for the Stretch zone questions will be estimates only and that they will need to use what they know already to make reasonable estimates for each distance. For example, the distance from the bus to the mall is 8 km and the trip takes 10 minutes. It takes half as long to get from sport stadium to school so you can estimate that it will also be half the distance, 4 km.

Differentiated outcomes	
All students	should use the timetables to write statements about time intervals with support.
Most students	will use the timetables to calculate time intervals.
Some students	may create their own problems relating to the timetable involving, for example, finding time differences.

Answers

Student Book page 190

Bus 1: 2 hours 45 minutes

Bus 2: 3 hours

Bus 3: 3 hours 15 minutes

Bus 4: 2 hours 45 minutes

Bus 5: 3 hours 15 minutes

Bus 6: 3 hours 5 minutes

Unit 10 Statistics

Practice Book page 147

bus station	08:30	09:20	10:10	11:00	11:50	12:40	13:30	14:20
mall	08:40	09:30	10:20	11:10	12:00	12:50	13:40	14:30
sports stadium	08:48	09:38	10:28	11:18	12:08	12:58	13:48	14:38
school	08:53	09:43	10:33	11:23	12:13	13:03	13:53	14:43
railway station	08:56	09:46	10:36	11:26	12:16	13:06	13:56	14:46
Park	09:13	10:03	10:53	11:43	12:33	13:23	14:13	15:03

Stretch zone: Mall to sports stadium 6 km, sports stadium to school 4 km, school to railway station 2.5 km, railway station to park 15 km.

10C Timetables

Explore Student Book page 191 • Practice Book page 148

Specific learning focus
- Read timetables using the 24-hour clock.

Global skills
- **Creative skills:** problem solving
- **Real-world skills:** presenting information/interpreting information

Key vocabulary
- timetable, time interval, duration, time difference

Resources
- copy of a bus timetable for the nearest local city, copy of timetables of flights or access to the internet
- globe or map
- mini whiteboards and markers

Language support
Show students different ways that hours, minutes and seconds are recorded in abbreviated form, for example: hr, hrs, mins, min, secs. Explain that you would still read these as hours, minutes and seconds.

 Introductory activity

Students should work in mixed-attainment groups of about three or four for this activity. Give each group a copy of a bus timetable for the nearest local city. As a group, on a whiteboard they should create a journey that takes in several key places around the city that can be travelled to on the bus. When each group has written their journey, they give it to another group to follow. Each group checks the answers to the problems they created.

 **Main activity**

Ask students to look at the activity on page 191 of the Student Book. Read the introductory text together, which sets up the context. Ask questions such as *Can anyone tell me where Bangkok is? What about Mumbai?* Show students both cities on a globe or map. Share facts about both cities and countries, with a focus on numerical data, for example population, area, monthly rainfall of a particular month or temperatures on this date.

Discuss with students how the timetable is arranged, for example each row represents one flight. *How is this flight timetable different from a bus timetable? How is it the same?* Explore with students how to calculate the flight durations. For example, if a flight leaves at 13:28 and arrives at 15:42, then it is easy to count on the hours (13 to 15 = 2 hours) then count on the minutes (28 to 42 = 14 minutes), giving a flight time of 2 hours 14 minutes.

Ask students to work out a flight time that leaves at 12:45 and arrives at 15:35. They might count on from 12:45 to 14:45 as 2 hours, then count on another 50 minutes to bring them 15:35. You might explain to them that they could have counted on 3 hours from 12:45 to 15:45 which is 3 hours, then taken off 10 minutes, as 15:35 is 10 minutes before 15:45.

They should now work in pairs as they complete the tasks on page 191 of the Student Book.

Ask questions to support students' learning, for example:
- *In what order are the flights organised?*
- *What do you notice about all the departure times?* (they are all multiples of five minutes)
- *What is the time difference between 11:55 and 19:05? How did you work that out?*

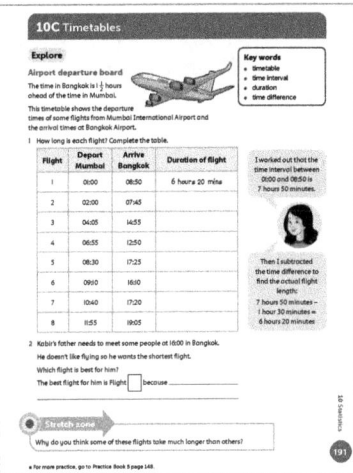

Differentiated outcomes	
All students	should work out the duration of flights by counting on or back along a 24-hour time number line.
Most students	will use the timetables to calculate flight durations accurately.
Some students	may create two-step problems relating to the timetable and involving finding time differences, and describe flight durations in minutes only and fractions of an hour as well as in hours and minutes.

Differentiation

Supporting: Help students to work out the flight times by counting on and back along a 24-hour time number line.

Consolidating: Ask students to share their strategies for finding the flight durations and say the durations in minutes only as well as in hours and minutes.

Extending: Ask students to create, and solve, two-step problems relating to the timetable and involving finding durations.

Stretch zone: *Why do you think some of these flights take much longer than others?*

Students may say due to stopovers in other countries.

You could also ask students to explore other timetables from the internet or newspapers.

Reflection time

Invite students to share their strategies for finding the lengths of the flights as well as their answers to question 2. Ask any students who wrote their own word problems to share them with the class to solve.

Practice Book: Students complete Practice Book page 148. They can do this directly after the Main activity, as homework, or as the focus of a separate mathematics session to help students consolidate their learning and build fluency.

Students will need access to the internet or to printed timetables of flights. Prior to starting the activity, you could print a selection of timetables for students. You could extend this activity by asking students to find out numerical facts about the departure and arrival cities or countries.

Answers

Student Book page 191

1. Flight 1 (Provided): 6 hours 20 minutes

   Flight 2: 4 hours 15 minutes

   Fight 3: 9 hours 20 minutes

   Flight 4: 4 hours 25 minutes

   Flight 5: 7 hours 25 minutes

   Flight 6: 5 hours 30 minutes

   Flight 7: 5 hours 10 minutes

   Flight 8: 5 hours 40 minutes

2. Kabir's father should take flight 2 because it is the shortest flight.

   Some flights are longer because they might stop in a different country on the way.

Practice Book page 148

Check that students have written three good questions about their particular timetable.

Stretch zone: Check that students have given good explanations for their choice of timetable improvement.

Unit 10 Statistics 245

10D Probability

Discover Student Book page 192 • Practice Book page 149

Specific learning focus
- Describe the occurrence of familiar events using the language of chance or likelihood.

Global skills
- **Creative skills:** investigating
- **Real-world skills:** presenting information
- **Interpersonal skills:** communication

Key vocabulary
- probability, chance, event

Resources
- three large sheets of paper: one with 'Impossible', one with 'Possible' and one with 'Certain' written on it
- reusable adhesive or sticky tape
- A4 pieces of paper with a variety of events (one per student) written on them such as:
  - The sun will set tonight.
  - It will rain in the next year.
  - I will be late for school tomorrow.
  - Somebody in the class will have a birthday in August.
  - Somebody in school will have a birthday today.
  - I can be in two places at the same time.

Language support
For Reflection time, pair up confident English speakers with less-confident speakers. The more confident one from each pair should report back to the whole class. Ask students to read out their responses so you can support them in developing the language of probability. You could help write the answers for less-confident English speakers for the final two questions so that they have good models of English in their Student Book.

 **Introductory activity**

Attach the 'Impossible', 'Possible' and 'Certain' sheets to the board. 'Impossible' should be on the left as the students look at the board to represent a 0–1 number line. Students should take it in turns to come up to the front of the class and attach their '**event**' on one of the sheets explaining why they are making this decision. As they place the 'events', model the language of **probability** listed on page 192 of the Student Book to further describe the likelihood of each event.

 **Main activity**

Ask students to discuss in pairs any other phrases they use in their native language to describe the likelihood of events. For example, in English, a phrase such as '**chance** would be a fine thing' means 'very unlikely', and 'fat chance' means little to no chance. Collect these phrases together and ask each pair to come up with an event of their own and write it on a piece of paper. They should then take it in turns to place it on the board.

Explain that probabilities are often given a value between 0 and 1, where 0 is for something impossible and 1 is for something certain. You can also use 0 to 100 in percentages in the same way. Ask students to look at the events they have sorted as a class and see whether they can give a percentage likelihood to each. For example, if something is equally likely as not, then it is 50% or $\frac{1}{2}$.

Look at the list of words in the table on page 192 of the Student Book. Explain that the words are ordered from least to most likely. Ask students to work in pairs to come up with events that match each word. They also add two events and their likelihoods to the list.

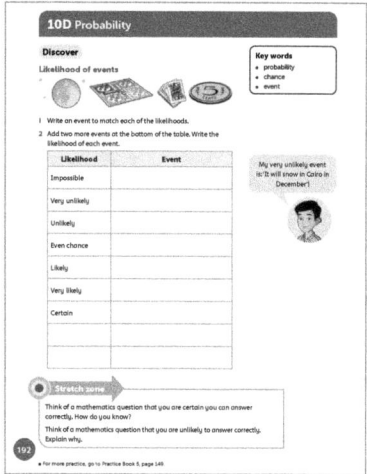

Differentiation

Supporting: Students should refer to the probability scale. Link each word to a familiar example.

Consolidating: Ask students to share a couple of examples for each word with you. Can they tell you what would need to change, for example, to make an example that was unlikely become very likely?

Extending: Ask students to justify their decisions about placing events on the probability scale.

Stretch zone: *Think of a mathematics question that you are certain you can answer correctly. How do you know? Think of a mathematics question that you are unlikely to answer correctly. Explain why.*

Students can use any mathematical topic from which to draw their questions. Check that they can explain why for both their choices.

Unit 10 Statistics

Reflection time

Rearrange students so that they are with a new partner. They should discuss which events they chose and the likelihood of them happening. Take feedback from pairs after they have had time for discussion.

Practice Book: Students complete Practice Book page 149. They can do this directly after the Main activity, as homework, or as the focus of a separate mathematics session to help students consolidate their learning and build fluency.

Tell students to use a dictionary to look up any vocabulary they are unfamiliar with.

Differentiated outcomes	
All students	should use simple statements of likelihood with support.
Most students	will use a range of vocabulary of probability.
Some students	may understand the probability scale.

10D Probability

Explore Student Book page 193 • Practice Book page 150

Specific learning focus
- Describe the occurrence of familiar events using the language of chance or likelihood.

Global skills
- **Creative skills:** investigating
- **Self-development skills:** reflecting on learning

Key vocabulary
- frequency, likelihood, probability, even chance

Resources
- coins
- spinners 1–6, digit cards 1–6

Language support
Add a probability scale, similar to the one on page 149 of the Practice Book, to your working wall. Include idiomatic expressions (for example: I have no doubt that …, chances are …, don't hold your breath) and examples of events in speech bubbles, which students can then place correctly on the scale.

Answers

Student Book page 192

Students choose their own events to match the phrases about likelihood, so answers will vary. Check that they seem reasonable. While students are working, ask them to explain why they thought a particular event was impossible (and so on).

Practice Book page 149

Students should have written a statement using the probability vocabulary, similar to those given as examples on the page.

Stretch zone: Students should have marked their events on a probability scale.

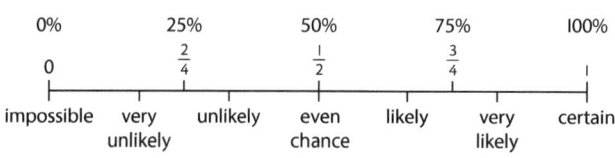

 Introductory activity

The Introductory activity should be very brief so that the focus of the lesson is the two experiments. Ask pairs to discuss what they think is the chance that a coin will land on a chosen side after being tossed a short way up into the air. Explain that this is called 'flipping a coin' and that in countries such as the UK, US and Canada the different sides of coins are called 'heads' and 'tails' and you would 'call heads or tails'. Use local currency and the local terminology, if you prefer. Students are likely to say that the chances are even or '50–50'.

Explain that as part of the Main activity they will do an experiment to investigate the **probability** of getting heads or tails when flipping a coin.

 Main activity

Students should work in mixed-attainment groups of four. They should record their findings on page 193 of the Student Book.

For the first experiment, each student should toss the coin 5 times while the others record the results. This will give them 20 coin tosses to record.

They should do the same with the spinners. This time, each student spins the spinner 5 times – this gives a total of 20 rolls. Before students write the sentences on page 193 of their Student Book, encourage the group to discuss their findings.

Unit 10 Statistics 247

While students are working, ask them to explain their answers to the questions on page 193. They can also answer one of the two questions in the speech bubbles and explain their reasoning. Ask, *Can you think of an event, using a coin, that is impossible? Can you think of an event, using a spinner, that is certain?*

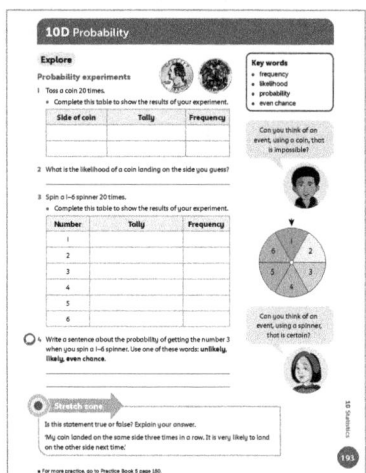

Differentiation

Supporting: Model simple probability vocabulary for students.

Consolidating: Ask students to use a range of vocabulary and explain what they are doing as they conduct each experiment.

Extending: Ask students to justify their probability estimates for each event from the data they collected.

Stretch zone: *Is this statement true or false? Explain your answer.*

'My coin landed on the same side three times in a row. It is very likely to land on the other side next time.'

False. Students should recognise that on each coin flip, both sides are equally likely, whatever happened previously, provided that the coin is a fair one.

 Reflection time

Draw the two tables from questions 1 and 3 on the board. Collate the results from the whole class into these tables. Take feedback from the groups, drawing on the sentences they have written.

The results of the experiments won't mirror the actual probabilities exactly, but the results from the whole class may be closer than from each individual group.

Ask students questions that might be answered from the tables, for example: *What is the probability of getting 'heads' and spinning a 3? How might you work this out?*

Practice Book: Students complete Practice Book page 150. They can do this directly after the Main activity, as homework, or as the focus of a separate mathematics session to help students consolidate their learning and build fluency.

Students will need digit cards 1–6 or they can make them themselves.

Differentiated outcomes	
All students	should record the outcomes of probability experiments with support.
Most students	will record the outcomes of probability experiments in frequency tables.
Some students	may use the outcomes of probability experiments to make statements about combined events.

Answers

Student Book page 193

Answers will vary as students record their results from the coin and spinner experiments.

Practice Book page 150

Check that students have completed the frequency table and bar chart for their data.

Stretch zone: Students may have found that 7 occurred the most, but this is not always going to happen, it is just the most likely. This is because of all the possible outcomes, more outcomes result in a total of 7 than any other total.

10 Statistics

Connect Student Book page 194

Big idea

I can use frequency tables to record data. I can present this data using different charts, including line graphs. I can read timetables and use a timetable to plan a journey.

Global skills

- **Creative skills:** investigating
- **Real-world skills:** presenting information
- **Interpersonal skills:** communication/teamwork
- **Self-development skills:** reflecting on learning

Key vocabulary

- time interval, line graph, continuous data, *x*-axis, *y*-axis

Resources

- water
- glasses or tumblers
- thermometers
- ice
- squared paper

Language support

Encourage students to refer to their glossaries and the classroom display to remind them of the meaning of key vocabulary as well as how to use it.

 Introductory activity

Review with students their learning about line graphs and what they can be used for from earlier in the unit. They should recall that line graphs are drawn from a series of data over a period of time. Recall with students the key ideas about discrete data and continuous data.

- Discrete data is for things that can be counted, for example shoe size, test scores.
- Discrete data is represented on, for example, bar graphs and pictograms.
- Continuous data is for things that can be measured, usually over a period of time.
- Continuous data is represented on a line graph.

Ask students to think of examples of things that happen over time that they could represent on a line graph. Examples might include, for example, the growth of a plant or the increasing weight of a new baby.

 **Main activity**

Explain that students are going to be carrying out an experiment where they will be measuring and recording the temperature of a glass of water over a period of time. Put students into small groups and ask them to think of questions they may have about the experiment, directing them to page 194 of the Student Book. Prompt them to consider their method by asking, for example:

- *How are you going to keep time?*
- *How accurately will you read the temperature (nearest degree or half degree?)*

Explain that once they have gathered their data in their frequency table, they will present it in a line graph. Discuss the intervals they might use on their horizontal *x*-axis. Ask students to suggest the scale, perhaps 2 cm for each 5-minute interval. Discuss the points between the intervals. Establish that the water temperature can still be changing even when it is not being measured. Agree that this is continuous data because something is happening all the time during the 40-minute period. Ask students to work in their groups on the experiment.

As students carry out the experiment, ask them questions about their data collection and recording, for example:

- *How will you measure your time intervals?*
- *How did you decide what scale to use for your x-axis and y-axis?*
- *How can you estimate the temperature between the measured points?*

Finally, students will need to review their data and write two statements about their data in their notebook, which they will share with the class.

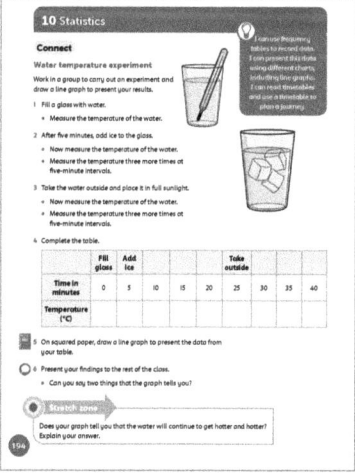

Differentiation

Supporting: Assign students a specific and appropriate role in the group to ensure their participation.

Consolidating: While students are working in their group, ask them to explain to you what they are doing.

Extending: Ask more-confident students to help others with the measuring and graph plotting.

Unit 10 Statistics 249

Stretch zone: *Does your graph tell you that the water will continue to get hotter and hotter? Explain your answer.*

Students should be able to comment on the slope of the graph and relate this to temperature rising or falling.

 Reflection time

Bring the groups together as a class to discuss the experiment and compare line graphs. Ask each group in turn to share two things the graph told them and to show their line graph. You may find that most graphs will be similar, but not exactly identical. Ask students why this might be. Ask them about the points where the line graph changed, for example when the ice was added and when the water was put in the sunlight. Ask them to predict the temperature of their water at points between the measurements and explain how they decided.

Differentiated outcomes	
All students	should measure, record and graph the data with support.
Most students	will measure, record and graph the data.
Some students	may measure, record and graph the data and use it to answer questions about what happened.

Answers

Student Book page 194

Check that students have recorded the water temperature in the table and drawn a line graph to represent the data. They should be able to make statements about the water temperature from the line graph, for example:

- 'The temperature went down quickly when the ice was added.'
- 'The temperature went up steadily in the sunshine.'

---

10 Statistics

Review Student Book page 195 • Practice Book page 151

Student Book

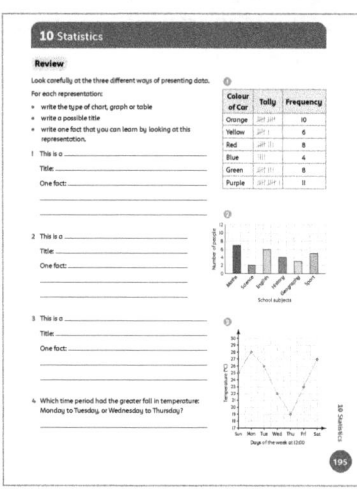

Global skills:
- **Creative skills:** problem solving
- **Real-world skills** interpreting information
- **Interpersonal skills:** communication/teamwork
- **Self-development skills:** reflecting on learning

With young students, assessment activities are most effective when carried out as an everyday classroom activity. Students should have access to examples of different types of charts and graphs, so they can refer to these to support them.

Extend by asking a range of questions about the data and using it to answer their original question, for example:

- *Which colour was the most popular?*
- *How many more people chose X than Y?*
- *How could you get a more accurate graph of temperatures?*

Answers

Student Book page 195

1. Frequency table

   It shows numbers of cars of particular colours. Example of information: there are 7 more purple cars than blue.

2. Bar chart

   It shows favourite subjects. Example of information: there are 7 people who like mathematics.

3. Line graph

   It shows the temperature over a week. Example of information: the coolest day is Thursday.

4. Wednesday to Thursday

Practice Book

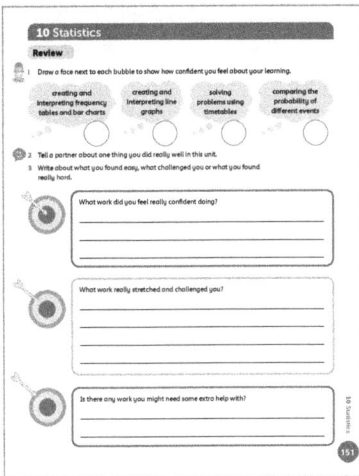

With students in upper primary years, it is appropriate to complete this Practice Book review as a whole-class discussion. You may choose to keep a record of the class discussion or a copy of the Review page for your own records. The Review provides an opportunity for students to reflect on their learning from the unit, to discuss any areas of mathematics that they feel went particularly well, and any areas that they feel less confident about. Ensure that all students have a copy of the Student Book as a reminder of the areas of mathematics that they have worked on in this unit.

Allow students plenty of time for discussion before asking them to complete the Practice Book page individually, and then, if appropriate, to share their responses with the rest of the class. If students complete this self-assessment at home, encourage them to discuss this with adults.

Make a note of areas that students still feel unsure about. As they are working, remind them to use the classroom display or work they have done and key vocabulary to help them. Reassure them that they will become more used to the vocabulary as they continue to collect data, represent it on different types of chart and use them to interpret it.

Additional material

There are additional end-of-unit assessment available on the *Oxford Owl for School* website.

Glossary

acute angle	an angle that measures between 0° and 90°
common factor	numbers that are factors to more than one other numbers, e.g. the common factors of 18 and 24 are 2, 3 and 6
common multiple	numbers that are common to two or more times tables, e.g. the common multiples of the two and three times tables include 0, 6, 12,
composite number	whole numbers that are not prime numbers are composite numbers
composite shape	a two-dimensional shape made up of basic two-dimensional shapes
convert	changing something from one thing into another. You can use conversion graphs and tables when converting between units
coordinates	the numbers on both axes that combine to give a particular point, e.g. (2,5)
cube number/ cubed (1^3, 2^3, etc.)	a number multiplied twice by itself, e.g. $2 \times 2 \times 2$
currency	the money system of a country, e.g. the currency of the US is dollars
decimal number	e.g. 0.05, 1.676, 22.7
decimal place	4.235 has three decimal places. Rounded to one decimal place, it is 4.2
denominator	the bottom number of a fraction. It tells you how many equal parts the quantity or shape has been divided into
diagonal	joining two opposite corners of a square, rectangle or other straight-sided shape
dividend	the number to be divided, e.g. in 24 divided by 6, 24 is the dividend
divisibility rules	quick checks to see whether one number will divide exactly into another
divisor	the number being divided into another number, e.g. in 24 divided by 6, 6 is the divisor
equivalent	fractions that are worth the same. When you simplify a fraction, the new fraction is equivalent to the original fraction
factor	e.g. 1, 3, 5 and 15 are the factors of 15.
factor pair	numbers that multiplied together give a particular number, e.g. 1 and 18 are a factor pair of 18.
formula	(plural: formulae) a rule that can be written in words or using letters and symbols
hundredth	one of 100 equal parts. You can find a hundredth of a shape, quantity or number
imperial	a type of measure, e.g. miles are an imperial measure
improper fraction	a fraction with a numerator that is larger than its denominator. It is a fraction that is worth more than 1
irregular	shapes with sides of different lengths
likelihood	the possibility or chance of an event happening
linear number sequence	a series of numbers that go up in equal size jumps, e.g. 1, 5, 9, 13, 17
line graph	a graph that shows continuous data over a period of time
metric	a type of measure, e.g. kilometres are a metric unit of measure
mixed number	this is made up of a whole number and a fraction
multiple	e.g. 300, 800, 2700 are multiples of 10
negative number	numbers below 0 are all negative numbers
numerator	the top number of a fraction. It tells you how many equal parts there are
obtuse angle	an angle that measures between 90° and 180°

outcome	when we perform an experiment, there are various possible outcomes, e.g. when we roll a dice there are six possible outcomes: 1, 2, 3, 4, 5, 6	**protractor**	a measuring tool to measure angles
		quotient	the product of a division problem
		ratio	a way of comparing one quantity to another
parallel	lines that are the same distance apart no matter how long they are. They can never cross each other. They can be straight or curved	**rectilinear shape**	a shape made up from rectangles
		reflection	a mirror image of a shape across a mirror line
		reflective symmetry	when one half of a shape is the symmetry reflection of the other half
percentage	out of a hundred (per cent). A percentage is another way of writing a fraction that has a denominator of 100. The symbol for per cent is %	**reflex angle**	an angle over 180°
		regular	shapes with sides that are all the same length
perpendicular	lines or faces that meet at right angles	**Roman numerals**	e.g. XXVI, CCXII
		scalene triangle	a triangle with no sides the same length. All its angles are different sizes
pounds	pounds are an imperial unit for measuring mass		
power of 10	10, 100 and 1000 are examples of powers of 10	**square number/ squared (1^2, 2^2, etc.)**	a number that is multiplied by itself. Whole numbers, fractions and decimals can be squared
prime factor	a factor that is a prime number, e.g. the prime factors of 15 are 5 and 3 because $3 \times 5 = 15$ and 3 and 5 are both prime numbers		
		tenth	one part of a whole divided by 10
prime number	a number that is divisible by only itself and 1	**thousandth**	one part of a whole divided by 1000
probability	the chance of something happening. We often write probability as a fraction. Words we can use when talking about probability include: chance, likelihood, odds	**timetable**	a schedule that tells you the times of buses, trains, classes, etc.
		translation	the movement of a shape without turning it, e.g. moving it 3 squares right and 7 squares up on a grid
proper fraction	a fraction with a numerator that is smaller than its denominator. It is a fraction that is worth less than 1	**volume**	the amount of three-dimensional space something takes up
protractor	a piece of equipment used to measure the size of angles	***x*-axis**	the horizontal axis of a graph
		***y*-axis**	the vertical axis of a graph
proportion	the number of items of a particular type in a group of items, e.g. 10 counters, 4 are yellow, 6 are green. The proportion of yellow counters is 4/10		